PROCESS CONTROL

**Structures
and
Applications**

PROCESS CONTROL

**Structures
and
Applications**

Jens G. Balchen
Kenneth I. Mummé

VAN NOSTRAND REINHOLD ELECTRICAL/COMPUTER SCIENCE AND
ENGINEERING SERIES

VAN NOSTRAND REINHOLD COMPANY
New York

Printed in the United States of America

Van Nostrand Reinhold Company Inc.
115 Fifth Avenue
New York, New York 10003

Van Nostrand Reinhold Company Limited
Molly Millers Lane
Wokingham, Berkshire RG11 2PY, England

Van Nostrand Reinhold
480 La Trobe Street
Melbourne, Victoria 3000, Australia

Macmillan of Canada
Division of Canada Publishing Corporation
164 Commander Boulevard
Agincourt, Ontario M1S 3C7, Canada

16 15 14 13 12 11 10 9 8 7 6 5 4 3 2 1

Library of Congress Cataloging-in-Publication Data
Balchen, Jens.
    Process control.
    Bibliography: p.
    Includes index.
    1. Process control.   2. Control theory.
I. Mummé, Kenneth I.   II. Title.   III. Series.
TS156.8.B35 1988   660.2'81   86-32608
ISBN 0-442-21155-4

PREFACE

This book is intended to help those concerned with the design and operation of systems in the chemical-processing industries so that they may come to understand the relationships between unit operation processes, their dynamic characteristics, and the various strategies needed to control them. The book also provides practical, specific ideas for such control.

We believe that this book will be useful to many different types of readers. Chemical engineering students will discover that unit operations require a wide variety of control structures to maintain steady-state operation. Control engineers will learn that unit operations often have design and operating constraints that are incompatible with expected "best" control strategies. Process design engineers will recognize that system performance can depend on dynamic behavior, which in turn may depend on seemingly innocuous factors such as size and plant layout. Many readers will come across useful control structures of which they may have previously been unaware. Finally, this book presents its material in such a way that engineers of various disciplines will better appreciate each other's problems in designing efficient and dynamic manufacturing processes.

The text is arranged so that readers already versed in unit operations can skip the introductory, descriptive material relating to those operations. Those readers may wish to proceed directly to the review of process control and then move on to the chapters in which process control strategies are discussed with respect to specific unit operations as well as complete processes. On the other hand, those readers well versed in control theory but not especially familiar with the various unit operations can find basic information about the latter and skip the introductory material. We should point out, however, that our treatment of unit processes emphasizes dynamic properties and behavior, which is different from the traditional approach.

Following an introductory chapter, Chapter 2 is a review of process control theory and structures, ranging from simple feedback systems to optimal control of multivariable systems. The review is organized so that the reader may enter the discussion at any appropriate level. Chapter 3 investigates basic unit operations, their dynamics, and their dynamic models. Chapter 4 presents control methods for various elementary functions such as level, flow, pH, pressure, and temperature.

In Chapter 5, we apply these elementary controls in the various groups and structures needed to control unit processes such as heat exchangers, separation processes, boilers, evaporators, drying processes, crystallizers, and distillation processes. Finally, in Chapter 6, we combine these unit processes into complex industrial systems.

Obviously, a single book cannot discuss more than a few complex systems. One purpose of this book, however, is to provide knowledge of the dynamics and control of the unit processes that constitute the building blocks of large systems. Using the techniques described here, the engineer can combine processes according to his own needs.

Much of the material in this book was developed for use in an upper-level control engineering course at the Norwegian Institute of Technology, Trondheim, Norway. Most of the students in this course had backgrounds in electrical engineering. In order to apply control-engineering methods to the processing industries, these students had to learn the fundamentals and dynamics of the elements of unit operations. It was in this environment that we discovered a serious need for a book that would provide a link between control engineering and process engineering, while recognizing the skills, training, and experience of both.

We hope that this book — with its organized review of the fundamentals of process control and unit operations, its relation of control theory, structures, and practical applications to the unit operation processes, and its ultimate application of all of this to complex industrial systems — will prove to be useful, both theoretically and practically, to students, process design engineers, and operating engineers. We hope that we have provided at least one small step in linking the control of processes to their initial design.

Jens G. Balchen
Division of Engineering Cybernetics
Norwegian Institute of Technology
Trondheim, Norway

Kenneth I. Mummé
Department of Chemical Engineering
University of Maine

CONTENTS

PROCESS CONTROL

Structures
and
Applications

Chapter 1
INTRODUCTION

1.1. THE FORM AND OBJECTIVES OF THE BOOK

This book is an attempt to bridge the gap between the enormous amount of material that has been developed in control theory in the last 40 years and the extensive practical experience and applications that have provided the basis for the development of industrial process control. It has long been recognized that this gap is much too wide. It has probably been due to the separation of theory and practice in education and management practices, as well as poor interdisciplinary communication of results and experiences.

Over the years there have been many attempts to explain the reasons for the problem and to propose solutions (Foss, 1973; Lee and Weekman, 1976), but little progress has been made. One reason for this is that during a formal program of study there is not enough time to gain insight into the many problem areas that must be considered in developing good control schemes for complicated processes.

In practice, the usual solution is to assemble a team of engineers with different backgrounds who can each contribute to the ultimate solution according to their individual skills and experience. Such an approach will gradually develop broader insight. A team of this type must be competent in at least two broad areas: process, chemical, and mechanical engineering on the one hand, and instrumentation, control, and on-line computer techniques on the other. In addition, the team must have access to auxiliary support in such areas as electronics, economics, administration, and so on. The basic difficulty may be that the education of engineers in the area of control and process computer methods provides little time for consideration of process systems, and the education of engineers in process systems (chemical engineers, for example) usually allows little time for consideration of advanced control system techniques.

No new theory regarding process control or process systems will be developed in this book. The literature dealing with these matters is extensive, and readers can study it according to their specific interests. However, the book does contain a series of appendixes describing some of the important results of applications in the areas of control theory and process dynamics.

The book has certain limitations. It is not possible to give a satisfactory description of all of the processes that one finds in industry, or all the other technical activities that use automatic control, although the similarities between many processes make it possible, at least in principle, to transfer certain techniques from one industry to another. Another obvious limitation of the book is that numerical values can seldom be assigned to the control systems that are developed; instead, real control systems must be adjusted to account for the physical characteristics of the particular process installation to be controlled (residence time, flowrates, etc.). Readers will find information that will enable them to apply the principles discussed here to choose the *structure* on which control systems should be based for a variety of processes. To be able to do this, one must develop insight into both process technology and automatic control technology.

The book has been organized to be a useful reference; that is, it is not necessary to read the sections in sequence if one is familiar with the techniques of either process control or process engineering. Chapter 2 is an overview, not an extensive development, of the principles of automatic control theory, which progresses to more difficult material later in the chapter. Chapter 3 is a systematic review of many important types of unit operations and unit processes found in industry, described with respect to both their functions in industry and their dynamic and static characteristics. Chapters 4, 5, and 6 consider process control as a synthesis of Chapters 2 and 3. Chapter 4 includes a discussion of some of the elementary process control functions such as level, flow, and temperature control. Chapter 5 discusses the control of various unit operations such as heat exchangers, separation processes, boilers, distillation columns, and so on. Chapter 6 describes how large processes are designed to include both the process technology and the required control technology, using examples from the paper, oil, petrochemical, and fertilizer industries.

A substantial number of references to the literature have been included to provide readers with a starting point for more detailed study of areas of their particular interest.

1.2. THE REASONS FOR AUTOMATIC CONTROL

The development of industrial automation has had a significant influence on modern society, especially in Europe, North America, and Japan. The methods and problems of the processing industries are somewhat different from those of the consumer goods industries. Although all industries have the same goals, this book will concentrate on the problems and objectives of the process industries.

The most important reasons for the use of increasingly sophisticated theory and technology in industrial automation have been and will continue to be the following:

- To improve productivity.
- To improve product quality.
- To improve the working environment and influence the total environment.

Productivity in a manufacturing operation is a function of factors such as work force, capital, raw materials, and energy. Partial productivity can be defined with respect to any one of these factors that gives the amount of product or value increase per unit of each factor. For example, in processes that are labor-intensive, production per worker hour can be the decisive factor for the economy of the process. In other processes, the ratio of product per unit of raw material might be the most important economic factor.

Many processes are fundamentally unstable, or nearly so, and must be equipped with automatic control to assure stable operation. The reason for using automatic control in such cases is to modify the process to attain the above goals regarding productivity, quality, the workers' environment, and so on.

The requirements for the control of quality can take many different forms. The most common is that the factors that determine the quality of the product have the least possible variation around the target value. Often a variable cannot be allowed to exceed (or fall below) a specified value because were it to do so, the product would be unacceptable to the customer. Figure 1.2.1a illustrates a case in which the variable $y_i(t)$ has large variation due to process disturbances or poor control. To ensure that the value does not exceed the specification, the average value y_i is kept relatively low. In Figure 1.2.1b it is assumed that the process can be controlled more effectively, thereby reducing the variation of $y_i(t)$ and making it possible to operate with the average value much closer to the specified limit. In many industrial situations an increase of just a few percent (or even a fraction of a percent) in the average value of the variable can represent very large economic gains in terms of, for example, energy savings, raw materials use, or customer acceptance. Here we also recognize the strong connection between the concepts of quality and productivity.

Quality and productivity can also be specified with respect to optimization. For example, the maximization of profit, minimization of loss, or maximization or minimization of other functions may be used to describe various objectives. The most common form of optimization of industrial processes is *steady-state optimization*, in which the dynamic characteristics of the process are ignored.

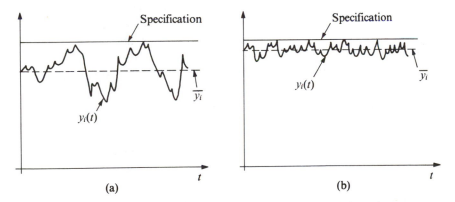

Figure 1.2.1. Reduction of the variance of a variable makes it possible to move the average value closer to the specification.

In these cases, a reference value (set point) is set in the various control systems to hold the process variables within the desired limits. *Dynamic optimization* is more complicated; it considers the dynamic behavior of the process, where it can be important to manipulate the process inputs during nonstationary conditions. In this book, dynamic optimization will be considered to some degree, but steady-state optimization will be detailed in a series of examples.

An important consequence of modern industrial control has been a significant improvement of the environment in which process operators must work. For example, operators have been freed from a large number of routine operations and detailed monitoring duties that can occur literally by the thousand in large industrial complexes. A modern petrochemical plant, for example, simply could not be operated without automatic control and instrumentation. With them, however, an operator in a central control room can easily monitor the process and make control adjustments when necessary. Certainly, good operator–process communication is important. This is a field in which many solutions have been proposed and in which many controversial questions remain to be settled.

From the point of view of the process operator, it is often an advantage to be able to control and monitor a process from a control room rather than to go out into process areas, which may be far away or may have parts that are dangerous or environmentally unpleasant. In practice, however, there is no general rule, as each process operating situation presents its own characteristic difficulties. Further, there are many processes in which it is important that the operator receive visual and audible information, where for one reason or another the process has not been completely instrumented. In such cases, the operator must augment the automatic systems by making personal inspections of the process from time to time. In contrast, many chemical processes are closed in the sense that little can be seen by the operator. In such a process, the need for manual monitoring and field inspections is usually relatively small, and the process can best be monitored and controlled from a central control room.

There is considerable disagreement over whether the working environment has been improved by the continuing increase in automation. The answer depends to some extent on the value one puts on the release of an operator from monotonous, routine work. If the use of advanced process control can lead to an understanding of how a complicated process functions, and to improvement of the system's design and operation, then the operator has the opportunity for a more meaningful and satisfying work experience.

For society as a whole, the growth of automatic control and process automation will increase the need for technical expertise in the areas of complex technology (of equipment and machines) and the complicated control solutions (control structure, measuring systems, information processing, etc.) that will be required. There is a risk, however, that industrial organizations (and society as a whole) will require increasingly qualified technical personnel and steadily less qualified operators. In the long run this is undesirable for many reasons. Therefore, a great deal of work is being done to find a way in which processes and systems can be designed and organizations and educational methods developed to provide a reasonable balance between the technical and economic production requirements and humanistic considerations.

A new important element in establishing the motivation for industrial automation is the relatively recent but serious concern for the natural environment. Since the 1960s, society has begun to demand a reduction of environmental pollution by industrial waste. The demand has become so strong that precise monitoring, and often control, of air and water quality has become necessary in many cases. This concern has not been detrimental to all industry, as the investment has often led to process improvements and significant savings of materials, manpower, and energy. These improvements have been especially important since the energy crisis of the 1970s and the resultant upheaval in the economic situation of many industries.

Thus there are many reasons for the use of automation in industrial processes. The most significant ones in the past were productivity and quality, but society has now made it clear that working conditions and the environment are also extremely important. An additional reason is the influence of competition. Most companies must continue to improve their manufacturing methods and many buy the means to do so from other companies, which also must continuously improve the systems they offer in order to remain competitive. An intense international competition has resulted, which is a major reason for the very rapid growth in the field of automatic control systems in the last 10 to 20 years.

1.3. WHICH TYPES OF PROCESSES ARE MOST SUITABLE FOR AUTOMATION?

There is no simple answer to the question of which types of processes are most suitable for automation. In principle, virtually all processes are candidates for the application of automatic control systems to help them meet the goals discussed earlier in this chapter. However, some processes are more suited to automatic control than others, and some have a greater need for it than others. Most often the question depends upon the size and design of the process. In the last 20 to 30 years it has become more usual to design processes so as to make them more suitable for automatic control. Many such processes were originally designed for manual control and are still influenced by that approach. Such processes, for example, might be designed to react very slowly to disturbances in order to reduce the need for quick operator reaction, thereby reducing the need for constant monitoring by the operator. With automatic control, a process can be designed to respond quickly, and such a process will usually require smaller equipment, shorter holdup times, and so forth. With the smaller equipment will come lower capital costs and better control of product quality.

The relationship between the process engineering and control engineering professions is not as good as it should be because of the educational dilemma discussed earlier. The difficulty shows up most clearly during the design stage of an industrial process.

The process aspects of the automation of an industrial process will be discussed in detail in Chapters 3 and 4, and no further comments will be made here except to point out that the control system and its associated equipment must be treated as an integral part of the process early in the process design stage. The separation of process design into various professional disciplines is undesirable, but it is perhaps partly due to the complexity of the process. A major effort must be made to avoid this sort of compartmentalization.

It would be unreasonable to try to collect examples of the solutions to all possible process control problems in one book. In this book, we will not discuss examples from typical servo systems (servomechanisms, electro-mechanical systems for position and velocity control, printing register control systems, machine tool control, etc.) or guidance systems (ships, platforms, aircraft, rockets, etc.). The major emphasis will be placed on industrial process control, which itself is so vast that only a limited number of examples can be considered.

A major purpose of this book is to provide examples that will be useful to the practicing engineer in the design of control systems in specific industrial situations. However, it still will be necessary to study specific process characteristics and relationships, which will lead to the development of a control problem solution and the eventual installation of the required equipment. The planning, design, installation, and testing of a complete automatic control system for a large industrial process are demanding and time-consuming jobs. The better the foundation one has in control theory, simulation techniques, and practical experience, the easier and less expensive the development of the final solution is. This book is intended to contribute to that foundation.

1.4. DEVELOPMENT OF THE VARIOUS TOOLS FOR DESIGN OF AUTOMATIC CONTROL SYSTEMS

Automatic control in industry was first recognized as a separate profession just after World War II. About the same time, chemical engineering experienced significant growth and development in the area of continuous processes. Developments in electrical engineering and electronics technology made possible the combination of chemical engineering and control engineering that is the main area of interest in this book.

The development of control theory can be traced approximately as follows:

1860–1940: Ordinary linear differential equations to describe simple processes and controllers.
1940–1960: Laplace transforms and frequency response methods to analyze single-variable, linearized systems.
1945–1960: Theory for discrete control systems based on the z-transform.

1955–1970: State space analysis for continuous and discrete multivariable systems. Optimal control theory.
1960–1975: Development of the theory for state and parameter estimation.
1960–1980: Methods for design of adaptive systems.
1965–   : Computer-aided design of multivariable control systems.

The development of a connection between general control theory and its practical application to industrial systems has been facilitated by the significant and wide-ranging possibilities offered by computer-aided design (CAD). There are, in the mid-1980s, many computer programs that offer interactive computer use for analysis of time series data, design, and simplification, reduction, and simulation of process models, development and testing of control strategies, and complete simulation of large systems under realistic conditions. Several systems that have been reported are: CYPROS (Tyssö, 1980), GEPURS (Mitchell and Gauthier, 1981), CSSL-IV (Nilsen, 1984), ND-TRAN (Uhran and Davisson, 1984), and CATPAC (Bunz and Gutschow, 1985). Development of systems in this area continues at a high rate. The rapid and continuing development of computer equipment technology steadily improves the capabilities and usefulness of graphic displays of results and the handling of larger, more detailed systems at higher speeds. These systems will give the practicing engineer easy access to the most advanced control theory.

1.5. PLANNING AND DESIGN OF LARGE INDUSTRIAL CONTROL SYSTEMS

The planning, design, and construction of a large modern industrial complex is a comprehensive task that can last several years and involve many professional disciplines. This also applies to automation of the technical processes; and because control is so closely related to the process units, it is especially important that the process engineers and the control engineers maintain good communication and cooperation. This has not always been the case. The result can easily be that the process engineer underestimates or misunderstands the control problem and prescribes a control system structure that cannot perform satisfactorily. Few modern industrial plants are exact copies of plants that were built earlier, even though individual processes might be quite similar. For example, even when two papermachines are built from the same plans, the machines, when erected, do not behave exactly alike. In practice it would be very rare for a complex, expensive process to be duplicated intentionally. Usually, important modifications in product or process are incorporated into later versions of the same equipment. Therefore, one cannot expect the control schemes of such processes to be directly transferable, and the design of the control scheme must be evaluated for each example of a complex process.

That a significant gap exists between available control techniques and usual practice is due not only to the fact that practicing engineers are frequently unaware of the many possibilities available, but also to pressure to use simple and generalized control system solutions. Design, maintenance, operation, understanding, and many other related conditions, in addition to basic principles and theoretical possibilities, must all be considered when one is choosing a control system design.

Large industrial corporations often have in-house groups for instrumentation and process control engineering, which can be assigned to the development of new projects. Such groups have the opportunity to apply extensive long-range planning and effort to the special processes and control problems within the company.

Nevertheless, many large companies that have their own process and control engineering groups hire outside engineering companies to plan and design new systems in those areas where they do not have satisfactory in-house competence. In this case, the control system must be included as an integral part of the total process design. The companies' own process and control engineers will participate in these projects to some extent to be sure the outside engineering firm meets the needs and wishes of the company in terms of control system performance and maintenance routines, and to gain insight into the system. In many such cases the company specializing in instrumentation and control systems will deliver the system installed and ready to run, a so-called turn-key installation. Prior to passing operating responsibility to the customer, these companies will spend some time operating, adjusting, and testing the newly installed system. After a certain period, the resident control engineers will have the opportunity to evaluate the system. Unfortunately, it often becomes apparent that the outside firm has provided a system that is not satisfactory, and the resident control engineers must spend a great deal of time and effort to improve the new control system. This problem, which unfortunately is not uncommon in industry, is difficult and time-consuming, but development of computer-aided-design packages for both processes and control systems can be expected to improve this situation dramatically.

Chapter 2
AUTOMATIC CONTROL:
THEORY AND STRUCTURES

2.1. INTRODUCTION

The theoretical basis for the study of the *behavior* of control systems (analysis) and for the *design* of control systems (synthesis) can be traced well back into history, perhaps as far back as the "South Pointing Chariot" of China in 900 B.C. However, it was not until the period following World War II that the field of automatic control could be said to have become a profession in its own right. To be recognized as a profession, a field of activity usually must develop through education, performance, recognition by national and international engineering societies, and the creation of new scientific journals devoted to the specialty. Today there is so much activity in the theory and application of automatic control (literature, conferences, etc.) that it is impossible for one to be knowledgeable in all areas. Although the number of engineers and scientists interested in the field has grown very rapidly in the postwar period, the growth of significant knowledge and technology has been characterized by quantum jumps caused by important specific and infrequent individual contributions. The digital computer, and perhaps in the future the development of computer-aided design for control systems, should help to generate a more uniform growth curve in the theory and application of automatic control.

This chapter presents an overview of the principles of automatic process control. We will not discuss the very large body of knowledge that applies specifically to the design of servomechanisms, but has less significance to process control. This chapter summarizes the most important concepts and techniques of basic process control theory. Further discussion can be found in one or more of the standard texts (Gibson, 1963; Luyben, 1973; Deshpande and Ashe, 1981; Balchen, 1984a; Stephanopolous, 1984; Smith and Corripio, 1985; Franklin, Powell, and Emami-Naeini, 1986).

We will present a discussion of how feedback and feedforward can be used for the control of most elementary single variable (SISO) processes, and how these structures can be further extended to cascade control, parallel control, and control based upon models derived from first principles such as mass and energy balances.

Multivariable processes, which have more than one input manipulated variable (drive), more than one output variable, and usually significant interactions between variables, can, in the simplest form, often be considered as groups of independent systems, each controlled by a system based upon the theory used for single variable processes. It is also possible to handle such systems with one of many available multivariable techniques. One such method is decoupling, which reduces the interactive coupling between variables and uses a diagonal control structure. Another method, often referred to as modal control, is based on a multivariable feedback coupling that places the closed loop eigenvalues at the desired locations in the complex plane. The method that has received the most attention, perhaps, is based upon the use of optimal control theory and leads to a multivariable feedback system that minimizes (usually) a quadratic objective functional. In cases where it may not be possible to measure all of the state variables of the system, it is possible to estimate them by means of a state estimator such as the Kalman filter.

In systems where the process parameters change over a wide range, the control system is often unable to operate satisfactorily. When this is the case, it may be reasonable to use an adaptive control scheme. A typical example of such a process is one that is highly nonlinear.

Because it is possible to influence only the *future* behavior of a process, and because this behavior is a result of both past and future disturbances, one could improve control if one could *predict* the nature of the most important disturbances. A system that does this is known as a *predictive* control system.

A wide range of control equipment can be used for the practical implementation of the various control structures mentioned above. For example, each structure could be realized by the use of analog equipment consisting of either pneumatic or electronic elements. Another alternative is to perform the functions with digital equipment, either individual microprocessors or a larger central computer operating in the real-time mode. Regardless of which physical realization is used, it is the control system's mathematical *structure,* rather than hardware, that is common to all and is of the greatest importance from the point of view of the objectives of this book.

In order to describe control system structures, it is convenient to use block diagrams and process flowcharts on which symbols represent the various elements. Unfortunately, different professions and areas of engineering use different sets of symbols. Therefore, there is no standard set that is comfortable to all. In this book we will try not to be dogmatic about notation and symbolism, but we will try to use symbols that are clear and self-explanatory. Naturally, we will use a common set when possible, and because this is a book about process control structures, we will generally use the symbols recommended by the ISA (Instrument Society of America, 1981).

In the discussion of control structures, one often hears of "new" structures as alternatives to "conventional" structures. In the literature one can find many structures that have been developed for specific important industrial systems. These discussions usually include arguments attempting to show that the new structure is superior to the old.

It probably is not possible to define a "best" structure in general, as "best" depends upon many things, not the least of which is the purpose of the process. For example, a technical definition of "best" might be based on an *objective functional* for the process, which includes the economics of the process and the cost of raw materials as well as the process behavior. It might also be argued that less technical aspects should be considered. For example, one could argue that operators dislike control systems they do not understand, the process should use "some other" variables for control, the measuring points should be changed, and so on. Therefore, an evaluation of what is "best" should be made in such a manner that the conditions applicable to the different control structures are similar. Unfortunately, this is difficult to do, so one can reasonably expect that the "best" structure discussion will continue well into the future!

Because this is a book about structures in process control, it is natural that we examine both conventional structures and others that can be based on either theoretical developments or practical experience. We will attempt as far as possible to provide an objective discussion of the pros and cons as they relate to the many different problems and solutions presented here.

Although the literature regarding structures is about as old as industrial process control itself, there has been remarkable systematic growth in the field since the mid 1960s (Buckley, 1964; Shinskey, 1979; Lee and Weekman, 1976; Govind and Powers, 1978; Morari, Arkun, and Stephanopoulous, 1980). One should expect this growth to intensify still further as systems for computer-aided design become more common in engineering firms and in the process industries' own development groups. This will make it possible to test and compare more theoretically based solutions with other structures. The most practical means of comparing various structures usually is simulation.

Although there are many simulation techniques, in this book we refer only to mathematical simulation, or mathematical models. Since it is virtually impossible to separate mathematical modeling from the design of modern control structures, a word about modeling is appropriate.

Many systems can be modeled on an essentially deterministic basis; that is, system or process behavior can be closely described by material and energy balances or other parameters that may be rather easily determined. Other systems that are not well defined or understood may have to be modeled on the basis of input–output data, which lead to a "black box" model of system behavior. Most models fall someplace between these extremes. In any case, a good model must describe the matters of interest (free of extraneous components that may increase complexity to the point of making the model useless), and it must be just precise enough to meet the objectives (for control, a useful model can often be quite simple and yet very effective). Development of useful models requires a skilled combination of process understanding, mathematical facility, and engineering sense. Discovering process errors, modifying processes, testing control algorithms, and so on, on the basis of competent modeling is far more efficient than doing so after the process has been built!

Of all the methods that provide a systematic derivation of a control system structure, the method referred to as optimal control is without doubt the most generally applicable. This method requires a mathematical model of the process, and that the desired process behavior be described quantitatively in a set of one or more objective functions. Thus, the problem is to formulate a model and an objective function(s) leading to a control system that provides a practical optimum with respect to system complexity, maintenance, function, and so forth. Simulation is necessary when making the transition from the theoretical to the practical optimal solution.

Most of the work reported in the literature on "optimal control" uses quadratic object functions, mainly because it leads to a linear control system for a linear process. However, objective functions that include process and plant economics are more realistic in the case of process control, and one should expect considerable development in this direction in the future (Balchen and Aune, 1966; Balchen, 1984b).

One of the frontiers in control research is the development of "expert systems." These systems use combinations of the many sophisticated control techniques now available to provide control algorithms that have the ability to function under abnormal operating conditions. These schemes must be able to recognize such conditions and decide what corrective action is necessary.

As a final caveat, let us emphasize that regardless of the control structure or algorithm used, it must be robust; that is, it must function properly despite process and/or modeling errors. Further, it must do so over the full range of expected process operation.

2.2. SINGLE VARIABLE PROCESS WITH FEEDBACK

The most elementary and important structure in automatic control is shown in Figure 2.2.1a. This is a process equipped with a measuring element that gives a

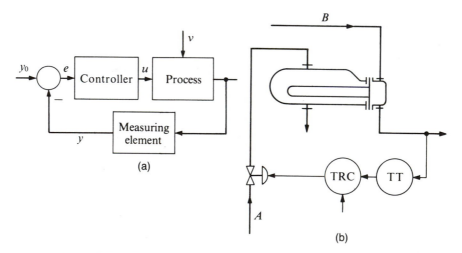

Figure 2.2.1. Two methods of diagramming an elementary process.

measurement signal y, where the process is forced by a *manipulated variable u.*
The measurement value is compared with a *set point* (reference) and the difference, the *deviation e,* drives the controller, which produces the value of the
variable *u.* This system is characterized by *feedback.* In Figure 2.2.1b, the
same system is shown in terms of the symbols typically used in process control.
The figure shows a heat exchanger where medium *A* (steam, for example) is
used to heat medium *B,* which flows through the tube inside the shell containing medium *A.* The problem here is to control the temperature of medium *B*
leaving the exchanger by controlling the flow of medium *A* by means of the
control valve. The temperature of medium *B* is measured and sent by means of
a temperature transmitter (TT) to the temperature recorder controller (TRC).
The simplest way to analyze the system in terms of its control performance is
to describe the components by means of transfer functions (Luyben, 1973;
Balchen, 1984a; Stephanopoulous, 1984).

The process response, if there is no feedback, can be described by the
equation:

$$y^{wo}(s) = h_u(s)u(s) + h_v(s)v(s) \tag{2.2.1}$$

where:

$h_u(s) =$ the process transfer function between the control variable *u* and the
measurement *y*

$h_v(s) =$ the process transfer function between the disturbance *v* and the measurement *y*

When the process does have feedback, the transfer function relating the response to the measurement will be:

$$y^w(s) = \frac{h_0(s)}{1 + h_0(s)} y_0(s) + \frac{h_v(s)}{1 + h_0(s)} v(s)$$

$$= M(s)y_0(s) + N(s)h_v(s)v(s) \tag{2.2.2}$$

where:

$$h_0(s) = h_c(s)h_u(s) : \text{loop transfer function} \tag{2.2.3}$$

$$M(s) = \frac{h_0(s)}{1 + h_0(s)} : \text{set point tracking ratio} \tag{2.2.4}$$

$$N(s) = \frac{1}{1 + h_0(s)} : \text{control ratio} \tag{2.2.5}$$

For the controller to be effective when the system is disturbed by *v(s),* the
control ratio N(s) must be numerically small. This means that $|h_0(s)| \gg 1$. Cor-
respondingly, the closer the *tracking ratio M(s)* is to 1, the better the process
will be able to follow changes in set point. In order to analyze the frequency

response of a linear (or linearized) system, we replace the operator s in the above equations with $j\omega$, where ω is the frequency of an impressed sinusoid. The functions $h_0(j\omega)$, $M(j\omega)$, and $N(j\omega)$ can be displayed graphically where their amplitudes and phase angles are plotted with respect to frequency. It is customary to present the amplitude as:

$$20 \log |h(j\omega)| = |h(j\omega)| \text{ db} \qquad \text{(db = decibel)}$$

and the phase angle as:

$$\angle h(j\omega) \qquad \text{(degrees or radians)}$$

with both plotted as functions of $\log j\omega$. This plot is known as a Bode plot (or Bode diagram). Figure 2.2.2a shows a Bode plot in which the amplitude and phase angle for $h_0(j\omega)$ are plotted against $\log j\omega$ for a hypothetical case. Figure 2.2.2b shows the corresponding phase angle for $h_0(j\omega)$. Figures 2.2.2c and 2.2.2d illustrate the amplitudes and phase angles for $M(j\omega)$ and $N(j\omega)$ for the same system.

The Bode plots in Figure 2.2.2 can be constructed either by relatively simple sketching techniques or on the basis of complete computation. The techniques are well described in most textbooks on elementary process control (Luyben, 1973; Balchen, 1984a).

The complex functions $M(j\omega)$ and $N(j\omega)$ are most conveniently determined graphically by means of a Nichols chart. Figure 2.2.3 shows how $M(j\omega)$ is determined from a plot of $|h_0(j\omega)|$ (dB) vs. $\angle h_0(j\omega)$ (degrees). The figure contains two sets of contours, one for the determination of $|M(j\omega)|$ (dB) and one for $\angle M(j\omega)$ (degrees). The frequency ω is a parameter along the locus describing $h_0(j\omega)$. Thus, for every ω one can find values of $|M(j\omega)|$ (dB) and $\angle M(j\omega)$ (degrees).

A locus of the $h_0(j\omega)$ of the system shown in Figure 2.2.2 is plotted on Figure 2.2.3.

The function $N(j\omega)$ of equation (2.2.5) is determined from a similar chart that in fact is the Nichols chart of Figure 2.2.3 mirror-imaged around both the $\angle h(_0)$ and $|h_0|$ axes. This chart, often referred to as the "inverted Nichols chart," is shown in Figure 2.2.4. The $h_0(j\omega)$ of Figure 2.2.2 is plotted as an example.

The most common form of control function (algorithm) is the so-called *PID controller*. This designation is often used even if the controller has only the P mode (proportional control), the PI modes (proportional + integral controller), or the PD modes (proportional + derivative controller) rather than all three PID modes (proportional + integral + derivative controller). The controller required depends upon the objectives of the control system and the transfer functions of the process to be controlled. It is important to know how many integrating elements the *process* has in order to determine whether the controller should have integral action.

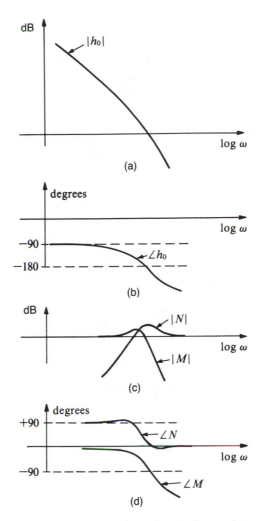

Figure 2.2.2. Frequency response functions characterize an elementary control loop.

In the transfer function

$$h_u(s) = \frac{K(1 + a_1 s + \dots)}{s^p(1 + b_1 s + \dots)}$$

p is the number of integrations.

The transfer function for a PID controller has the form:

$$h_R(s) = K_p \frac{1 + T_i s}{T_i s} \frac{1 + T_d s}{1 + \alpha T_d s} \qquad (2.2.6)$$

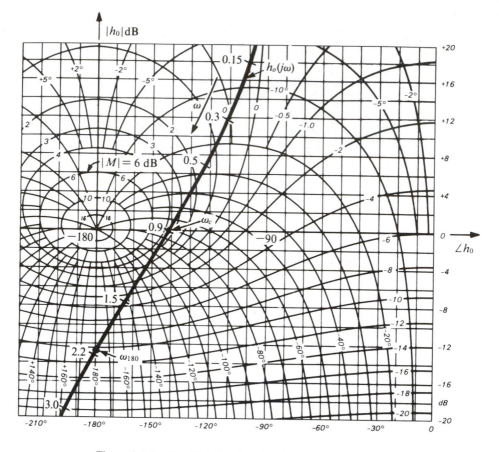

Figure 2.2.3. The Nichols chart for determination of $M(j\omega)$.

The factor $(1 + \alpha T_d s)$ in the denominator (where $\alpha < 0.1$) is there because one usually cannot achieve derivative action for more than about one decade along the frequency axis.

Tuning of PID controllers (the determination of the controller parameters K_p = gain = 1/proportional band, T_i = integral time, and T_d = derivative time) can be done with the help of any of several different methods. Among the common techniques are the Ziegler-Nichols method and one of "organized trial and error" (Luyben, 1973).

Other forms of controller transfer functions can be developed, especially if the controller is to be realized in discrete form. In the discrete case the transfer functions must be expressed in either the z or w domain [$h_c(z^{-1})$ or $h_c(w)$]. These transforms are discussed in the literature (for the z operator see, for example, Luyben, 1973; for the w transform, see Appendix A).

The control solution shown in Figure 2.2.1 is satisfactory in most simple cases, especially when the process dynamics are such that $\angle h_u(j\omega) > -180°$

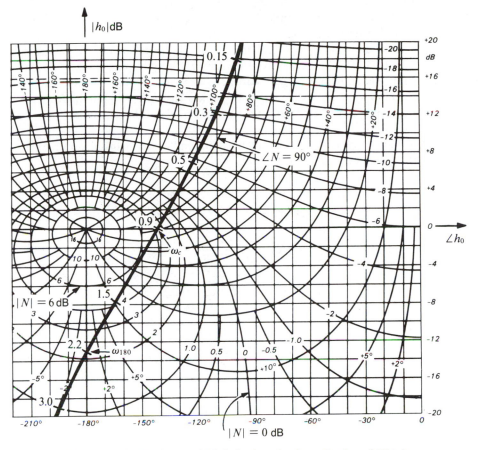

Figure 2.2.4. The inverted Nichols chart for determination of $N(j\omega)$.

over the range of frequencies in which the control system is to be effective. When this condition is met, one can expect satisfactory control with a simple feedback controller. If this condition is not met, then feedback control alone will not be satisfactory. In the latter case, one must use other techniques, such as feedforward control (see section 2.3), move the measurement positions within the process, or make changes in the process itself to improve the speed of response.

Most processes subject to automatic control have one or more external disturbances for which the control system must provide compensation or correction. Some of these can be very slow, others very rapid. The response of the controller and the system depends, of course, upon how the disturbances affect the process. Equations (2.2.1) and (2.2.2) clearly show the suppression of disturbances to processes with feedback control. The control ratio, $N(s)$, expresses the ratio between the response *with* feedback to the response *without* feedback. Many processes, however, are *self-regulating* (disturbances are highly attenuated) and require very little automatic control except, perhaps, at very low frequencies (which represent long response times).

The process shown in Figure 2.2.1 is very simple. A detailed model of virtually any real industrial process, even a relatively simple one, is likely to be very complicated. Therefore, the control engineer usually faces a situation in which a control system must be designed on the basis of a simplified description (model) of a complex process. One could say about this situation:

It is often better to find an acceptable solution than to be unable to find the perfect solution.

For single-variable systems, one can often use one of the following transfer functions as a satisfactory approximation of a linearized dynamic process:

$$h_1(s) = \frac{Ke^{-\tau s}}{s^p(1 + T_1 s)} \tag{2.2.7}$$

$$h_2(s) = \frac{Ke^{-\tau s}}{s^p(1 + T_1 s)(1 + T_2 s)} \tag{2.2.8}$$

$$h_2(s) = \frac{Ke^{-\tau s}(1 + T_3 s)}{s^p(1 + T_1 s)(1 + T_2 s)} \tag{2.2.9}$$

$$h_4(s) = \frac{K(1 - e^{-\tau s})}{(1 + T_1 s)\tau s} \tag{2.2.10}$$

$$h_5(s) = \frac{K \cdot e^{-\tau s} \cdot T_2 s}{(1 + T_1 s)(1 + T_2 s)} \tag{2.2.11}$$

The factor s^p that appears in three of the above expressions will often be equal to 1; that is, $p = 0$. This is the case when the process has no integration, and it will usually apply in the following discussion.

Of the above forms, h_1 is the simplest and is perhaps the most often used. When $p = 0$, the function is characterized by a dominant time constant T_1, a steady-state gain K, and an equivalent delay (transport lag, distance velocity lag, dead time, etc.) τ. This function is known as the *First Order Plus Dead Time* (FOPDT) model. The apparent transport delay will often actually consist of a combination of physical phenomena. However, the process might be distributed spatially, in which case it actually takes time for the transmission of a transient signal. The equivalent transport delay τ can seldom be predicted on the basis of the physical situation. Therefore, it usually must be determined experimentally.

The advantage of the transfer function $h_1(s)$ is that it is very simple and contains only three parameters (K, T_1, and τ). It is usually easy to obtain values for these parameters, either by rough calculation or by treating experimental data with analytical tools ranging from simple graphical methods to sophisticated mathematical estimation techniques.

The transfer function $h_2(s)$ is somewhat more complicated than $h_1(s)$ and is useful if the process has two time constants of about the same value. As with $h_1(s)$, $h_2(s)$ has an equivalent transport delay and a steady state gain K. This model is usually known as the *Second Order Plus Dead Time* (SOPDT) model, when $p = 0$.

The transfer function $h_3(s)$ is used when the process can be characterized by two first order dynamic phenomena that have a strong two-sided interaction with each other (expressed by the term $[1 + T_3 s]$ in the numerator). For this model, five parameters must be determined.

The factor $e^{-\tau s}$, which appears in these three transfer functions, can, when τ is large, be responsible for most of the phase angle $\angle h(j\omega)$ for the corresponding frequency response for each transfer function. Because $\angle e^{-j\omega\tau} = -\pi$ when $\omega = \pi/\tau$, a control system having only feedback cannot possibly be effective when the frequency is higher than $\omega = \pi/\tau$.

The transfer function $h_4(s)$ can be used in certain situations, for example, the description of a countercurrent heat exchanger, but it usually produces the same result as the transfer function $h_2(s)$ with $p = 0$.

The transfer function $h_5(s)$ is characterized by a zero at the origin in the s plane. This means that the steady-state response is zero (the phenomenon is transient). This situation can occur in certain processes; for example, in connection with mass balances on the plates of a distillation column. It is important to recognize a zero at the origin.

Two examples will illustrate the use of the models given above.

▽ EXAMPLE 2.2.1

Assume that the process dynamics can be represented by the transfer function:

$$h_{u1}(s) = \frac{Ke^{-\tau s}}{1 + T_1 s}$$

If τ/T_1 is not less than about 0.1, it will not be satisfactory to use a simple proportional controller for the process. Further, use of derivative control to compensate for the phase shift caused by the dead time will not be very effective. The simplest controller algorithm that can be used in this case is the PI controller having the transfer function:

$$h_c(s) = K_p \frac{1 + T_i s}{T_i s}$$

We then have the loop transfer function:

$$h_0(s) = \frac{K_p K(1 + T_i s)e^{-\tau s}}{T_i s(1 + T_1 s)}$$

We see that in this case a suitable integration time for the controller will be $T_i = T_1$. This situation corresponds to a process that can be characterized by a transport delay and a controller having pure integration.

For this system we can easily find the frequency ω_{180} where the phase angle is $-180°$. Then:

$$\angle h_0(j\omega_{180}) = -\frac{\pi}{2} - \omega_{180}\tau = -\pi$$

which gives:

$$\omega_{180} = \pi/2\tau$$

If we want this control loop to have a gain margin of $\Delta K = 2 = 6$ db, we must choose the gain parameter such that:

$$K_p K = (\pi/4)(T_1/\tau)$$

This gives the crossover frequency:

$$\omega_c = \omega_{180}/2 = \pi/4\tau$$

which is the frequency at which $|h_0| = 1 (= 0$ db$)$.

The upper limit of the control system bandwidth is determined by the transport delay and the fact that the controller gain is proportional to the ratio T_1/τ. The development given above leads to $|N(j\omega)|_{max} \approx 7$ db. This is acceptable in most cases, but it depends somewhat on the frequency distribution of the disturbances. The resonance peak can easily be reduced by slightly reducing K_p.

We can now draw the important conclusion that the control system can only be effective [that is, $|N| < 1 (= 0$ dB$)$] for frequencies lower than approximately $\omega_c = \omega/4\tau$. A corresponding expression for the control ratio in which the controller will be effective can be seen to asymptotically approach $|N(j\omega)| \approx \frac{4}{\pi}\omega\tau$. If we want the control system to suppress a disturbance by a factor of 10, then the control ratio must be $|N| \leq 0.1 (-20$ dB$)$. This means that the frequency of the disturbance must be $\omega \leq (\pi/40)(1/\tau)$. If the disturbance is sinusoidal with a period of $T_v = 2\pi/\omega$, then the control ratio will be:

$$|N(j\omega)| \approx 8 \cdot \frac{\tau}{T_v}$$

Therefore, if we want a control ratio of $N = 0.1$, the period of the disturbance T_v must be greater than 80 times the delay time τ!

This discouraging result clearly shows how poorly suited a simple feedback
△ controller is for processes having a significant transport delay.

▽ EXAMPLE 2.2.2

If the process transfer function is of the type given by equation (2.2.8), with $p = 0$, the result will be essentially the same as that found in the previous example if $T_2 \ll T_1$. However, if T_2 is of the same order of magnitude as T_1 (let us say $T_2 \geq 0.1\ T_1$), the result obtained with a PI controller will be somewhat different.

If we use a PID controller having the transfer function given by equation (2.2.6), we could choose $T_i = T_1$ and $T_d = T_2$. This will then give essentially
△ the same result as that obtained in example 2.2.1.

The following observations sum up the behavior of elementary feedback controllers applied to a single-variable process:

- The process dynamics determine how effective the control system can be, especially if there is a significant transport delay in the process.
- If the control system must have good accuracy at low frequencies, it must have integral action.
- Derivative action can, to some degree, eliminate the influence of short time constants in the process (e.g., those related to the measuring elements).
- The final closed loop system control ratio $N(j\omega)$ must be viewed in relation to the relative amplitudes of the disturbances over the frequency spectrum. This means that in order to determine the amplitude of response at any single frequency, one must multiply the disturbance amplitude at that frequency by the value of $|N(j\omega)|$.
- If simple feedback control does not give satisfactory control of a single-variable process, then a more complicated structure such as feedforward or cascade control may be necessary.

The systems shown in Figure 2.2.1b are examples of the simplest type of control with feedback. The control variable (u) consists of the heat flow in a medium A that flows through the control valve on the primary side of the heat exchanger. However, it is the opening of the valve that really is controlled, and the actual energy flow depends upon several factors. If A is a liquid, the flow of the liquid will depend upon the pressure drop across the valve. This in turn depends upon the flow resistance of the heat exchanger and the pressure ahead of the valve. If the flow leaves the heat exchanger at constant pressure, atmospheric for example, then the net upstream pressure will represent a potential disturbance to the flow. In addition to the liquid flow, the liquid temperature will influence the net heat flow to the heat exchanger. Therefore, temperature also is a process disturbance. A third possible disturbance, which can be significant in some cases, is a change in specific heat of the liquid. The secondary medium will be heated (or cooled) as a result of heat transfer from (or to) medium A. The object of the control system could be to maintain a constant temperature of medium B leaving the heat exchanger. A very important process disturbance would be a change in the heat content of medium B entering the heat exchanger. The heat content of this stream is a function of the mass flow, temperature, and specific heat. All of these must be considered to be important process disturbances. In addition the stream velocity of medium B affects the heat flow into and out of the system because the heat transfer coefficient be-tween the exchanger tubes and medium B is related to the degree of turbulence in the flow. The same is true of the flow on the primary side of the exchanger.

In addition to the disturbances within the system, the process will be influenced by the temperature of the surrounding environment and the heat transfer coeffi-cient between the heat exchanger and the environment, a factor that depends on the degree of insulation of the exchanger system. There is also usually a long-term disturbance in the form of a gradual change in the values of the heat trans-fer coefficients, due primarily to scale buildup on the walls of the tubes.

We can now summarize the many disturbances that can affect even this simple process. A convenient form for expressing this list of process distur-bances is a vector, which we designate as v:

v_1: Pressure in the primary medium A ahead of the control valve.
v_2: Pressure in the primary medium A downstream of the control valve.
v_3: Specific density of the primary medium.
v_4: Specific heat of the primary medium.
v_5: Temperature of the primary medium.
v_6: Volume flow of the secondary medium (B).
v_7: Specific density of the secondary medium.
v_8: Specific heat of the secondary medium.
v_9: Temperature of the secondary medium entering the heat exchanger.
v_{10}: Temperature of the environment.
v_{11}: Heat transfer coefficient.

Each of these disturbances can have different levels of significance, depend-ing upon the role of the heat exchanger in the total process. Some of the distur-bances can change very quickly (v_1, v_2, v_6), some more slowly (v_5, v_9, v_{10}), and some very slowly (v_{11}). Some of them represent a linear additive effect (v_5, v_9, v_{10}), and others are multiplicative and therefore nonlinear in their effect (v_1, v_2, v_3, v_4, v_6, v_7, v_8, v_{11}). A linear analysis of the influence of the factors that are multiplicative must be done by means of linearization around some stationary operating point. Usually, the accuracy of such a linearization is satisfactory.

The control structure shown in Figure 2.2.1b has the temperature measuring device placed in the secondary medium (B) at the point where the flow leaves the heat exchanger. The temperature controller (TRC: temperature recorder-controller) that operates the control valve will sometimes not be "clever" enough to give satisfactory control performance. The system can be significantly im-proved by the use of techniques that will be discussed in later sections of this chapter. However, if one decides to try this solution, it is likely that a PID con-troller will be the best choice for the following reasons:

• Integral action is necessary in order to get good accuracy under stationary conditions (low frequencies).

- Derivative action is desirable in order to counteract the dynamics of the temperature measuring element (see example 2.2.2).

For a countercurrent heat exchanger of the type shown in Figure 2.2.1b, it is often advantageous to use the transfer function given by equation (2.2.10), in which the term

$$\frac{1 - e^{-\tau s}}{\tau s}$$

reflects the fact that the secondary medium remains in the heat exchanger for a certain time τ. At low frequencies, however, this transfer function can be approximated as:

$$h_4(s) = \frac{K(1 - e^{-\tau s})}{(1 + T_1 s)\tau s} \approx \frac{K}{(1 + T_1 s)(1 + T_2 s)} \qquad (2.2.12)$$

where $\tau/2 < T_2 < \tau$.

2.3. FEEDFORWARD CONTROL

Feedforward control can be a very effective means of improving the dynamic response of a control system when simple feedback is not satisfactory. Feedforward control assumes that it is possible to measure some of the major disturbances that drive the process away from the desired state. If one knows how a process will respond to a disturbance, one can, in principle, generate a control signal that will compensate for the predicted response to a disturbance before it occurs, thereby holding the process at the desired state. In order to design an effective feedforward control system, one must carefully study the dynamics of the process; that is, develop a mathematical model of the process. Sometimes this is quite simple; other times, it is virtually impossible!

An important advantage of feedforward over feedback is that there are no related stability problems, which often are the limiting factors in feedback systems. Feedforward control can be used in both single- and multivariable systems.

Let us consider a single-variable process that can be linearized and described by the simple model given in equation (2.2.1). This model can be generalized to take account of the fact that the process can have many disturbances:

$$y(s) = h_u(s)u(s) + \mathbf{h}_v^T(s)\mathbf{v}(s)$$

$$= h_u(s)u(s) + \sum_{i=1}^{p} h_{vi}(s)v_i(s) \qquad (2.3.1)$$

Note that if we have just one disturbance, this expression reduces to that of equation (2.2.1). Although we will examine the case for a single disturbance, the solution for the case of multidisturbances can be solved by adding feedforward controls for each disturbance.

If we want to use feedforward control of the control variable $u(s)$ to compensate for the disturbance $v(s)$ so as to force the response $y(s) \rightarrow 0$ in Figure 2.3.1, we must use the feedforward transfer function:

$$h_{Fi}(s) = -\frac{h_v(s)}{h_u(s)} \tag{2.3.2}$$

The transfer function $h_{Fi}(s)$ is referred to as the "ideal feedforward" function.

If we are not able to realize the "ideal feedforward" transfer function, and there may be many reasons why we cannot, we must insert the actual function $h_F(s)$ into the equation for the response:

$$
\begin{aligned}
y(s) &= h_u(s)u(s) + h_v(s)v(s) \\
&= (h_u(s) \cdot h_F(s) + h_v(s))v(s) \\
&= \left(-\frac{h_F(s)}{h_{Fi}(s)} + 1\right)h_v(s)v(s) \\
&= (g_0(s) + 1)h_v(s)v(s) \\
&= L(s)h_v(s)v(s) \tag{2.3.3}
\end{aligned}
$$

where $L(s)$ refers to the *feed forward control ratio* and has the same significance in the feedforward system as the *control ratio* $N(s)$ has in the feedback system. From the value

$$g_0(s) = -\frac{h_F(s)}{h_{Fi}(s)}$$

which expresses the deviation between the actual feedforward function and the ideal, one can (for example, with a Nichols chart) easily determine $L(s)$.

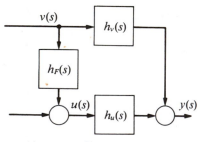

Figure 2.3.1. Feedforward in a single-variable process.

If one interprets feedforward in the frequency domain, the value of $|L(j\omega)|$ is a measure of the reduction of response to a disturbance at different frequencies. Figure 2.3.2 is a graphical determination of $|L|$ as a function of $|g_0|$ and $\angle g_0$ (Balchen, 1984a). We see that in order to reduce the response to 0.1, that is, $|L| = 0.1 \ (= -20 \text{ dB})$, the matching of the actual feedforward to the ideal must be within 0.5 to 1 dB in amplitude and within an angle of less than 5°. This is often difficult to obtain.

For combined feedforward and feedback control of the total process, illustrated in Figure 2.3.3, we have a response given by:

$$y(s) = M(s)\,y_0(s) + N(s) \cdot L(s)h_v(s)v(s) \qquad (2.3.4)$$

where:

$$M(s) = \frac{h_0(s)}{1 + h_0(s)} = \text{tracking ratio}$$

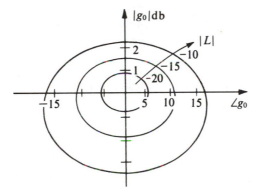

Figure 2.3.2. Graphical determination of feedforward control ratio.

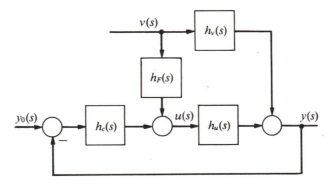

Figure 2.3.3. Combination of feedforward and feedback control.

$$N(s) = \frac{1}{1 + h_0(s)} = \text{control ratio}$$

$$L(s) = (g_0(s) + 1) = \text{feedforward control ratio}$$

$$h_0(s) = h_c(s)h_u(s) = \text{loop transfer function}$$

$$g_0(s) = -\frac{h_F(s)}{h_{Fi}(s)} = \text{feedforward accuracy}$$

Examining equation (2.3.4) in the frequency plane, we see that the most important contribution of feedforward is to reduce the response to those disturbances that cannot be effectively suppressed by feedback; that is, frequencies that are higher than the cross-over frequency ω_c. Control can be significantly improved if the feedforward is adjusted such that $|L|$ is small in the frequency range $0.1\,\omega_c < \omega < 10\,\omega_c$. *There is no particular advantage to feedforward at very low frequencies where feedback is most effective ($|N| \ll 1$).*

Feedforward is especially easy to implement if the transfer function of the control variable $h_u(s)$ has a shorter transport delay (and therefore less phase shift) than the disturbance transfer function $h_v(s)$. If the opposite situation exists with respect to the relative transport delay times, then it is not possible to use feedforward control.

There have been many successful applications of feedforward control to both single- and multivariable systems. Several of them will be described in following sections of this chapter. Many control schemes that have come to be known by various special terms are actually special cases of feedforward control. Typical of this is ratio control, which will be discussed in the next section.

In our discussion of feedforward we have assumed that the process is linear and that the disturbances are additive to the process. In many practical situations this is not the case, including the case of the heat exchanger that was discussed in section 2.2. To illustrate, consider a simplified system having much in common with the heat exchanger discussed earlier. Figure 2.3.4 shows a water-heating system in which an electric current supplies heat at a rate Q to

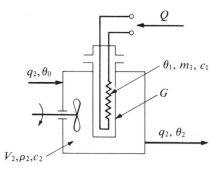

Figure 2.3.4. Ideal electric water heater.

heat water in a tank. Let the water flowrate through the system be q_2. The temperature of the water is θ_0 at the inlet and θ_2 at the outlet. The temperature of the heating element is θ_1. The mass of the heating element is m_1, and the specific heat is c_1. The water in the system has a volume V_2, density ρ_2, and specific heat c_2. The heat transfer coefficient between the heating element and the water is G. Assume that the tank is perfectly mixed and perfectly insulated from the surrounding environment. We can then write the following differential equations for the system:

$$\dot{\theta}_1 = \frac{1}{m_1 c_1}(Q - G(\theta_1 - \theta_2)) \qquad (2.3.5)$$

$$\dot{\theta}_2 = \frac{1}{V_2 \rho_2 c_2}(G(\theta_1 - \theta_2) + q_2 \rho_2 c_2(\theta_0 - \theta_2)) \qquad (2.3.6)$$

If the flow q_2 and the temperatures θ_0 and θ_2 can be measured, then an approximate feedforward strategy can be developed. The quantities q_2 and θ_0 are the only disturbances to this system, in which θ_2 is to be held constant. For θ_2 to be constant, it must be true that $\dot{\theta}_2 = 0$. Furthermore, we may assume that m_1, c_1, and $\dot{\theta}_1 \approx 0$. Using these values in equations (2.3.6) and (2.3.5), we get the heat balance:

$$q_2 \rho_2 c_2(\theta_2 - \theta_0) = G(\theta_1 - \theta_2) = Q \qquad (2.3.7)$$

This equation tells us that if we multiply the measurement of the flow q_2 by the temperature change $(\theta_2 - \theta_0)$ and the constant factor $\rho_2 c_2$, we will get the power necessary to maintain an undisturbed heat balance. This is a typical example of feedforward control.

In the system shown in Figure 2.3.5, this principle is used in combination with an ordinary temperature controller based on measurement of θ_2. If the mathematical model used in equation (2.3.5) is exact, the feedforward scheme will give very good results. If not, then the feedback will compensate for the resulting errors, especially if the controller includes the integral mode (PI or PID).

This control structure can easily be extended to more complicated processes, but at this point we will limit our discussion to feedforward as applied to control of the most important disturbances to the heat exchanger shown in Figure 2.2.1b.

Figure 2.3.6 shows the heat exchanger equipped with additional instrumentation. Consider the case where both heat exchange media, A and B, are liquids. The amount of heat delivered by medium A can be determined by multiplying the measured temperature difference ΔT between the inlet and the outlet by the measured flow of A. For simplicity, assume that the flow measurement is linear. The product is fed to a controller (EC, enthalpy controller) that operates the control valve.

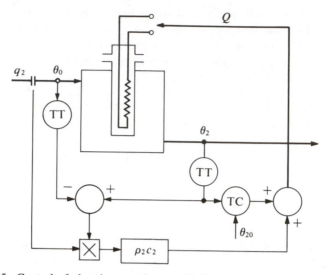

Figure 2.3.5. Control of electric water heater with feedback and feedforward based on energy flow in primary and secondary sides.

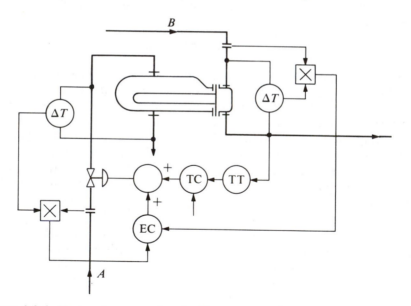

Figure 2.3.6. Heat exchanger equipped with temperature feedback control and energy flow feedforward control.

Similar instrumentation of the secondary side of the exchanger permits computation of the heat flow in medium B. This gives a direct measure of the heat needed from the primary side and can be used as a reference (set point) for the controller EC. This technique gives rapid feedforward control of all of the most important disturbances in the system.

Since we want to control the outlet temperature of the secondary medium (B), we now use the measurement of that temperature as the signal to a temperature controller (TC), which adds its output to the total signal to the control valve. This is the same as in the elementary feedback control system shown in Figure 2.2.1b.

In reality the system shown in Figure 2.3.6 contains a *cascade* control loop, which will be discussed in section 2.5. With this we will significantly reduce the effects of other important process disturbances such as a change in the stream pressure on the primary side. The change in the flowrate will immediately be detected by the flow meter. This sends a signal that ultimately is used to operate the control valve. The only remaining uncontrolled disturbances are the temperature changes of the environment surrounding the process and the changes within the exchanger, such as those due to tube fouling (scale buildup). An ordinary temperature control system will compensate for these changes because they are very slow compared with the other process dynamics.

The system shown in Figure 2.3.6 is obviously more complicated than the system of Figure 2.2.1b. We now have six measuring elements instead of one, and two controllers instead of one, plus two multipliers and three add–subtract elements. The additional equipment is expensive in itself, but perhaps even greater are the costs of installation and maintenance. Therefore, it is important to balance the control advantages of this system against the disadvantages of increased complexity and cost.

2.4. RATIO CONTROL

Ratio control is a special control structure closely related to feedforward control. The purpose of a ratio control system is to maintain a desired ratio between two or more process variables. For example, a process might require a certain ratio of the mass flows of two components in order to maintain the stoichiometric balance in the system. Or, the system might be nonreacting, such as a ratio of dye to raw material in a papermaking or textile process. Because this situation is very common in industry, ratio control is one of the most frequently used special control solutions.

Figure 2.4.1 illustrates a very simple application of ratio control. Two streams (A and B) must be metered at a certain ratio so that the mixture has the desired characteristics. For example, stream A might be an acid that is neutralized by a base (alkali) B in order to produce a neutral mixture having a pH of 7. Stream A varies independently, perhaps as a result of previous processing, and stream B

is to follow stream A at a fixed ratio. This can be done according to Figure 2.4.1 if the linear measurement of the flowrate of A, multiplied by a constant factor k, is used as the set point for a flow controller (FC) in the B branch. Control of the flow in the B branch is realized by means of the linear measurement of the flow q_B and the control valve. The relation between the two flows is determined by $q_B = k \cdot q_A$.

It should be recognized that this scheme, while providing excellent ratio control, will not necessarily give the desired product unless the compositions of streams A and B are constant. If these compositions vary, it is necessary to provide a control system based on measurement of the final product by means of some type of analytical instrumentation. For example, one can use a pH measurement to provide a signal to a controller (AC) that can adjust the factor k in the ratio controller. Such a scheme gives a precise and quick response to variations in the component concentrations of the feed streams.

Figure 2.4.2 shows a similar system, but in this case there is a long delay before the mixture can be analyzed. This makes it difficult for the feedback system to control rapid variations in the concentrations of streams A or B. Therefore, the concentration of each feed stream is measured prior to mixing by means of analyzers A_1 and A_2. The product of the concentrations and the corresponding flowrates gives a measure of the flow of the active components that are to be held at the required ratio. When these disturbances are handled in this manner, the need for rapid control response to the measurement from A_3 decreases substantially.

An important application of ratio control is in combustion processes. The very important ratio between fuel and air will be discussed in detail in section 4.7.

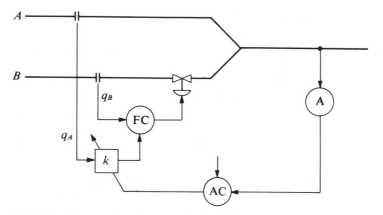

Figure 2.4.1. An elementary ratio control with updating from analysis of the mixed streams.

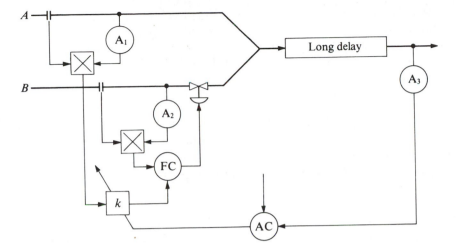

Figure 2.4.2. Ratio control including measurement of concentration of the feed streams and updating from analysis of the mixture.

2.5. CASCADE CONTROL

The principle of cascade control is shown in Figure 2.5.1. Here we have a control variable that has a series of several measurements (three in this case) with cause-and-effect relationships. In many processes there can be a long time lapse between a control variable change and the resulting effect on the output variable (here, y_3) being controlled. Often in such cases it is possible to measure some intermediate variable that can be used in an inner control loop, as shown in Figure 2.5.1. The two inner loops shown in the figure include dynamic feedback elements $h_{T1}(s)$ and $h_{T2}(s)$. The objectives of these loops are to shape the process dynamics. This can be done because the effect of the feedback on the

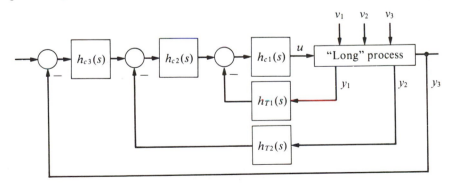

Figure 2.5.1. Cascade control in block diagram form.

system is that it behaves as the inverse of the feedback. This principle is used often in the design of tracking systems (servomechanisms). In process control it is more common to let $h_{T_1}(s)$ and $h_{T_2}(s)$ equal 1, thereby removing the transient feedback. Cascade control has much in common with state feedback, which will be discussed in sections 2.8 through 2.13.

Some of the important advantages of cascade control are:

- Disturbances early in the process will be quickly suppressed by the inner control loops and will not be carried on further through the system.
- Parameter variations and nonlinearities in the early part of the process will be suppressed by the inner control loops, making it possible to achieve much better control in later parts of the process.

We have used cascade control in some of the previous examples. The system in Figure 2.3.6 is a cascade controller in that the set point of the controller EC, which controls the heat content of the primary stream, is controlled on the basis of the heat content in the secondary stream. In Figure 2.4.1 we have cascade control of part of the ratio control system, in that the set point for controller FC is controlled. A similar system is seen in Figure 2.4.2.

Figure 2.5.2 illustrates a quite common cascade control system. The system consists of a valve positioner VC in cascade with flow controller FC, which in turn is in cascade with temperature controller TC. We have seen the latter two in previous examples. The valve positioner is a self-contained servomechanism designed to make the valve move precisely in response to the output from the flow controller (FC).

Many control valves, in fact most of those used in the process industries, are driven by motors in which the motion of a pneumatically driven diaphragm is opposed by a coil spring. Because of dry friction, the stem movement caused by the relative balance of forces between the diaphragm and the spring may not be precise or smooth, especially in the case of larger valves. As a result, the

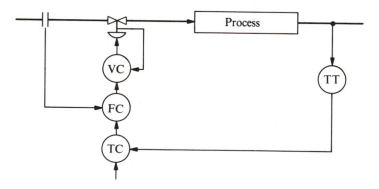

Figure 2.5.2. Elementary cascade structure in process control.

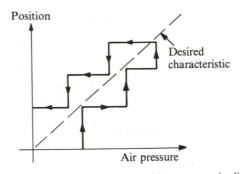

Figure 2.5.3. Hysteresis in a control valve with a pneumatic diaphragm motor.

valve stem motion may exhibit hysteresis or discontinuities. (See Figure 2.5.3.) The problem is often aggravated by the effects of the stream pressure in the valve body. The valve positioner compares the actual movement of the valve stem with the signal supplied to the valve motor, and its internal servomechanism forces agreement between the two. Valves with positioners cost more than ordinary pneumatic valves, but the advantages in terms of performance can be very important. As mentioned above, this is especially true for large valves, but it can be important for any valve that requires tight stem packing because of high stream pressure.

Additional examples of cascade control will be given in Chapters 4 and 5.

A detailed description of the dynamic behavior of cascade control systems can be found in most elementary control texts.

2.6 PARALLEL CONTROL

In many processes more than one control variable can be used to influence a process variable. These variables can often affect the controlled process in different ways. For example, one might cause a very rapid response of the output, while another might cause a longer-term but more precise response. Therefore, it can be advantageous to use more than one control variable (in this example, two) in order to obtain both quick and precise process control. To accomplish this, we use two controllers in *parallel*, both operating from the same process deviation, but each manipulating a different control variable. This arrangement is shown in Figure 2.6.1. If controller $h_{c1}(s)$ is dominated by the proportional mode, and controller $h_{c2}(s)$ is dominated by the integral mode, the parallel effect will be that of proportional + integral control.

A typical version of parallel control is shown in Figure 2.6.2. A portion of a stream of cold liquid is passed through a heat exchanger of relatively large volume. The remainder of the stream bypasses the exchanger and is then mixed with the heated portion. The final temperature, of course, will be between that of the feed and that of the liquid leaving the heat exchanger. The control valve

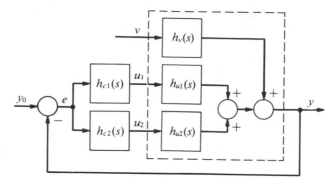

Figure 2.6.1. Block diagram of a parallel control system.

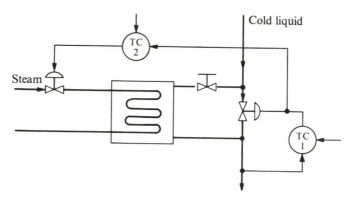

Figure 2.6.2. Parallel control of a heat exchanger.

on the primary side of the exchanger allows only a slow change in temperature on the secondary side. Using a measurement of the temperature of the final mixed stream, a parallel temperature control scheme can be designed using temperature controller TC1, which can be either a P or a PD controller. The output of this controller, which is porportional to the controller deviation, is fed to temperature controller TC2, which controls the valve in the primary flow. This controller must have integral action and can be a PI controller.

It is desirable to operate the control valve in the parallel (bypass) branch at approximately 50% of its maximum value. When the output from TC1 has this value, the input to TC2 will be zero, and the valve on the primary side will not move. Therefore, we can say that the objective of TC2 is to adjust the temperature of the secondary side of the exchanger in such a manner that the valve in the parallel branch always returns to its midpoint. Rapid variations in the temperature of the feed stream will be quickly suppressed by this system, and the temperature of the mixed stream will be able to follow fast variations in the setpoint of TC1. This system can also be designed to use feedforward and/or cascade control, but they will not be considered here.

The fact that only one of the controllers in an ordinary parallel control can have integral action can best be understood by consideration of Figure 2.6.1. Assume that the two process transfer functions are:

$$h_{u1}(s) = \frac{K_1}{1 + T_1 s}$$

$$h_{u2}(s) = \frac{K_2}{1 + T_2 s}$$

Now assume that both controllers are pure integral controllers:

$$h_{c1}(s) = \frac{1}{T_{i1} s}$$

$$h_{c2}(s) = \frac{1}{T_{i2} s}$$

This leads to the block diagram given in Figure 2.6.3. Here, δ_1 and δ_2 represent noise inputs to the integrators. If δ_1 and δ_2 are constants, then the conditions for equilibrium for the system are that $e(t = \infty) + \delta_1 = 0$ and $e(t = \infty) + \delta_2 = 0$. This means that equilibrium can occur only if $\delta_1 = \delta_2$, a situation that is not possible. Therefore, the system will be driven to infinity (in practice, saturation). If one of the integrators is replaced with proportional control, the problem disappears.

Parallel control, as described above, has many well-known applications. One is the speed control (frequency control) of turbines used to drive alternating current power generators that are connected together in the same power network. In order to operate these generators at the same frequency, their individual controllers all use the same control deviation, as shown in Figure 2.6.3. Therefore, only one of the turbine controllers can be equipped with integral control, and all of the others must be proportional controllers (Balchen, 1984a; Stephanapoulos, 1984).

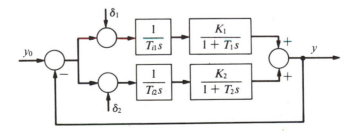

Figure 2.6.3. Parallel control with two integral controllers.

2.7. MODEL-BASED CONTROL

In previous sections we have often referred to the use of mathematical models of process behavior in order to generate suitable control variables. This approach will be used again many times, especially in section 2.12. First let us consider several solutions that can be applied to single-variable systems, but which can also be expressed in multivariable form.

In the examples of section 2.3, we used energy balance calculations as the basis for feedforward control. This principle can also be applied to mass and material balances, momentum balances, and so on. However, sometimes it is difficult to develop a material balance on a dynamic process that has many state variables. If we use the stationary, or steady-state, condition of the process as a basis and consider the net material balance between the input and output of the process (that is, if we ignore the dynamics), it is usually easy to develop expressions for use with feedforward control, as discussed in section 2.3. On the other hand, if we must consider the dynamic material balance, the problem might be treated more easily as a dynamic optimization problem of the type to be discussed in sections 2.10 and 2.15.

The principles for development of feedforward control based on an energy balance given by equations (2.3.5) and (2.3.6) can be generalized on the basis of a process described by the following differential equation:

$$\dot{x} = f(x, u, v) \tag{2.7.1}$$

where x = state vector, u = control vector, and $f(\cdot)$ = vector of nonlinear functions. If we assume that the process can be approximated as steady state $\dot{x} = 0$, and that the steady state is characterized by $x = x^0$, then we can find the control vector that makes the right side of equation (2.7.1) zero. Or:

$$u = p(x^0, v) \tag{2.7.2}$$

The solution given in equation (2.7.2) requires that x^0 and v be available measurements (or calculated values), and that the dimension of u be dim $u = r \le$ dim $x = n$. If $r = n$, there will be as many unknowns as there are equations. If $r < n$, there will be more equations than unknowns, and it will be possible to calculate some of the elements in x^0 [$(n - r)$elements]. That is exactly what was done in equation (2.3.7).

The solution of nonlinear equations of the type given in equation (2.7.1) often requires the use of iterative methods that will not be discussed here. In relatively simple cases where the number of variables r is three or less, one can usually find a simple method of solution. The $(n - r)$ elements of the state vector x^0 that must be known in order to solve equation (2.7.2) must be obtainable either from measurements or from a so-called *state estimator* (see section 2.13).

Equation (2.7.2) describes how the control vector must be manipulated on the basis of the disturbances (v) and the steady-state condition (x^0) so that the process states do not change ($\dot{\mathbf{x}} = \mathbf{0}$). Because the computation of *ideal feedforward control* cannot be expected to be perfect, adjustment of the control vector must also be based on *feedback control,* which uses the actual measurements from the process. One of the main reasons why feedforward cannot be perfect is the fact that perfect knowledge of the various elements in the disturbance is not available. Discussion of this principle applied to the dynamic case ($\dot{\mathbf{x}} \neq \mathbf{0}$) will be presented in section 2.10.

As we saw in equation (2.3.4), the response of a process having both feedforward and feedback control will consist of two terms, one having to do with the process disturbances and the other having to do with changes in process set point (y_0). We will develop a control structure that is especially effective in improving the tracking ability in a single-variable system having a significant transport delay. This is a special case of the method described in section 2.13. The problem is illustrated by Figure 2.7.1, where a control (u) influences a process consisting of two cascade-coupled components having the transfer functions $h_1(s)$ and $h_2(s)$, respectively. The difficulty is due to the fact that a transport delay in $h_2(s)$ will be inside the feedback path if the measurement is taken as indicated by the dashed line in Figure 2.7.1b. In this case, a simple feedback controller would not be very effective.

On the other hand, if one could measure the response of the first part of the process (x in Figure 2.7.1a), simple feedback would be effective because there is assumed to be no significant transport delay in $h_1(s)$ and therefore little phase angle at the frequencies of interest. Under our assumption that x cannot be measured directly, we can use an *estimator* scheme such as that shown in Figure 2.7.1c. This system estimates the values of x based on a model in parallel with the actual process. This estimator is related to the more general *state estimators* that will be discussed in section 2.13. The estimator generates an estimate ($\hat{v}$) of the actual disturbance (v), which we assume affects the process at the same point as the control variable. The estimate is generated by feedback through the transfer function $k(s)$ where the input to $k(s)$ is the difference between the actual process output (y) and the output of the model ($\hat{y}$).

In the system shown by the solid lines in Figure 2.7.1b we have:

$$x = \frac{h_c h_1}{1 + h_c h_1} y_0 + \frac{h_1}{1 + h_c h_1} v \qquad (2.7.3)$$

where for convenience we have chosen to use the notation $x(s) = x$, $h_c(s) = h_c$, and so on. Similarly, we can describe the system of Figure 2.7.1c by:

$$x = \frac{h_c h_1}{1 + h_c \tilde{h}_1} y_0 + \frac{h_1}{1 + h_c \tilde{h}_1} (1 + (\tilde{h}_1 - \overline{h}_1) h_c) v \qquad (2.7.4)$$

$$\tilde{h}_1 = \frac{1}{\beta_1} (1 + \hat{M} Y) h_1 \qquad (2.7.5)$$

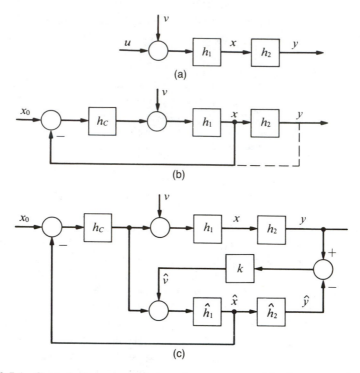

Figure 2.7.1. Control of a process having a large transport delay by means of a model-based "estimator."

$$\overline{h}_1 = \beta_2 \hat{M} h_1 \qquad\qquad (2.7.6)$$

$$\beta_1 = \frac{h_1}{\hat{h}_1} \qquad\qquad (2.7.7)$$

$$\beta_2 = \frac{h_2}{\hat{h}_2} \qquad\qquad (2.7.8)$$

$$Y = \beta_1 \beta_2 - 1 \qquad\qquad (2.7.9)$$

$$\hat{M} = \frac{k\hat{h}_1\hat{h}_2}{1 + k\hat{h}_1\hat{h}_2} \qquad\qquad (2.7.10)$$

If we succeed in making $\hat{h}_1 = h_1$ and $\hat{h}_2 = h_2$, we will have $\beta_1 = 1$ and $\beta_2 = 1$ and $Y = 0$. Then $\hat{h}_1 = h_1$ and $\overline{h}_1 = \hat{M}_1 h_1$.

When we substitute these results into equation (2.7.4), we will see that the first term in equation (2.7.4) is identical to the first term in equation (2.7.3), and that the second term of equation (2.7.4) differs from the second term of equation (2.7.3) by the factor:

$$(1 + (1 - \hat{M})h_1 h_c) = 1 + \hat{N} h_1 h_c \qquad (2.7.11)$$

This means that the system shown in Figure (2.7.1c) will behave exactly the same way as the ideal system with respect to a change in the set point y_0. However, the response to a disturbance (v) will differ by the factor given in equation (2.7.11). This factor indicates that for the case of a disturbance there is no particular advantage to the scheme shown in Figure 2.7.1c over that given by simple feedback as shown by the dashed line in Figure 2.7.1b. The reason is that the quickest way to detect a change in the disturbance (v) is to measure y. The feedback loop that consists of h_1, h_2, and k must have about the same characteristics as the system consisting of h_1, h_2, and h_c, as shown by the scheme that includes the dashed line in Figure 2.7.1b. Obviously, this development assumes that it is not possible to use feedforward from v because v is not measurable.

A complete development of equation (2.7.11) is given in Appendix B.

The system described above has much in common with the "Smith Predictor" (Smith, 1957), which was further developed by others (Moore, Smith, and Murrill, 1970; Alevisakis and Seborg, 1974).

Let us now consider another example of the application of a model-based estimator to the design of a single-variable control system.

▽ EXAMPLE 2.7.1

Silo 2 in Figure 2.7.2 can contain materials such as liquids, powders, or granules that must be metered out on a precise mass flow basis (w_2). This is done

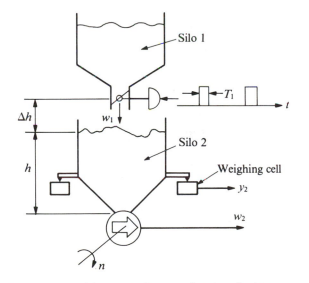

Figure 2.7.2. Equipment used in a control system for mass feed rate.

by means of a discharge mechanism (pump, conveyor, vibrator, etc.) operating at a velocity n. The silo is supplied from an upstream storage unit, such as another silo, at a mass flow rate w_1 that is very high compared with the outflow rate w_2, possibly in the form of a sudden batch or pulse. The problem is that the discharge system is imprecise, and the mass flow w_2 will depend not only on the velocity n, but also on some physical characteristic of the material (μ), the density of the material (ρ), the level in the silo (h), and the general condition (s) of the material handling mechanism. In most cases it is difficult to directly measure the mass flow ($\hat{w}_2$). Therefore, we must use a modeling approach and devise an estimator for the mass flow w_2. The estimate is based upon knowledge of the empty weight of the silo, an estimate of the impulse effect of the material dropping into the silo, and the measured total weight. The weight is measured by accurate load cells. A block diagram of the system is shown in Figure 2.7.3. Clearly, we must develop a mathematical model that will express the mass flow w_2 as a function of the other variables in the process. We can reasonably assume that such a model will have the form:

$$w_2 = \rho \cdot n \cdot f(\mu, h, s) \tag{2.7.12}$$

where:

n = velocity of the discharge system

ρ = density of the material

μ = measurable physical characteristics of the material

h = level of the material in the silo

s = a performance measure of the material transport system

It is not difficult to derive a reasonable function $f(\cdot)$ for specific materials and equipment.

Let us designate the measured weight of the silo and its contents plus the shock due to the falling mass by y_2 and the estimated value of the same variable by $\hat{y}_2$. Similarly, let the estimated weight of the material in the silo be $\hat{y}_1$. The estimator is based on the following simple models:

$$\hat{y}_2 = \hat{y}_1 + \hat{y}_{shock} + \hat{y}_{tare} \tag{2.7.13}$$

$$\hat{y}_{shock} = k_3 \hat{w}_1 \sqrt{h_0 - \hat{h}} \tag{2.7.14}$$

$$\hat{y}_{tare} = \text{constant} \tag{2.7.15}$$

$$\dot{\hat{y}}_1 = \hat{w}_1 - \hat{w}_2 \tag{2.7.16}$$

$$\dot{\hat{w}}_1 = 0 \tag{2.7.17}$$

$$\dot{\hat{s}} = 0 \tag{2.7.18}$$

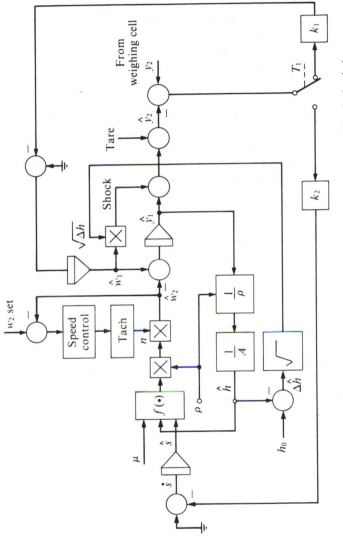

Figure 2.7.3. Block diagram of system for mass feed rate based on model of physical relations.

Equation (2.7.14) is a model of the shock force caused by the falling mass. In equation (2.7.17) we assume that the flow into the silo is constant; that is, either zero or a fixed value (pulse height). Equation (2.7.18) assumes that the condition of the discharge equipment changes very little or very slowly. Therefore, $\hat{w}_1$ and $\hat{s}$ will each be generated by an integrator in the estimator as shown in Figure 2.7.3. The solution of equation (2.7.16) will also be generated by an integrator. The value of $\hat{h}$ needed in equation (2.7.12) is found by dividing the estimated mass contents of the silo ($\hat{y}_1$) by the known material density (ρ) and the cross-sectional area (A) of the silo. The estimated total weight ($\hat{y}_2$) is compared with the measured total weight (y_2), and the difference can be used to update either $\hat{s}$ or $\hat{w}_1$. In Figure 2.7.3, $\hat{w}_1$ is updated during the time period T_1, which is the short period during which the material is transferred from silo 1 to silo 2. The quantity $\hat{s}$ is updated during $T - T_1$; that is, when one of the switches in the diagram is open, the other is closed. As a result, the estimate of the mass flow out of the silo, $\hat{w}_2$, will be adjusted such that the difference between the estimate and the measured value will be zero. Once the estimate has been found, it can be subtracted from the desired value ($w_{2\text{ref}}$), and the difference can be used by an integral controller to control the speed (n) of the discharge system.

The advantage of this system, which is based only on the control of the discharge system, is that it can be very accurate. It should be kept in mind, however, that overall system accuracy is limited by the accuracy of the load cells.

The system described above can be simplified somewhat if one replaces s in equation (2.7.18) with the entire unknown function $f(\cdot)$ of equation (2.7.12). In this case, the model will not use all of the information available because the characteristics of the material have now been lumped into one unknown; but, in
△ many cases, this solution will be satisfactory.

An especially simple form of model-based system is one in which the control structure is selected according to a comparison of the magnitudes of a number of process variables. Many important problems in industrial control can be solved in this way:

- Protection of the equipment
- Multiple instrumentation (majority vote)
- Choice of instrument location
- Reconfiguration of control

Figure 2.7.4 shows a chemical reactor with four temperature measurements placed at different points in the vertical direction. By means of a selector the highest temperature at any time can be chosen for use in the system that controls the cooling of the reactor. According to standard selector symbolism, when the arrow points to the right, the highest value is used; when it points to the left, the lowest is used (Shinskey, 1979). The location of the maximum

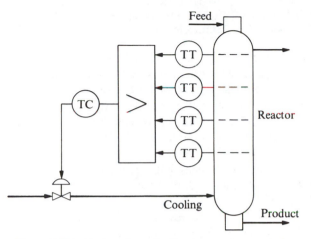

Figure 2.7.4. System for selective temperature control.

temperature can change because of variation of the reactant concentration, catalyst activity, or production rate of the reactor. The system in Figure 2.7.4 will
prevent overheating of the reactor.

Figure 2.7.5 shows a system in which the product of a reactor is analyzed by
two instruments, either of which might fail. The selection problem here is to
control the reactants on the basis of the analyzer which provides the largest
output value. This will give the safest operation because if a possible failure of
an analyzer causes it to increase the signal, the system will stop the supply
of reactant.

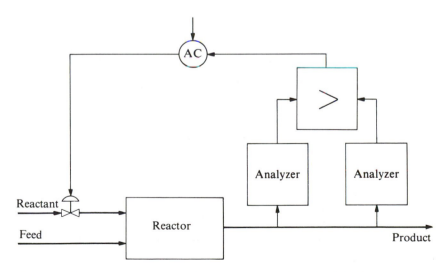

Figure 2.7.5. Selective control on the basis of two identical analytical instruments.

2.8. MULTIVARIABLE CONTROL WITH THE DIAGONAL CONTROLLER

Many industrial processes are *multivariable*; that is, they have multiple control variables (**u**) and multiple measurements (**y**). There are many approaches to the control of such systems, and the theoretical basis for the analysis and synthesis of these systems has been well developed (Athans and Falb, 1966; Gould, 1969; Rosenbrock, 1969; Balchen, Fjeld, and Solheim, 1970, 1978; Anderson and Moore, 1971; Fisher and Seborg, 1976; MacFarlane, 1980; Ray, 1981).

The simplest and by far the most common means of multivariable control is to equip a process with as many individual control systems as there are control variables, and then treat each of these systems as a single-variable system. This simple approach might be called the "ostrich" method, in that it fails to recognize the real world. It ignores cross-couplings, or interactions, between variables. This solution is shown in the block diagram of Figure 2.8.1. It is assumed that in this system there are as many control variables as there are measurements, and that it has been determined which measurement and control variable will interact in each single control loop.

The structure shown in Figure 2.8.1 can often give satisfactory control, and it is the most common multivariable structure because it is the simplest one can imagine. Further, it is easy to operate and maintain. On the other hand, this control solution can be totally unsatisfactory for a number of reasons, typical of which are stability problems and dynamic interactions, which can cause severe oscillations of the variables.

In this section we will concentrate on the following problems:

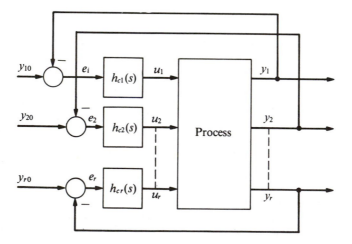

Figure 2.8.1. Control of a multivariable process by means of independent single-variable controllers; a "diagonal controller."

- Analysis of linearized multivariable systems by means of transfer matrices.
- Choice of the control–measurement variable pair for each control loop (pairing).
- Determination of the degree of interaction (cross-couplings) in the system.

After any necessary linearization, the multivariable process shown in Figure 2.8.1 can be described by the following model:

$$\mathbf{y}(s) = H_u(s)\mathbf{u}(s) + H_v(s)\mathbf{v}(s) \qquad (2.8.1)$$

If the system is described in the state space by the equations:

$$\dot{\mathbf{x}}(t) = A\mathbf{x}(t) + B\mathbf{u}(t) + C\mathbf{v}(t) \qquad (2.8.2)$$

$$\mathbf{y}(t) = D\mathbf{x}(t) \qquad (2.8.3)$$

then the transfer function matrices will be:

$$H_u(s) = D(sI - A)^{-1}B \qquad (2.8.4)$$

$$H_v(s) = D(sI - A)^{-1}C \qquad (2.8.5)$$

The individual transfer functions that make up the *transfer function matrix* will have the following relationship:

$$H(s) = \begin{bmatrix} h_{11}(s) & h_{12}(s) & \cdots & h_{1r}(s) \\ h_{21}(s) & h_{22}(s) & \cdots & h_{2r}(s) \\ \vdots & \vdots & & \vdots \\ h_{r1}(s) & h_{r2}(s) & \cdots & h_{rr}(s) \end{bmatrix} \qquad (2.8.6)$$

The transfer functions $h_{ij}(s)$ have the form

$$h_{ij}(s) = \frac{t_{ij}(s)}{n_{ij}(s)}$$

where $t_{ij}(s)$ and $n_{ij}(s)$ are polynomials in s, and they will all have the same denominator $n_{ij}(s)$. Some of the roots of $n_{ij}(s)$, that is, the poles of the transfer functions $h_{ij}(s)$, can be canceled by the roots of the numerator polynomial $t_{ij}(s)$. This means that not all of the poles necessarily appear in all of the transfer functions. The physical explanation of this is that the multivariable process can contain many subprocesses that have especially weak couplings.

If the transfer function matrices $H_u(s)$ or $H_v(s)$ result from a series of experiments in which each input–output pair is observed individually, the matrices will often appear to have more poles than the system should theoretically have. This will have no significance in most applications of the model described here.

A control matrix has the general form

$$H_c(s) = \{h_{cij}(s)\} \tag{2.8.7}$$

but in the case of the diagonal controller we will have:

$$h_{cij}(s) = \begin{cases} h_{ci}(s), & i = j \\ 0, & i \neq j \end{cases} \tag{2.8.8}$$

This is a major reduction in the complexity of the control matrix, in that the number of controller parameters that must be determined and adjusted will be reduced by a factor r (the number of control variable and measurement pairs), where it is assumed that each controller has the same number of parameters.

It is usually quite easy to pair the various measurements with their proper control variables by considering their physical relationships. For example, energy in gives temperature out, flow in gives level out, and so on. Sometimes, however, the couplings between variables are such that it is difficult to determine the most logical pairings. In these cases, it may be helpful to examine the forms of the transfer functions. In general there are $r!$ possible pairs, but only some are realistic.

If we look at the contribution of the control variables to the response of one particular measurement variable, we see:

$$y_i(s) = h_{i1}(s)u_1(s) + h_{i2}(s)u_2(s) + \cdots + h_{ir}(s)u_r(s) \tag{2.8.9}$$

We can compare the various inputs on the right side of equation (2.8.9) in order to see the frequency response using relative values of the variations in the control variables (for example, 10% of the normal variation range). We find:

$$|y_i(j\omega)| \leq (|h_{i1}(j\omega)| + |h_{i2}(j\omega)| + \cdots |h_{ir}(j\omega)|)\,\Delta u \tag{2.8.10}$$

where Δu is the normalized change in the control signal.

Most of the frequency response characteristics for the terms in the equations above can be categorized by the curves shown in Figure 2.8.2, where amplitude and phase are plotted as functions of frequency. In the broad sense, there are three types of systems that produce the three different types of amplitude characteristics shown in the figure. In a *Type 1* system where the transfer function can have a number of poles and zeroes, and perhaps a transport delay, the gain at low frequencies approaches a constant value. A *Type 2* system is characterized by the presence of one or more pure integrations in the transfer function. A *Type 3* system is dominated by a zero at the origin, which means there will be very little response at low frequencies.

The phase angles are shown in the lower part of Figure 2.8.2. Curve 1a is typical of systems that have no transport delay, whereas curve 1b shows the

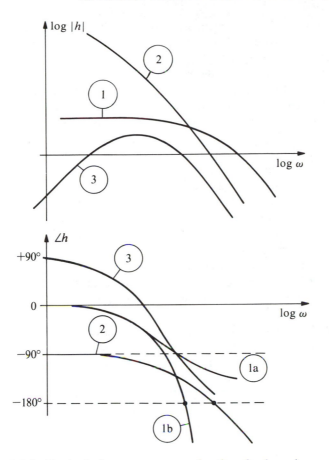

Figure 2.8.2. The basic frequency response functions for dynamic processes.

severe phase shift a transport delay adds to the system. Phase curve 2 is typical of systems having pure integrations, and it starts at an integer multiple of $-90°$ at zero frequency. The multiple corresponds to the number of pure integrators. Phase curve 3 starts at $+90°$ at zero frequency and is the curve for a Type 3 system.

In many cases with low cross-coupling, a set of simple rules can be used when pairing measured variables with control variables in a diagonal multivariable control system:

1. If any transfer function in equation (2.8.9) is of Type 2 (contains a pure integrator), the input and output of that transfer function must be taken as a pair.
2. A pair that has a transfer function of Type 3 (a zero at the origin) can be used only for transient control.

3. If all of the transfer functions are of either Type 1 or Type 3, one should normally choose a pair corresponding to Type 1. If several of the transfer functions are of Type 1, the pairing should proceed as follows: The transfer function that gives the greatest steady-state gain $[|h_{ij}(0)|]$ and/or the greatest value of ω_{180} $(\angle h_{ij}(j\omega_{180}) = -180)$ should be chosen as a pair.
4. If there are multiple transfer functions of Type 2, the input and output of the transfer function that has the highest gain and/or the largest value of ω_{180} should be used as a pair.

The basic rules given above will usually be satisfactory, but there are cases that require further analysis. These cases are caused primarily by cross-couplings that become active when feedback loops are closed between measurements and control variables. Furthermore, many practical considerations do not appear in the transfer functions but can be very important in the choice of the input–output pairs.

After the pairings have been chosen, the process transfer function matrix $H_u(s)$ should have the dominant transfer functions on the diagonal. Therefore, the controller transfer function matrix of equation (2.8.7) may be chosen to be a diagonal matrix.

The following is a technique for testing the choice of pairing with respect to the effect of cross-coupling (interaction) when the feedback loops are closed:

Consider the system of Figure 2.8.3, in which there are no disturbances. We can consider any one of the control loops, for example, the one with control variable u_1 and a measurement y_1. Further, let this be an open system, but assume that all the others in the process are closed through their controllers. We want to find the effective loop transfer function between u_1 and y_1. A simple outline of the procedure will be given here, but a detailed development is given in Appendix C.

The transfer function between the control and the measurement for open loop number i will be:

$$h_{ii}(s)\,(1 - \Delta_i) \tag{2.8.11}$$

where the subscripts indicate that h refers to control variable u_i and measurement y_i and:

$$\Delta_i = \frac{1}{h_{ii}}\mathbf{h}^{iT}(H_u^i)^{-1}M^i\tilde{\mathbf{h}}^i \approx \frac{1}{h_{ii}}\mathbf{h}^{iT}(H_u^i)^{-1}\tilde{\mathbf{h}}^i \tag{2.8.12}$$

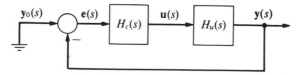

Figure 2.8.3. Block diagram for a multivariable control system.

The above notation is completely defined in Appendix C.

A multivariable system is often dominated by two control loops. In this case, the correction factor in equation (2.8.11) will be:

$$(1 - \Delta_i) = 1 - \frac{h_{12}h_{21}}{h_{11}h_{22}} M^i \approx 1 - \frac{h_{12}h_{21}}{h_{11}h_{22}} \qquad (2.8.13)$$

If more than two control loops dominate the system, then equation (2.8.12) must be used to compute the correction factor. From equation 2.8.13 we see that if $h_{12}(s) = 0$ and/or $h_{21}(s) = 0$, the mutual coupling between the parts of the system will not influence the design of the controllers in the individual system parts; that is, each controller can be designed individually. This does not mean that the various sections of the system do not influence each other, but it does mean that these influences will have no effect on stability. This can be illustrated by some examples.

▽ EXAMPLE 2.8.1

Figure 2.8.4 shows an especially simple case of a system with two control variables, two measurements, and strong coupling between the two system parts. Two liquids, A and B, having about the same flowrates and controlled by the two valves, are mixed together. Some property of the mixture is measured by an analytical instrument that is relatively slow because of either a large time constant or a transport delay. If the two valves have about the same speed and ability to affect the value of the total flow, then a flow controller could be applied to either valve A or valve B with about the same results. However, if the flowrate of stream A were significantly greater than that of stream B, which

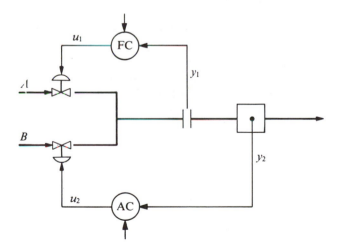

Figure 2.8.4. Multivariable process with two single-variable controllers.

could be the case if stream B had a very high concentration relative to A, then the structure shown in Figure 2.8.4 would be preferable.

If the flows through the two valves are q_A and q_B respectively, the total mixed flow will be:

$$q_C = q_A + q_B \tag{2.8.14}$$

If the concentrations of the materials of interest in the two streams are c_A and c_B, the concentration of the mixed stream will be:

$$c_C = \frac{q_A c_A + q_B c_B}{q_A + q_B} \tag{2.8.15}$$

From this we find:

$$\frac{\partial q_C}{\partial q_A} = 1 \tag{2.8.16}$$

$$\frac{\partial q_C}{\partial q_B} = 1 \tag{2.8.17}$$

$$\frac{\partial c_C}{\partial q_A} = \frac{(c_{A0} - c_{B0})q_{B0}}{(q_{A0} + q_{B0})^2} \tag{2.8.18}$$

$$\frac{\partial c_C}{\partial q_B} = \frac{(c_{B0} - c_{A0})q_{A0}}{(q_{A0} + q_{B0})^2} \tag{2.8.19}$$

If c_A is negative and c_B is positive, we have the situation that occurs when an acid is mixed with a base. The mixture could be neutral. In the above equations the subscript 0 refers to nominal values.

The two control valves are sized so that when either control signal (u_1 or u_2) is 100%, the corresponding valve will pass twice the nominal value; that is, $2q_{A0}$ or $2q_{B0}$. We assume the motors of the two valves will have first-order transfer functions characterized by time constants T_1 and T_2, respectively.

The measuring system for the mixed stream flow is assumed to have a time constant T_3 and a steady-state gain of 1, and measurement of the concentration of the mixture, as determined by the analyzer, is assumed to have a time constant T_4 and a steady-state gain of 1.

We can now study two different situations:

Case 1:

Let $c_A = -5$ and $c_B = +5$. The desired concentration $c_{C0} = 0$, and the desired mixed stream flow rate $q_{C0} = 100 = q_{A0} + q_{B0}$. Equation (2.8.15) gives:

$$c_{C0} = 0 = \frac{-5q_{A0} + 5q_{B0}}{q_{A0} + q_{B0}}$$

This gives $q_{A0} = q_{B0} = 50$. Substituting this into equations (2.8.18) and (2.8.19), we get:

$$\frac{\partial c_C}{\partial q_A} = \frac{(-5 - 5)50}{100^2} = -0.05$$

$$\frac{\partial c_C}{\partial q_B} = \frac{(5 + 5)50}{100^2} = +0.05$$

We can now determine the four transfer functions of this process:

$$\frac{y_1}{u_1}(s) = h_{11}(s) = \frac{2q_{A0}}{1 + T_1 s} \cdot \frac{\partial q_C}{\partial q_A} \cdot \frac{1}{1 + T_3 s} \qquad (2.8.20)$$

$$\frac{y_1}{u_2}(s) = h_{12}(s) = \frac{2q_{B0}}{1 + T_2 s} \cdot \frac{\partial q_C}{\partial q_B} \cdot \frac{1}{1 + T_3 s} \qquad (2.8.21)$$

$$\frac{y_2}{y_1}(s) = h_{21}(s) = \frac{2q_{A0}}{1 + T_1 s} \cdot \frac{\partial c_C}{\partial q_A} \cdot \frac{1}{1 + T_4 s} \qquad (2.8.22)$$

$$\frac{y_2}{u_2}(s) = h_{22}(s) = \frac{2q_{B0}}{1 + T_2 s} \cdot \frac{\partial c_C}{\partial q_B} \cdot \frac{1}{1 + T_4 s} \qquad (2.8.23)$$

Now assume that

$$T_2 = 2T_1, \qquad T_3 = 0.1T_1, \qquad \text{and} \quad T_4 = 20T_1$$

Putting these values into the transfer functions gives:

$$h_{11}(s) = \frac{100}{(1 + T_1 s)(1 + 0.1T_1 s)} \qquad (2.8.24)$$

$$h_{12}(s) = \frac{100}{(1 + 2T_1 s)(1 + 0.1T_1 s)} \qquad (2.8.25)$$

$$h_{21}(s) = \frac{-5}{(1 + T_1 s)(1 + 20T_1 s)} \qquad (2.8.26)$$

$$h_{22}(s) = \frac{5}{(1 + 2T_1 s)(1 + 20T_1 s)} \qquad (2.8.27)$$

Because this is a 2 × 2 system, the expression from equation (2.8.12) will be:

$$\Delta_1 = \frac{h_{12}h_{21}}{h_{11}h_{22}} \cdot \frac{h_{C2}h_{22}}{1 + h_{C2}h_{22}} \qquad (2.8.28)$$

Using the appropriate numerical values:

$$\Delta_1 \cong -1 \cdot \frac{1}{(1 + 2T_1s)^2} \tag{2.8.29}$$

Therefore, the flow controller loop will have the process transfer function:

$$h_{11}(1 - \Delta_1) \approx \frac{100}{(1 + T_1s)(1 + 0.1T_1s)} \cdot \frac{2(1 + \sqrt{2}T_1s)^2}{(1 + 2T_1s)^2} \tag{2.8.30}$$

We see that the last term of equation (2.8.30) approaches 1 when s becomes large; that is, near the crossover frequency, which will be approximately $1/(0.1T_1)$ in this case. This means that we can completely ignore the interaction with respect to the flow control loop. For concentration control we have:

$$\Delta_2 = \frac{h_{12}h_{21}}{h_{11}h_{22}} \cdot \frac{h_{C1}h_{11}}{1 + h_{C1}h_{11}} \tag{2.8.31}$$

Using the numerical values, we get:

$$\Delta_2 \approx -1 \cdot \frac{1}{(1 + 0.1T_1s)^2} \tag{2.8.32}$$

The process transfer function for the concentration loop is:

$$h_{22}(1 - \Delta_2) \approx \frac{5}{(1 + 2T_1s)(1 + 20T_1s)} \cdot \frac{2\left(1 + \frac{\sqrt{2}}{2}0.1T_1s\right)^2}{(1 + 0.1T_1s)^2} \tag{2.8.33}$$

At the crossover frequency for the concentration control loop (approximately $1/2T_1$), the last term of equation (2.8.33) will have a value of about 2. This means we will have to consider the transfer function:

$$h_{22}(1 - \Delta_2) \approx \frac{10}{(1 + 2T_1s)(1 + 20T_1s)} \tag{2.8.34}$$

Therefore, the gain of the concentration controller must be reduced to half of what it would have been had there been only a single control loop.

Case 2:

Consider the situation when the numerical values are:

$$c_{A0} = -5, \quad c_{B0} = +100$$

$$q_{C0} = 100 = q_{A0} + q_{B0}$$

and:

$$c_{CO} = \frac{-5q_{A0} + 100q_{B0}}{q_{A0} + q_{B0}} = 0$$

This gives:

$$q_{A0} = 95.24$$

$$q_{B0} = 4.76$$

If we substitute these values into equations (2.8.16) through (2.8.19), we get:

$$\frac{\partial q_C}{\partial q_A} = -1, \qquad \frac{\partial q_C}{\partial q_B} = 1$$

$$\frac{\partial c_C}{q_A} = -0.05, \qquad \frac{\partial c_C}{\partial q_B} \approx 1$$

Inserting these values into equations (2.8.20) through (2.8.23) gives:

$$h_{11}(s) = \frac{190.5}{(1 + T_1 s)(1 + 0.1T_1 s)} \tag{2.8.35}$$

$$h_{12}(s) = \frac{9.5}{(1 + 2T_1 s)(1 + 0.1T_1 s)} \tag{2.8.36}$$

$$h_{21}(s) = \frac{-9.5}{(1 + T_1 s)(1 + 20T_1 s)} \tag{2.8.37}$$

$$h_{22}(s) = \frac{4.76}{(1 + 2T_1 s)(1 + 20T_1 s)} \tag{2.8.38}$$

Equivalent to case 1, we now find:

$$\Delta_1 = \frac{h_{12}h_{21}}{h_{11}h_{22}} \cdot \frac{h_{C2}h_{22}}{1 + h_{C2}h_{22}} \tag{2.8.39}$$

which gives:

$$\Delta_1 \approx -0.1 \frac{1}{(1 + 2T_1 s)^2} \tag{2.8.40}$$

From this we find that the transfer function for the flow process is now:

$$h_{11}(1 - \Delta_1) \approx h_{11} \tag{2.8.41}$$

This means that the concentration loop now has a negligible effect on the flow control. This is due, of course, to the low flow in branch B.

We can also write:

$$\Delta_2 = \frac{h_{12}h_{21}}{h_{11}h_{22}} \cdot \frac{h_{C1}h_{11}}{1 + h_{C1}h_{11}} \tag{2.8.42}$$

which, after substitution of the numerical values, gives:

$$\Delta_2 \approx -0.1 \cdot \frac{1}{(1 + 0.1T_1s)^2} \tag{2.8.43}$$

From this we find that the transfer function for the concentration control loop is:

$$h_{22}(1 - \Delta_2) \approx h_{22} \tag{2.8.44}$$

This means that the controller on the concentration loop has little effect on the flow control loop because the latter is much faster than the concentration con-
△ trol system.

We will return to more examples that use isolated control loops in multivariable processes in later sections.

The reason why an overwhelming majority of industrial processes use this type of multivariable control is that this structure is easy to design, tune (adjust), and maintain. If it also performs nearly as well as a more complicated system, the practicing engineer naturally will use it. The problem facing the engineer who must make this decision is that one does not often know the advantages and disadvantages that might be associated with a more complicated system. These questions will be discussed further in the next sections.

2.9. DECOUPLING

A major characteristic of a multivariable system, as we have just seen, is the strong coupling between the subsystems, which causes the individual controllers to interact with each other. One way to eliminate, or at least reduce, this effect is to provide the multivariable system with artificial cross-couplings that eliminate the real ones. The goal is to arrive at a diagonal multivariable system. Doing this by means of artificial cross-couplings is called *decoupling*.

The easiest way to understand and develop a decoupler is to start with a transfer matrix that is the result of a suitable pairing of variables. Such a matrix will have the "dominant" transfer functions on the diagonal. It is necessary that the dimension of the control vector be the same as that of the measurement vector; that is, $r = m$.

The decoupler can be regarded as a part of the controller. Therefore, the controller can be thought of as consisting of two parts — the decoupler, which has a transfer function matrix $H_{C2}(s)$ and is closest to the process, and the controller, which will have a diagonal structure and a transfer function matrix $H_{C1}(s)$. This is shown in Figure 2.9.1. Between the two parts of the controller we have a new set of variables (**m**) that can be thought of as new control variables for the modified process characterized by the transfer function matrix:

$$H_u(s)H_{C2}(s) = H_u(s)_m \tag{2.9.1}$$

The purpose of the decoupler $[H_{C2}(s)]$ is to make $H_u(s)_m$ a diagonal matrix, or at least to weaken the transfer functions that are off the diagonal and represent the strong interactive terms in the system.

Because the matrix $H_u(s)$ was obtained after a satisfactory pairing of variables, it is logical (and usual) to try to form $H_{C2}(s)$, such that:

$$H_u(s)_m = \text{diag } [H_u(s)] = [\text{diagonal elements in } H_u(s)]$$

According to equation (2.9.1) this can be done by forming a decoupler as:

$$H_{C2}(s) = H_u^{-1}(s)H_u(s)_m = H_u^{-1}(s) \text{ diag } (H_u(s)) \tag{2.9.2}$$

This equation should be used as a guideline rather than an inflexible demand on the system because many other considerations enter into the design of a practical decoupling scheme (decoupling algorithm). For example, it may be advantageous to use a simple algorithm that might give a less perfect but nevertheless satisfactory decoupling. Also, if $H_u(s)$ contains elements of the "non-minimum-phase" type, the realization of $H_u^{-1}(s)$ may be a problem. In this case, an approximation must be used.

It is easy to develop the transfer function for a decoupler for a 2 × 2 "minimum phase" system. We have:

$$H_{C2}(s) = \begin{bmatrix} 1 & -\dfrac{h_{12}(s)}{h_{11}(s)} \\ -\dfrac{h_{21}(s)}{h_{22}(s)} & 1 \end{bmatrix} \cdot \dfrac{1}{1 - \dfrac{h_{12}(s)h_{21}(s)}{h_{11}(s)h_{22}(s)}} \tag{2.9.3}$$

The last term of equation (2.9.3) is a scalar function. If we neglect this term (set it equal to 1), the result will be a diagonal matrix, but it will not satisfy equation (2.9.2). Nevertheless, this is the solution most used in practice.

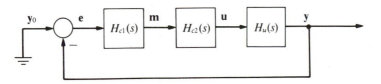

Figure 2.9.1. Decoupling of a multivariable process.

In this case the modified process transfer matrix is:

$$
H_u(s)_m = \begin{bmatrix} h_{11}\left(1 - \dfrac{h_{12}h_{21}}{h_{11}h_{22}}\right) & 0 \\[2ex] 0 & h_{22}\left(1 - \dfrac{h_{12}h_{21}}{h_{11}h_{22}}\right) \end{bmatrix}
\tag{2.9.4}
$$

If the last term in equation (2.9.3) is not neglected, we will have a more complicated decoupler, but the terms on the diagonal of equation (2.9.4) will be somewhat simpler. The transfer functions for the modified process of equation (2.9.4) remind one of those developed in equations (2.8.11) through (2.8.13). From the point of view of the controller, this decoupling has not significantly simplified the problem.

When we form the first part of the controller, characterized by the diagonal matrix $H_{C1}(s)$, the problem is reduced to one of finding a number of completely independent single-variable control systems. In the case of a 2×2 system, the modified transfer functions will be given by either the elements of the diagonal of the original matrix or the diagonal elements of equation (2.9.4).

In spite of the simplicity expressed in equation (2.9.2), this method of decoupling is not as widely used in practice as one might expect. The reason for this is that the resulting system is still more complicated than one that uses the method of multiple single-variable controllers.

Many methods have been developed for the systematic design of control transfer matrices that will provide at least an approximation of a diagonalized system. One such method is the "inverse Nyquist array" introduced by Rosenbrock (1969). Another, but related, method was presented by Mayne (1973).

"Multivariable frequency response analysis" includes a number of different techniques that in addition to those mentioned above includes the "generalized Nyquist stability criterion" developed in MacFarlane and Postlethwaite (1977). This area is very well described in two papers by MacFarlane (1979, 1980).

Consider now the following situation:

Suppose one has a so-called LQG (Linear Quadratic Gaussian) controller consisting of a combination of an optimal controller based on the quadratic objective functional [described in section 2.11, equation (2.11.9)] and optimal state estimation provided by a Kalman filter (described in section 2.13). Under stationary conditions, this can be described by the block diagram of Figure 2.9.2. The following matrices appear in this diagram:

- X_s—the stationary solution of the covariance matrix for the state estimate, given by the Riccati equation:

$$
\dot{X}(t) = AX(t) + X(t)A^T + CVC^T - X(t)D^TW^{-1}DX(t) = 0 \tag{2.9.5}
$$

- R_s—the stationary solution of the Riccati equation that determines the optimal feedback, given by:

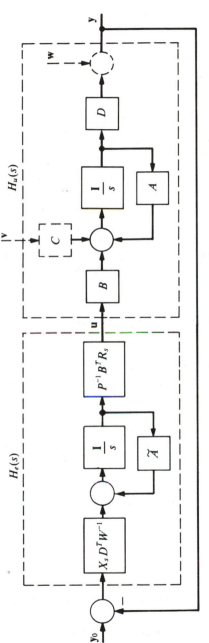

Figure 2.9.2. Block diagram of multivariable process controller based on optimal control and state estimation (LQG controller).

$$\dot{R}(t) = -R(t)A - A^TR(t) + R(t)BP^{-1}B^TR(t) - Q = 0 \quad (2.9.6)$$

$$\tilde{A} = A - X_sD^TW^{-1}D - BP^{-1}B^TR_s \quad (2.9.7)$$

where:

A = system matrix for the process
B = control matrix for the process
C = disturbance matrix for the process
D = measurement matrix for the process
V = covariance matrix for the process disturbances
W = covariance matrix for the measurement noise
P = weighting matrix for the control vector in the objective functional
Q = weighting matrix for the state vector in the objective functional

We see that the controller in Figure 2.9.2 has dynamics that contain information regarding the process dynamics, the disturbances, the measurements, the characteristics of the system noise, and the objectives partly derived from the process dynamics. It is easy to realize this controller because it contains only three matrices and a number of integrators equal to the order of the process. This type of controller has significant decoupling of the control loops.

2.10. FEEDFORWARD IN MULTIVARIABLE SYSTEM

The reasoning used with respect to feedforward in the case of single-variable systems in section 2.3 can be applied to feedforward control of multivariable systems. The demands on feedback control can be reduced if one is able to establish a feedforward relationship between those disturbances that can be measured and the various control variables that can, at least in part, suppress the response to those disturbances. In this case it should then be much simpler to build a feedback controller that gives satisfactory performance.

Consider again the transfer matrix given by equation (2.8.1.):

$$\mathbf{y}(s) = H_u(s)\mathbf{u}(s) + H_v(s)\mathbf{v}(s) \quad (2.10.1)$$

We see that the ideal feedforward from the disturbance vector to the control vector is:

$$H_{Fi}(s) = -H_u^{-1}(s)H_v(s) \quad (2.10.2)$$

where it is assumed that $H_u(s)$ has an inverse. This implies that $\mathbf{y}$ and $\mathbf{u}$ have the same dimension.

If one cannot realize an ideal feedforward matrix, but instead has an actual feedforward matrix $H_F(s)$, one gets:

$$\mathbf{y}(s) = H_v(s)[G_0(s) + I]\mathbf{v}(s) = H_v(s)L(s)\mathbf{v}(s) \quad (2.10.3)$$

where:

$$G_0(s) = -H_{Fi}^{-1}(s)H_F(s)$$

This result is completely analogous to that found in equation (2.3.3). The matrix $L(s)$ can be defined as the *feedforward control ratio matrix*.

The field of feedforward multivariable control has been studied mostly in connection with the control of distillation columns because of the large number of variables and the complexity of the dynamics in these systems. It is especially difficult to control a distillation column that is part of a series of other processes because those processes often generate severe disturbances to the distillation column. Feedforward and ratio control are important means of maintaining the proper material balances in the system. This will be discussed further in the section on distillation in Chapter 5.

In practice, multivariable feedforward control will usually be designed as a number of single-variable feedforward systems. As discussed in section 2.3, feedforward can lead to multiplicative effects such as those described in Figure 2.3.6. A multivariable version of this, which appears in equations (2.3.5) and (2.3.6), can also be derived in some cases. For the case of a nonlinear state space process model of the form

$$\dot{\mathbf{x}} = \mathbf{f}(\mathbf{x}, \mathbf{u}, \mathbf{v}) \tag{2.10.4}$$

feedforward control can be determined in the following way:

We want to find a control vector $\mathbf{u}$ that drives the process with a desired rate of change of the process states $\dot{\mathbf{x}}_0$. From equation (2.10.4) we find:

$$\mathbf{u} = \mathbf{h}(\dot{\mathbf{x}}_0, \mathbf{x}, \mathbf{v}) \tag{2.10.5}$$

assuming that the value given by equation (2.10.5) exists.

We can establish feedback control by, for instance:

$$\dot{\mathbf{x}}_0 = G(\mathbf{x}_0 - \mathbf{x}) \tag{2.10.6}$$

where $\mathbf{x}_0$ is a vector of the desired process states, and G is a control matrix. This solution is especially simple if the process model has the form

$$\dot{\mathbf{x}} = \mathbf{F}(\mathbf{x}, \mathbf{v}) + B(\mathbf{x})\mathbf{u} \tag{2.10.7}$$

where $B(\cdot)$ is a matrix. Then:

$$\mathbf{u} = B(\cdot)^{-1}(\dot{\mathbf{x}}_0 - \mathbf{F}(\mathbf{x}, \mathbf{v})) \tag{2.10.8}$$

if $B^{-1}(\cdot)$ exists. We can check the result by applying equation (2.10.6) to equations (2.10.7) and (2.10.8) to get:

$$\dot{\mathbf{x}} = \mathbf{F}(\mathbf{x}, \mathbf{v}) + B(\cdot) \cdot B(\cdot)^{-1}(G(\mathbf{x}_0 - \mathbf{x}) - \mathbf{F}(\mathbf{x}, \mathbf{v}))$$

$$= G(\mathbf{x}_0 - \mathbf{x}) \tag{2.10.9}$$

Equation (2.10.9) shows that the feedforward of equations (2.10.5) and (2.10.8) has made the feedback simple.

A model of the form of equation (2.10.7) can be expected in processes having interaction between mass and energy balances. Such a case exists in the production of a metal (X) by electrolysis of a molten metal chloride (XCl_2). This process has two significant state variables: the concentration of the chloride (x_1) and the temperature (x_2).

The material balance that determines the rate of change of x_1 is influenced by the inflow of chloride and the conversion of chloride to metal resulting from the electrolysis current (I). The chloride is fed to the process in two forms, as pellets (u_1) and as a molten liquid (u_2) at a temperature near that of the process (x_2).

The energy balance that determines the rate of change of x_2 is influenced by the heat generated by the electrolysis, the heat losses, and the ratio of solid to molten chloride. (The latter is especially important because heat is needed to melt the pellets.)

A simplified mathematical model of this process has the form

$$\dot{x}_1 = -f_1(x_1, x_2)I + b_{11}(u_1 + u_2) \tag{2.10.10}$$

$$\dot{x}_2 = f_2(x_2, I) - b_{21}u_1 + b_{22}u_2 \tag{2.10.11}$$

where $f_1(\cdot)$ expresses the dependence of "current efficiency" on the concentration and temperature, and $f_2(\cdot)$ expresses the heat generation and loss with respect to concentration, temperature, and current.

Equations (2.10.6) and (2.10.8) can be applied directly to yield expressions for u_1 and u_2:

$$u_1 = \frac{1}{b_{11}(b_{22} + b_{21})}(b_{22}g_1(x_{10} - x_1) - b_{11}g_2(x_{20} - x_2) + b_{22}f_1(\cdot)I + b_{11}f_2(\cdot)) \tag{2.10.12}$$

$$u_2 = \frac{1}{b_{11}(b_{22} + b_{21})}(b_{21}g_1(x_{10} - x_1) + b_{11}g_2(x_{20} - x_2) + b_{21}f_1(\cdot)I - b_{11}f_2(\cdot)) \tag{2.10.13}$$

These expressions constitute the control algorithm based upon internal feedforward and external feedback.

2.11. OPTIMAL CONTROL

The term *optimal control* refers to systems designed to generate control variables that maximize (or minimize) a scalar objective functional that includes the process states and control variables. This problem has been widely discussed in the control literature since about 1957–60 (Pontryagin et al., 1962; Athans and

Falb, 1966; Bryson and Ho, 1969; Balchen, Fjeld, and Solheim, 1970, 1978; Anderson and Moore, 1971).

One of the best-known methods for solution of this problem is the "Pontryagin maximum principle." A summary of the main points of this method follows.

A nonlinear process (with no disturbances, for the moment) can be described by:

$$\dot{\mathbf{x}} = \mathbf{f}(\mathbf{x}, \mathbf{u}) \qquad (2.11.1)$$

We wish to find the $\mathbf{u}(t) \in U$ which gives the maximum of the objective functional (optimal criterion):

$$J = S(\mathbf{x}(t_2)) + \int_{t_1}^{t_2} L(\mathbf{x}, \mathbf{u})\, dt \qquad (2.11.2)$$

A scalar function (the Hamiltonian) is introduced:

$$H(\mathbf{x}, \mathbf{p}, \mathbf{u}) = L(\mathbf{x}, \mathbf{u}) + \mathbf{p}^T \mathbf{f}(\mathbf{x}, \mathbf{u}) \qquad (2.11.3)$$

The optimal control is the $\mathbf{u}(t)$ that at any time maximizes $H(\cdot)$ when $\mathbf{x}(t)$ and $\mathbf{p}(t)$ are given by solution of the equations:

$$\dot{\mathbf{x}} = \frac{\partial H(\cdot)}{\partial \mathbf{p}} \qquad (2.11.4)$$

$$\dot{\mathbf{p}} = -\frac{\partial H(\cdot)}{\partial \mathbf{x}} \qquad (2.11.5)$$

The boundary conditions for equations (2.11.4) and (2.11.5) depend on how $\mathbf{x}(t_2)$ is specified. Usually the boundary conditions are given as:

$$x(t_1) = \text{given} \qquad (2.11.6)$$

$$\mathbf{p}(t_2) = \left.\frac{\partial S(\cdot)}{\partial \mathbf{x}}\right|_{t=t_2} \qquad (2.11.7)$$

The solution of the optimization problem by this method requires the solution of a two point boundary value problem. There are several ways of doing this reported in the literature. In all cases this demands a real-time computation, which must be done rapidly by means of some iterative computation scheme.

A condition for the successful application of the Maximum Principle is that one must have a good mathematical model of the system, and the objective functional equation (2.11.2) must express accurately what one wishes to accomplish. The literature reports many applications of this method that have provided remarkable improvements compared with what can be done by other techniques.

If the process in equation (2.11.1) also has a disturbance vector ($\mathbf{v}(t)$), the problem is best solved by first developing a mathematical model of the generation of the disturbance as a response to a vector of uncorrelated "white noise sources" through a dynamic process. The state vector ($\mathbf{x}$) for the process model is augmented by the disturbance state to form a super process having the state vector ($\tilde{\mathbf{x}}$). This vector replaces the original state vector ($\mathbf{x}$) in the equations above and leads to an optimal solution for the disturbed system.

The state vector ($\mathbf{x}$) or ($\tilde{\mathbf{x}}$) must be completely available in order to derive the optimal control. Therefore, it is often necessary to use a state estimator to calculate those state variables that cannot be directly measured. This leads to another area of great concern that has been developing in the control literature since about 1960. We will discuss this subject further in section 2.13.

An especially simple result is obtained if the process of equation (2.11.1) is linear and time-invariant, and if the objective functional of equation (2.11.2) is a scalar quadratic function. We then have what is known as the *linear quadratic problem* (LQ problem). If the process disturbances are white noise with zero mean, then the required control strategy will be just as if the process had no noise at all; that is, as if the system were deterministic! If the disturbances are not white noise, but can be described by a mathematical model of a system that generates equivalent disturbances from a white noise source, then the optimal control strategy can be derived by augmenting the process model with the disturbance model.

Assume we have white noise and the following process model:

$$\dot{\mathbf{x}} = A\mathbf{x} + B\mathbf{u} + C\mathbf{v} \qquad (2.11.8)$$

for which we will assume the optimization criterion:

$$J = E\left(\int_{t_1}^{t_2} (\mathbf{x}_0 - \mathbf{x})^T Q(\mathbf{x}_0 - \mathbf{x}) + \mathbf{u}^T P\mathbf{u})\, dt\right) \qquad (2.11.9)$$

where $\mathbf{x}_0$ is a constant or very slowly changing set point vector. The optimal control vector can be found by the algorithm

$$\mathbf{u}(t) = P^{-1}B^T R(t)\, (\mathbf{x}_0 - \mathbf{x}(t)) = G(t)\, (\mathbf{x}_0 - \mathbf{x}(t)) \qquad (2.11.10)$$

where:

$$\dot{R} = -RA - A^T R + RBP^{-1}B^T R - Q \qquad (2.11.11)$$

Equation (2.11.11) is known as the *Riccati matrix differential equation*. It can be shown that when $t \to \infty$, $R(t) \to R =$ constant, and we have the stationary case. The optimal control vector is:

$$\mathbf{u} = G(\mathbf{x}_0 - \mathbf{x}) \qquad (2.11.12)$$

This means that when we have feedback from the state variables, we have a controller that is a matrix of constant terms. This solution is shown in Figure 2.11.1. An alternative to the objective functional of equation (2.11.9) is to include a term that considers the velocity with which the control vector can change:

$$J = E\left(\int_{t_1}^{t_2} ((\mathbf{x}_0 - \mathbf{x}_1)^T Q_1(\mathbf{x}_0 - \mathbf{x}_1) + \mathbf{u}^T Q_2\mathbf{u} + \dot{\mathbf{u}}^T P\dot{\mathbf{u}})\, dt\right) \quad (2.11.13)$$

where the state vector of the process is $\mathbf{x}_1$, and the corresponding weighting matrix is Q_1. If we think of the new control vector as a new "state" such that $\mathbf{u} = \mathbf{x}_2$ and $\dot{\mathbf{u}} = \mathbf{m}$, then equation (2.11.13) can be written as:

$$J = E\left(\int_{t_1}^{t_2} (\mathbf{x}_0 - \mathbf{x}_1)^T Q_1(\mathbf{x}_0 - \mathbf{x}_1) + \mathbf{x}_2^T Q_2\mathbf{x}_2 + \mathbf{m}^T P\mathbf{m})\, dt\right)$$

$$= E\left(\int_{t_1}^{t_2} (\tilde{\mathbf{x}}_0 - \tilde{\mathbf{x}})^T Q(\tilde{\mathbf{x}}_0 - \tilde{\mathbf{x}}) + \mathbf{m}^T P\mathbf{m})\, dt\right) \quad (2.11.14)$$

where:

$$\tilde{\mathbf{x}} = \begin{bmatrix} \mathbf{x}_1 \\ \mathbf{x}_2 \end{bmatrix} \qquad Q = \begin{bmatrix} Q_1 & 0 \\ 0 & Q_2 \end{bmatrix}$$

This means that if the result of equation (2.11.12) is used, the optimal control vector will be generated by:

$$\mathbf{m} = G_1(\mathbf{x}_0 - \mathbf{x}_1) - G_2\mathbf{x}_2 \quad (2.11.15)$$

which is shown in Figure 2.11.2. This solution means that the control is generated by a first-order system where the velocity is determined mostly by the matrix G_2. The advantage of this solution over the original one [equation (2.11.1)] is that matrix Q_2 [which was matrix P in equations (2.11.9)–(2.11.11)] need not be nonsingular. This is fortunate because the weighting of the control matrix expressed in equation (2.11.9) is rarely realistic in process control. However, the form in equation (2.11.13), where possibly $Q_2 = 0$, is very realistic.

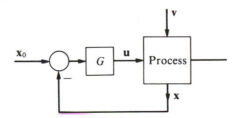

Figure 2.11.1. Multivariable control based on state feedback.

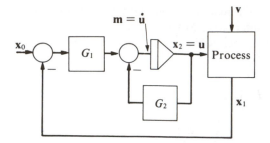

Figure 2.11.2. Multivariable control based on state feedback with consideration for the rate of change of the control vector.

The solutions shown in Figures 2.11.1 and 2.11.2 assume that all of the state variables in the process are measurable. This is not always the case. If we have fewer measurements than states, and/or if the measurements are masked by noise, then it might be possible to use a state estimator to compute the state variables that can be used for the feedback systems shown in Figures 2.11.1 or 2.11.2. This problem will be discussed further in section 2.13.

Even if the solution shown in Figure 2.11.2 is preferable to that of Figure 2.11.1, in the following derivations we will assume that the simplest solution is used. Applying this solution to a practical process control system is subject to the following conditions:

- A suitable process model must be developed.
- The process states must be measurable or obtainable from an estimator.
- The matrices Q and P in equation (2.11.9) must be given reasonable numerical values.
- The Riccati equation (2.11.11) must be solved.
- The final system dynamics [the eigenvalues of the matrix $(A - BG)$] must be acceptable.
- The stability margin of the resultant system must be acceptable; that is, one must be able to change any parameters or combinations of parameters in the process matrices A and B within realistic limits without the system becoming unstable ($\text{Re}\lambda_{A-BG} > 0$).

In many cases it is very difficult to develop a reasonable process model. In such cases this form of process control is unsuitable. In what follows, however, we will assume that the required model can be derived. In many applications all of the state variables can be measured or estimated, and we will assume that to be the case here. A few rather simple guidelines can help to establish the matrices of the objective functional, and these at least provide a starting point:

1. The matrices Q and P are first formed as diagonal matrices.

2. The elements on the diagonals of these matrices are chosen with magnitudes relative to the expected (or nominally acceptable) value of the corresponding variable. For example:

$$q_{11} = \frac{1}{(\Delta x_i)^2}$$

where Δx_i is a nominally acceptable variation relative to a reference of the state variable x_i. Correspondingly:

$$p_{ii} = \frac{1}{(\Delta u_i)^2}$$

It is difficult to associate a physical meaning with any terms that might lie off the diagonals of matrices Q and P, but these terms will affect the dynamic behavior of the process. There exists some theory regarding the relationship between the parameters of the weighting matrices and the resultant system eigenvalues (Solheim, 1976). This provides at least some help in formulation of these matrices. For systems having many state variables, the computations necessary for this formulation will require the use of a computer. There are several CAD (computer-aided design) programs that can do this.

Solution of the Riccati equation (2.11.11) also requires the use of a computer for most practical situations. Many algorithms and programs are available for this (Bierman, 1977; Laub, 1979; Tyssø, 1980).

The eigenvalues of the final system matrix $(A - BG)$, which results from the application of linear quadratic theory, will lie in the left half-plane, and the system will be stable. One cannot be sure, however, that the relative damping factors of the eigenvalues will be satisfactory, so they should be examined further. The relative damping factors are influenced above all by the weight put on the velocities of the control variables (the matrix P). It is also true that the behavior of a dynamic system is determined by the eigenvectors as well as the eigenvalues.

Even a system that has been designed on the basis of linear quadratic theory, and has its eigenvalues suitably placed in the complex plane when the process parameters have their nominal values, can become unstable owing to small variations in those parameters. A system with this characteristic is said to lack robustness. Since good robustness is a necessity for a practical system, it is important to find methods of selecting parameters that will provide the desired degree of robustness (stability margin). Work in this area is relatively recent, but several useful techniques have been developed. All of them require machine computation (Safonov and Athans, 1977; Lehtomaki, Sandell, and Athans, 1981; Doyle and Stein, 1981). A detailed account of the theory of "robustness" is given in Appendix E.

The advantage of using optimal control theory in designing a multivariable system is that it gives a direct answer to how one should coordinate the various control variables, depending on how the state variables develop.

Some methods for design of multivariable controllers are problematic with processes that display significant "non-minimum-phase" behavior. These processes have characteristics such as zeroes in the right half complex plane, "inverse response," dead time, and transport delay. This difficulty does not occur with the optimal control method based on a state space model. With this method, when the process tends to have inverse response, the only consequence is that the system matrices (*A* and *B*) will be changed somewhat, and the resulting system will be more sluggish. The optimal control method is a direct method of synthesis which when combined with the application of CAD, makes it possible to test different parameter values in the system matrices.

Several reasons may be suggested why the method has been slow to receive wide application to practical process control (Foss and Denn, 1976). One reason is that it requires a relatively good process model in order to assure robustness of the solution. Another is that computation and implementation of the solution require a computer, a requirement that is becoming less significant as the cost of computation continues to decrease. Many engineers think that such a system is much more difficult to adjust, or tune, than a system consisting of isolated single-variable control loops, because of the intense cross-coupling of virtually all of the control parameters. It has also been claimed that the systems that have been studied have not shown that optimal control theory provides significantly better control than other, simpler solutions. Most of these claims lack generality, however, and are based on specific experiences. There is no doubt that basic optimal control theory, with its extensions to handle typical process control problems, has proved to be a powerful tool for the designer (Rijnsdorp and Seborg, 1976; Ray, 1983; MacPherson, 1985).

2.12. MODAL CONTROL

One of the early methods proposed for design of multivariable feedback systems was based on "nondynamic" feedback from the state variables to the control variables with the objective of placing the system eigenvalues at the desired positions in the complex plane (Morgan, 1964; Gould, 1969; Rosenbrock, 1969; Balchen, Fjeld, and Solheim, 1970, 1978; Ray, 1981).

Consider the following mathematical process model:

$$\dot{\mathbf{x}} = A\mathbf{x} + B\mathbf{u} + C\mathbf{v} \qquad (2.12.1)$$

and assume that all of the states are measurable. That is:

$$\mathbf{y} = \mathbf{x} \qquad (2.12.2)$$

It is possible to find a negative feedback:

$$\mathbf{u} = -G\mathbf{x} \qquad (2.12.3)$$

such that we get:

$$\dot{\mathbf{x}} = (A - BG)\mathbf{x} + C\mathbf{v} \qquad (2.12.4)$$

where the eigenvalues of the matrix $(A - BG)$ lie at the desired positions.

If we formulate the modal matrix

$$M = (\mathbf{m}_1, \mathbf{m}_2, \ldots \mathbf{m}_n) \qquad (2.12.5)$$

where $\mathbf{m}_i$ is an eigenvector associated with the eigenvalue λ_i, we can then use the transformation

$$\mathbf{x} = M\mathbf{q}$$

where $\mathbf{q}$ is the modal state vector. If we apply this to equation (2.12.1), we get:

$$\dot{\mathbf{q}} = \Lambda\mathbf{q} + M^{-1}B\mathbf{u} + M^{-1}C\mathbf{v} \qquad (2.12.6)$$

where:

$$\Lambda = M^{-1}AM = \begin{bmatrix} \lambda_1 & & 0 \\ & \lambda_2 & \\ 0 & & \lambda_n \end{bmatrix} \qquad (2.12.7)$$

and λ_i are the process eigenvalues. For the sake of simplicity, assume that these values are distinct. Substituting equation (2.12.3) into equation (2.12.6), we get: get:

$$\dot{\mathbf{q}} = \Lambda\mathbf{q} - M^{-1}BGM\mathbf{q} + M^{-1}C\mathbf{v}$$
$$= (\Lambda - M^{-1}BGM)\mathbf{q} + M^{-1}C\mathbf{v} \qquad (2.12.8)$$

If we specify that the matrix $(\Lambda - M^{-1}BGM)$ be a diagonal matrix Λ_0 (it need not be so specified), we must find a matrix G that satisfies the equation

$$\Lambda - M^{-1}BGM = \Lambda_0 \qquad (2.12.9)$$

where Λ_0 contains the specified closed loop eigenvalues.

This problem is particularly easy to solve if matrix B is nonsingular (there are as many control variables as state variables). Then we have:

$$G = B^{-1}M(\Lambda - \Lambda_0)M^{-1} \qquad (2.12.10)$$

It is well known that if a process is controllable and if all the state variables are available, then one can find a feedback matrix G that will place the closed loop eigenvalues at any arbitrary positions in the complex plane. This means that it is not necessary that matrix B be nonsingular. However, it may not be easy to see how matrix G should be formed (Simon and Mitter, 1968).

It is not always obvious where the eigenvalues of a dynamic system should be placed. If one wanted the system to damp quickly, one would choose large eigenvalues with the desired damping ratio. In other situations, one might choose smaller eigenvalues in order to suppress disturbances with high frequencies. In general, we probably cannot say that larger eigenvalues will give better performance than small ones. Optimal control theory provides a rational answer to exactly this question. If fast response were required of a process that is itself quite slow, thus implying large displacements of the eigenvalues, then control variables with large magnitudes (high energy, fast mass transfer, etc.) would be needed. This is not always possible, so we see that one must consider the effects of both the process and the control variables.

There has been a great deal of developmental work in the use of modal control theory for the synthesis of control systems for *distributed* processes. One of the major problems here is to determine where to place the measuring element in order to get the best possible control of the process. This matter is considered in Balchen, Fjeld, and Olsen (1971) and Ray (1981).

2.13. MULTIVARIABLE CONTROL WITH STATE ESTIMATION

One of the most significant advances in modern control theory is the use of a state estimator to estimate (calculate) the value of a state variable that cannot be directly measured. A state estimator can also be used to extract an estimate of the real signal from a process measurement that is corrupted by noise. A state estimator is based on a mathematical model (programmed in a computer) that is operated in parallel with the actual process and is updated by feedback of the difference between the actual measurements and the estimated measurements. This is shown in Figure 2.13.1. Perhaps the best-known form of state estimator is the Kalman filter. Best known in the discrete form, this is actually an "optimal" estimator, in that its feedback matrix K is such that it minimizes the variance of the state estimate expressed as the trace of the covariance matrix (Kalman and Bucy, 1961; Maybeck, 1979–1982).

The most important characteristics of the Kalman filter are given by the expressions that follow.

The Continuous Kalman Filter

Assume that the process can be described by:

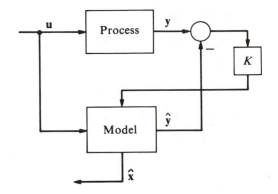

Figure 2.13.1. State estimator for multivariable process.

$$\dot{\mathbf{x}}(t) = A\mathbf{x}(t) + B\mathbf{u}(t) + C\mathbf{v}(t)$$

$$\mathbf{y}(t) = D\mathbf{x}(t) + \mathbf{w}(t)$$

The state estimate will be generated by the equation:

$$\dot{\hat{\mathbf{x}}}(t) = A\hat{\mathbf{x}}(t) + B\mathbf{u}(t) + K(\mathbf{y}(t) - D\hat{\mathbf{x}}(t))$$

where:

$$K(t) = X(t)D^TW^{-1}$$

and:

$$\dot{X}(t) = AX(t) + X(t)A^T + CVC^T - X(t)D^TW^{-1}DX(t)$$

In these equations:

$\mathbf{x}(t)$ = state vector for the process
$\mathbf{y}(t)$ = process measurement vector
$\hat{\mathbf{x}}(t)$ = estimate of the state vector
$X(t)$ = covariance matrix for $\mathbf{x}(t)$
V = covariance matrix for $\mathbf{v}(t)$
W = covariance matrix for $\mathbf{w}(t)$

The Discrete Kalman Filter

Assume that the process can be described by the discrete model:

$$\mathbf{x}(k + 1) = \Phi\mathbf{x}(k) + \Delta\mathbf{u}(k) + \Omega\mathbf{v}(k)$$

$$\mathbf{y}(k) = D\mathbf{x}(k) + \mathbf{w}(k)$$

The state estimate is now given by:

$$\bar{\mathbf{x}}(k + 1) = \Phi\hat{\mathbf{x}}(k) + \Delta\mathbf{u}(k)$$

$$\hat{\mathbf{x}}(k) = \bar{\mathbf{x}}(k) + K(k)(\mathbf{y}(k) - D\bar{\mathbf{x}}(k))$$

$$K(k) = \bar{X}(k)D^T(D\bar{X}(k)D^T + W)^{-1}$$

$$\bar{X}(k + 1) = \Phi\hat{X}(k)\Phi^T + \Omega V\Omega^T$$

$$\hat{X}(k) = (I - K(k)D)\bar{X}(k)$$

where:

$\mathbf{x}(k)$ = discrete process state vector
$\mathbf{y}(k)$ = discrete process measurement vector
$\bar{\mathbf{x}}(k)$ = *a priori* estimate of $\mathbf{x}(k)$
$\hat{\mathbf{x}}(k)$ = *a posteriori* estimate of $\mathbf{x}(k)$
$\bar{X}(k)$ = *a priori* covariance matrix of $\mathbf{x}(k)$
$\hat{X}(k)$ = *a posteriori* covariance matrix for $\mathbf{x}(k)$
V = covariance matrix for $\mathbf{v}(k)$
W = covariance matrix for $\mathbf{w}(k)$
$K(k)$ = Kalman filter gain matrix

Techniques, such as the Kalman filter, used for state estimation in linear systems can be extended to derive state estimators for systems that use non-linear, and therefore more accurate, dynamic models of real processes. This is of special interest in process control. The nonlinear state estimator shown in Figure 2.13.2 is based on a process model of the form:

$$\dot{\mathbf{x}} = \mathbf{f}(\mathbf{x}, \mathbf{u}, \mathbf{v}) \qquad (2.13.1)$$

$$\mathbf{y} = \mathbf{g}(\mathbf{x}, \mathbf{w}) \qquad (2.13.2)$$

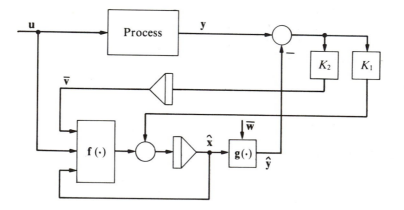

Figure 2.13.2. State estimator for nonlinear dynamic process.

It is assumed that the process disturbance vector (v) has an average value ($\bar{v}$) that changes slowly; that is, $\dot{\bar{v}} \approx 0$. Therefore, it can be modeled with a set of integrators and updated through a constant feedback gain matrix K_2. The gain matrix K_1 for the nonlinear estimator will usually be based on a linearization of the process around some nominal value or operating point. Then, linear estimation theory can be used as a first approximation. If the nominal values of the system state variables change very much, it may be necessary to make some of the elements of K_1 depend on average values of the state estimates.

A multivariable system that is developed on the basis of optimal control theory or modal control theory (as shown in Figure 2.11.1, perhaps), but which uses a state vector obtained from a state estimator, is shown in Figure 2.13.3. Matrix G, which is part of the solution, contains only "proportional" action (no integral action). The most common way to achieve what corresponds to integral action in a single-variable system is to model the disturbances (which are assumed to change slowly) by using integrators and feeding the values obtained back to the controller. It is easy to show that the largest allowable number of such integrators is equal to the number of measurements in the process. This number is also sufficient to obtain zero stationary control deviations (errors) of the various measured variables (Balchen et al., 1973).

A state estimator is useful for more than the design of multivariable control systems. When combined with instrumentation, it can help to give an operator a good overview of the process. For example, when combined with a suitable display, the state estimator can give an output plot of variables that are internal to the system and cannot be measured. The temperature profile and the reaction rate profile in a reactor are examples of this situation, and knowledge of the profile can be critical to successful operation of the reactor. It often happens that when an operator has a satisfactory estimate of the inner states of the process, control of the process can become so simple that manual control may be

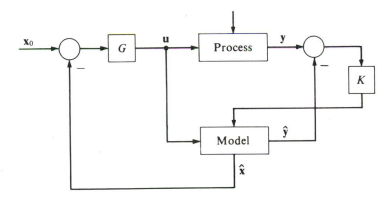

Figure 2.13.3. Multivariable control based on estimated process states.

sufficient. The concept of "state" itself often represents a clear description of the behavior of a dynamic process. Since most real processes that are subject to process control actually have many state variables (perhaps infinitely many), it is important to develop a model in which a reasonably small number of states describe the essential process phenomena. There has been, and continues to be, a great deal of research effort directed toward development of effective modeling principles in this respect. Many times a model based on first principles such as mass and energy balances leads to a very complicated model with many state variables. Other models, for example a model of temperature profiles in a distillation column, can provide a good process description with a very simple structure (Saelid, 1976; Gilles and Retzbach, 1980). There is no unique single structure of a mathematical (numerical) model for a given process. Many types of models can describe the same process, and the choice depends upon what one wishes the model to describe. Therefore, it is especially important to think carefully about the model structure problem when one wishes to derive the simplest possible system.

In section 2.7, we saw the use of a state estimator in a single-variable process with a large transport delay. Exactly the same, and an equally important, situation occurs in multivariable systems. However, in a multivariable system the various time delays can be placed in different positions in the model structure, and they can have many different values. Perhaps the simplest way to understand a multivariable system with many time delays is to use a discrete model in which the delays represent a number of "empty" time intervals in which one or more variables will be delayed by the number of time steps over which the variable can neither change nor be manipulated. A discrete state space model of a multivariable system with transport delays will be a higher-order system than one without delays, but it will not have more parameters. It is the number of parameters that really determines the complexity of the model. This can be illustrated by an example.

▽ EXAMPLE 2.13.1

Consider a discrete model of a linearized system described by:

$$\mathbf{x}(k + 1) = \Phi\mathbf{x}(k) + \Delta\mathbf{u}(k) + \Omega\mathbf{v}(k) \tag{2.13.3}$$

$$\mathbf{y}(k) = D\mathbf{x}(k - N) \tag{2.13.4}$$

We see from equation (2.13.4) that the elements of the measurement vector are all delayed by N time steps relative to the corresponding state variables. The corresponding transport delays are NT where T = the sampling interval.

This equation can be written to include a number of artificial state variables that account for the delay as follows:

$$\mathbf{x}_1(k + 1) = \Phi\mathbf{x}_1(k) + \Delta\mathbf{u}(k) + \Omega\mathbf{v}(k)$$

$$\mathbf{x}_2(k + 1) = \mathbf{x}_1(k)$$

$$\mathbf{x}_3(k + 1) = \mathbf{x}_2(k) \tag{2.13.5}$$

$$\mathbf{x}_{N+1}(k + 1) = \mathbf{x}_N(k)$$

$$\mathbf{y}(k) = D\mathbf{x}_1(k - N) = D\mathbf{x}_{N+1}(k) \tag{2.13.6}$$

A block diagram of equations (2.13.5) and (2.13.6) is given in Figure 2.13.4. We see that we have significantly increased the size of the state space in that we have N new state variables, each of which has the same dimension as the original state vector $\mathbf{x}_1$ (dimension n). We now have an extended state space of dimension $(N + 1)n$.

The extended system, which includes the transport delay, has the state vector

$$\tilde{\mathbf{x}}(k) = \begin{bmatrix} \mathbf{x}_1(k) \\ \mathbf{x}_2(k) \\ \cdot \\ \cdot \\ \cdot \\ \mathbf{x}_{N+1}(k) \end{bmatrix}$$

and a system matrix

$$\tilde{\Phi} = \begin{bmatrix} \Phi & 0 & 0 & 0 & \cdots & 0 \\ I & 0 & 0 & 0 & \cdots & 0 \\ 0 & I & 0 & 0 & \cdots & 0 \\ 0 & 0 & I & 0 & \cdots & 0 \\ 0 & 0 & 0 & I & \cdots & 0 \\ \cdot & \cdot & \cdot & \cdot & & \cdot \\ 0 & 0 & 0 & 0 & I & 0 \end{bmatrix}$$

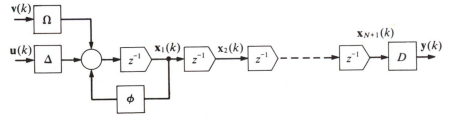

Figure 2.13.4. Block diagram of discrete process with delayed states.

$$\tilde{\Delta} = \begin{bmatrix} \Delta \\ 0 \\ 0 \\ \vdots \\ 0 \end{bmatrix}$$

We see that the essential information contained in the original system appears in the upper left-hand corner of the extended matrix $\tilde{\Phi}$. The computations that involve the rest of the terms are almost trivial, although they require some additional computer memory and time if N is large. An alternative form of the system is given by equations (2.13.3) and (2.13.4). This requires less computer storage and can be derived if one chooses the following new state space:

$$\mathbf{x}_1(k + 1) = \Phi\mathbf{x}_1(k) + \Delta\mathbf{u}(k) + \Omega\mathbf{v}(k)$$
$$\mathbf{x}_2(k + 1) = D\mathbf{x}_1(k)$$
$$\mathbf{x}_3(k + 1) = \mathbf{x}_2(k) \tag{2.13.7}$$
$$\mathbf{x}_{N+1}(k + 1) = \mathbf{x}_N(k)$$

$$\mathbf{y}(k) = x_{N+1}(k) \tag{2.13.8}$$

A block diagram representation of this system is shown in Figure 2.13.5, where the only important change is the fact that block D has been moved to the left, compared with Figure 2.13.4. We see that the system matrix now has the form

$$\tilde{\Phi} = \begin{bmatrix} \Phi & 0 & 0 & 0 \cdots & 0 \\ D & 0 & 0 & 0 \cdots & 0 \\ 0 & I & 0 & 0 \cdots & 0 \\ 0 & 0 & I & 0 \cdots & 0 \\ 0 & 0 & 0 & I \cdots & 0 \\ 0 & 0 & 0 & 0 \cdot\cdot I & 0 \end{bmatrix}$$

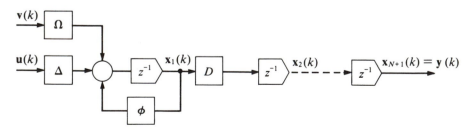

Figure 2.13.5. Block diagram of discrete process with delayed measurements.

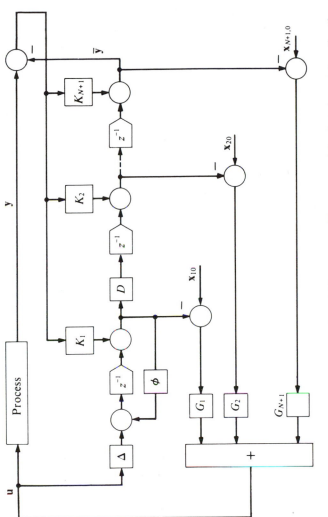

Figure 2.13.6. Discrete control of a multivariable process with transport delay based on state estimation.

It is clear that the new state vectors $x_2(k) \ldots x_{N+1}(k)$ have the dimension m, which is the dimension of the measurement vector. This will usually be less than dimension n. The result is that the new system matrix has a dimension of $n + nm$. It is true here, as before, that the model will contain no more parameters than the system without transport delay.

If the model in Figure 2.13.5 is used as a state estimator and the estimated states are used in a multivariable control system, the solution will be that shown in Figure 2.13.6. We see that in addition to the feedback from the original state variables, we have a "distributed" feedback from the delayed states. This also △ applies to the feedback in the state estimator.

Use of a state estimator in connection with multivariable control is especially effective when one has a good model of the physical phenomena involved in the system. It is then possible to estimate many process states on the basis of relatively few measurements. However, if one does not have good knowledge of the characteristics of the process, it is very difficult for an estimator to replace good measurements. In any case, good measurements are important in any form of control system, and it is often a matter of economics whether it is better to spend resources on detailed model building or improved instrumentation. Usually, the best approach is a compromise.

The question of how complicated a model must be in terms of order, nonlinearities, and so on, for use in a state estimator is a topic of considerable discussion. In this matter there are few fixed truths; many considerations that determine the answer change with time. One example is the fact that the cost of computing power must be considered before one can decide if a complicated model and state estimator should be implemented. Another consideration is whether the level of available knowledge of the process dynamics is sufficient to develop a complicated process model. As research and technology progress, these conditions can change, and what was once impossible can become possible.

It is important to recognize in this connection that it is possible when using a state estimator to use measurements other than those needed for direct control. There are many good examples of systems in which easy and/or inexpensive measurements are combined with a state estimator to give a better and cheaper result than could be obtained from more sophisticated instrumentation and no estimator. An estimator also provides *redundancy* in the system. For example, if one measuring instrument fails, control can be maintained by using the remaining measurements to update the state estimator that provides the control signals. This can be extremely important in critical processes.

Dynamic models of complicated processes will always contain a number of parameters that are poorly known or that change slowly for one reason or another. Therefore, we may need *automatic parameter estimation* just as we need state-variable estimation. Several effective methods for parameter estimation have been developed on the basis of theory closely related to the theory of state estimation. As a result, it is natural to implement parameter and state estimation in the same system (Sage and Melsa, 1971; Gustavsson, 1972, 1976; Åstrom and Eykhoff, 1974; Eykhoff, 1974; Goodwin and Payne, 1977; Ljung, 1978; Balchen, 1979a).

The concepts of *observability* and *identifiability* provide rules and guidelines for the limitations to the theory of state and parameter estimation. If the measurements provide insufficient information about a state or parameter, the estimator will not have good convergence. In fact, it may not converge at all! This problem can often be solved by adding new or different measurements, or moving the measurement to a different location in the system.

If often happens that the natural disturbances to a process are not the most efficient excitations with respect to parameter estimation. In such cases it may be necessary to create new disturbances in the system to "help" the parameter estimator. These extra disturbances must be planned in terms of timing, magnitude, and type such that the normal operation of the system is not disrupted more than necessary. There are many possibilities in this field, but the transition from theory to practical use simply has not yet been fully developed.

One difficulty that has inhibited the use of state and parameter estimation based on dynamic models with multivariable feedback is that tuning (adjustment) of the system may become complicated because the parameters in the estimator and the controller depend upon the parameters of the process. If multiple controllers are used in a multivariable process (independent control loops, diagonal control matrix, etc.), there will usually be no more than three parameters in each controller. These can usually be tuned one at a time. A correspondingly easy routine is necessary for tuning of the more complicated multivariable strategies. This has led to attempts to develop computing algorithms that make it possible to adjust multiple control parameters on the basis of a few key process parameters. There is still much to be done in order to develop acceptable, practical solutions in this area.

It is a *fundamental* of process control that it is possible to affect only the *future* behavior of a process. Since the process is dynamic, it is advantageous to know the future disturbances in order to suppress them by timely manipulation of the control variables. This is the purpose of modeling and estimation.

Prediction is thought to be difficult, but if one knows the phenomena that determine the future, it should be possible to predict the future with good accuracy. For example, it is easy to predict the arrival time of a train when one knows that the train is on the way, its route, and its average speed. It is also easy to predict the quality changes in a material flowing out of a pipe if one can measure the properties of the material somewhere upstream. A disturbance caused by harmonic oscillation can be estimated by means of a model (which is an oscillator) where frequency, phase, and amplitude are continuously updated. There are many successful practical examples of this sort (Saelld, Jenssen, and Balchen, 1983).

There has been a tendency toward two schools of thought in the choice of a model structure for use in a control system. One school believes that the model structure should be based on known physical phenomena that characterize the process; that is, a *first principles model*. Examples would be the known linear and nonlinear phenomena and the known process parameters. The other school tends toward the *black box approach,* which uses observed relations between the input and the output of the process to characterize a general, usually linear, transformation (transfer function) where all the parameters are unknown. The best method is probably a combination of the two in which a structure having physical meaning uses black box subsections. Dogmatic adherence to either school is not especially productive!

2.14. ADAPTIVE CONTROL

In the discussion of parameter estimation in section 2.13, it was mentioned that it is often necessary to adjust the parameters of a control system when the process changes for one reason or another. A system that does this automatically is known as an *adaptive* system. Because it is also sometimes necessary to automatically adjust the control parameters during initial startup of the process even if the process parameters are relatively constant, we also have what are known as *self-tuning controllers*. The similarity between these two systems is so strong that they can be treated as one.

Many adaptive control systems have been developed over the last 30 years or so, and several have appeared on the market. However, most of the developments in this area have been limited to adaptive controllers for single-variable systems (Åstrom and Wittenmark, 1973; Åstrom, 1977; Clarke and Gawthrop, 1981; Fjeld and Wilhelm, 1981; Isermann, 1982).

There are many ways to realize an adaptive system. The most important methods for single-variable processes are shown in Figure 2.14.1. It is important to distinguish clearly between the problems associated with the control algorithm and those connected with the adaptive action itself. For example, it is possible to adapt a relatively simple algorithm such as a PID controller, or, one could choose a special algorithm that requires adaption because it is close to singularity, or instability. With this, the question of robustness again enters the picture.

The literature contains very few practical examples of adaptive multivariable controllers based on parameter estimation in multivariable models. This is certainly due to the complexity of the problem. There is no doubt, however, that this subject will have much greater significance and utility once it has been developed to the point where it has the reliability necessary for practical application. A complex solution requires considerable computer hardware and software capability. At the moment, many practicing engineers believe that the probability of a failure someplace in the system is too high. The practical de-

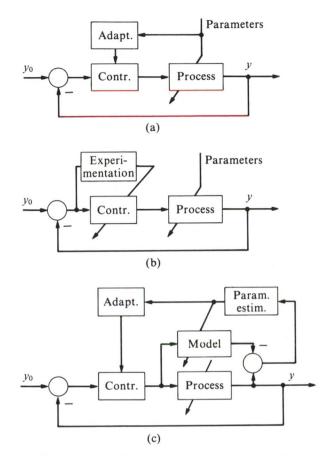

Figure 2.14.1. Three structures for adaptive control of single-variable processes.

velopment of a complex program system requires much time and manpower, and a potential user is not likely to expend the necessary development time and effort without being sure that the likely results, in terms of improved operating characteristics, improved quality, improved production, and so forth, offer a favorable investment–risk relationship.

Engineers disagree about what characteristics a control system must have in order to be called adaptive. One definition is that an adaptive system is one that optimizes (finds either a maximum or a minimum) an objective function with respect to the variations of the controller parameters. This is probably the most common definition. Another opinion holds that feedback from a process variable (perhaps as measured by a complicated device such as a chromato-graph, analyzer, etc.) used to adjust (trim) a feedforward control or ratio con-troller located earlier in the process is also an adaptive system. This type of system includes no direct form of optimization.

It is important to recognize that ordinary feedback systems and adaptive systems based on some form of feedback, such as parameter estimation with model adjustment, always have a potential stability problem. The practical control system designer must insist that systems have very high operating reliability, and stability plays a central role in this connection. This makes it easy to prefer systems that use feedforward as indicated in Figure 2.14.1b; they do not have inherent stability problems. Unfortunately, feedforward can be realized only when one has measurements of the most important disturbances and a reasonably good knowledge of how the controller parameters should change as a function of these disturbances. Since these requirements often cannot be met, one should not expect feedforward adaptive control to be precise. Further improvements in the adjustment of the controller parameters must be made with the help of feedback of the actual behavior.

Details of the most important algorithms for adaptive control of single-variable systems are given in Appendix D.

Examples of results that have been obtained with adaptive control of several different industrial processes are given in Chapters 4 and 5. An excellent recent survey of applications of adaptive control to process control is given by Seborg, Edgar, and Shah (1986).

2.15. PREDICTIVE CONTROL

It was stated in sections 2.11 and 2.13 that a control system can affect only the future behavior of a process. Therefore, in order to develop effective optimal control it is important not only to be able to predict which disturbances will affect the process in the future, but also how the process will respond to different control variables. A dynamic process model is concerned with exactly this point. It was also stated in section 2.11 that the solution of a real optimal control problem involves a two point boundary value problem. The amount of computation involved in this is usually relatively large, so it can be beneficial to look for simpler solutions. One possible simplification, especially in the case of batch processes (batch pulp digesters, batch reactors, etc.), is to assume certain simple patterns that the various control variables can follow (for example, piecewise constant, piecewise linear, etc.).

Figure 2.15.1 shows the principles of a structure that can be called predictive control. The system consists of a state and parameter estimation routine that works in real time. The parameters of the process model are continuously updated. A model with exactly the same structure and with updated parameters, but with a much smaller time scale, perhaps 100 or 1000 times faster than the real process, is made to produce responses to fast repetitions of different control actions (within given restrictions). To do this, one needs an optimization routine and a control-variable generator that can compute the appropriate control-variable values for perhaps two to five sample periods in the future.

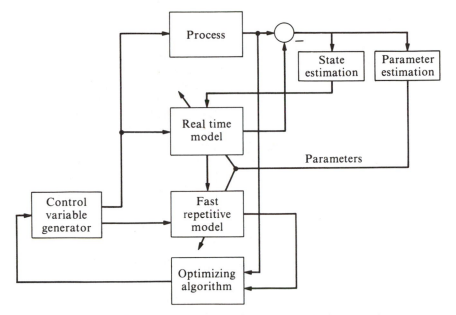

Figure 2.15.1. Block diagram of system for predictive control.

This control-variable behavior is chosen for the real process during the first period, and the same process is repeated in subsequent periods.

This very simplified realization of the two point boundary value problem is especially useful when the process has significant nonlinearities. Figure 2.15.2 illustrates the progress of a search of different possible control actions for the case of a single variable. Several industrial applications of this principle have been reported (Wells, Johns, and Chapman, 1975; Irving, Boland, and Nicholson, 1979; Balchen, 1980).

The literature in recent years has reported several control algorithms that are closely related to the predictive control technique described above, as well as the optimal control techniques outlined in section 2.11. These methods, based on predictions derived from an input–output model in the form of a multivariable impulse response matrix, bear such names as Model Algorithmic Control (MAC) (Rouhani and Mehra, 1982), Model Predictive Heuristic Control (Richalet et al., 1978), and Dynamic Matrix Control (DMC) (Cutler and Ramaker, 1980). Of these, we will describe only MAC.

The main steps in the MAC algorithm as applied to a single-variable system are shown in Figure 2.15.3. By means of a discrete impulse response model of the process (a model that must be established beforehand), one computes the predicted process response $y_p(t)$ an interval T into the future on the basis of the behavior of a control variable during the same interval as determined by an optimization procedure. We want $y_p(t)$ to follow a reference trajectory $y_r(t)$ that

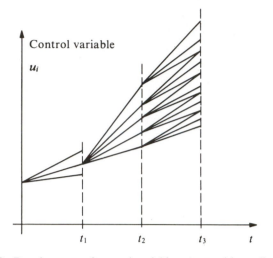

Figure 2.15.2. Development of control-variable pattern with predictive control.

takes $y_p(t)$ to a set point y_0. The optimization procedure finds the control variable trajectory that minimizes the objective function:

$$J_N = \sum_{p=1}^{N} (y_p(k + p) - y_r(k + p))^2 \cdot \omega_p \tag{2.15.1}$$

where ω_p is a weighting factor. The remaining notation in the above equation corresponds to that used in Figure 2.15.3.

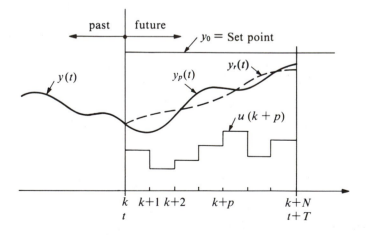

Figure 2.15.3. The variables in the MAC control algorithm.

The output $\mathbf{y}(t)$ of a multivariable dynamic system in the steady state (that is, the transients due to the initial conditions have died out) can be described by the convolution integral:

$$\mathbf{y}(t) = \int_{t_0}^{t} H(t - \tau)\mathbf{u}(\tau)\, d\tau \qquad (2.15.2)$$

For a system of the type

$$\dot{\mathbf{x}} = A\mathbf{x} + B\mathbf{u}, \qquad \mathbf{y} = D\mathbf{x} \qquad (2.15.3)$$

we get

$$H(t) = De^{At}B \qquad (2.15.4)$$

The discrete form of equation (2.15.2) is:

$$\mathbf{y}(k + N) = \mathbf{y}(k) + \sum_{p=1}^{N} H(N - p)\mathbf{u}(p) \qquad (2.15.5)$$

It is common to specify a reference trajectory $y_r(t)$ that approaches the set point exponentially from its current value. In the discrete form this leads to a formulation that for the scalar case is:

$$y_r(k + p) = y_0 - \alpha(y_0 - y(k + p + 1))$$
$$= \alpha y(k + p - 1) + (1 - \alpha) y_0 \qquad (2.15.6)$$

where $0 < \alpha < 1$ is a constant that determines the time constant of the exponential function, and $y(\cdot)$ is the measured output of the process.

MAC requires relatively high computing power for solution of the optimization problem [the minimization of J_N in equation (2.15.1)] and large memory for storage of the impulse response (between 10 and 50 points per response), but this will be less important as the cost of computation decreases. A major advantage of the method is that it is quite robust with respect to variations in the process parameters (a result of a high degree of redundancy). Marchetti, Mellichamp, and Seborg (1983) have presented a reformulation of the DMC algorithm that clearly illustrates the relationship of these methods to discrete optimal control. An excellent review of these results is presented in Deshpande (1985).

2.16. STEADY-STATE AND DYNAMIC OPTIMIZATION OF CONTINUOUS PRODUCTION PROCESSES

The desire to operate industrial processes optimally is as old as industry itself, but the ability to do this automatically is relatively new. The theoretical basis on which methods for automatic optimization have been developed has been available for the last 30 years or so. More important is the fact that the computing power required for operation of such systems has become available at an attractive price only during the last 10 years at most—perhaps actually within the last 5 years.

During process operation, one chooses between two types of optimization: *steady-state optimization* and *dynamic optimization*. The difference between the two is that dynamic optimization considers and uses dynamic process effects, while steady-state optimization is concerned with the process in the steady state. The theory of steady-state optimization is much simpler than that of dynamic optimization. How much can be gained by the use of the more complicated dynamic optimization depends upon whether the process and the states that affect the evaluation of process performance are significantly nonstationary (transient, or dynamic).

Solution of the dynamic optimization problem actually falls under the category optimal control, which was discussed in section 2.11.

Steady-state optimization can be illustrated by the simple example shown in Figures 2.16.1 and 2.16.2. For simplicity, these figures show a multivariable

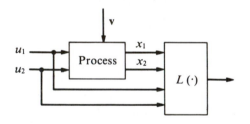

Figure 2.16.1. Multivariable dynamic process with calculation of objective function.

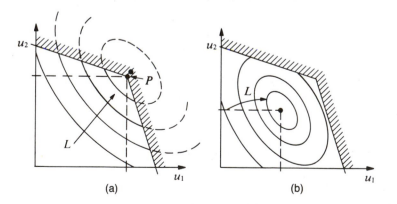

Figure 2.16.2. Steady-state objective function with constraints on control-variable space.

process having two control variables, two states, and multiple disturbances. The steady-state yield of this process is described by the objective function $L(\mathbf{x}, \mathbf{u})$, which can be evaluated at any time. Under stationary conditions, $\mathbf{x}$ will be a function of $\mathbf{u}$ and $\mathbf{v}$. Assume that we have a fixed value of the disturbance vector in the two-dimensional case. The value of the objective function produced as a function of the control variables is shown in Figure 2.16.2 for two different cases. We see here that the control variables have definite limits determined by the process. In this case, two different phenomena cause the limits indicated by the nearly straight lines in the $u_1 u_2$ plane. Notice in the figure to the left that the contours for constant values of L increase monotonically toward the corner labeled P. Under stationary conditions the optimal solution is apparently to force u_1 and u_2 up to the value indicated by P. In the figure to the right, we see a case in which the objective function L has a maximum in about the middle of the area of the figure not bounded by the two limit lines.

If we consider the disturbance vector $\mathbf{v}$, we can examine two different cases: (1) when $\mathbf{v}$ represents very slow phenomena, and (2) when $\mathbf{v}$ represents very rapid excitations that quickly affect $\mathbf{x}$. When $\mathbf{v}$ contains only slow components and the process itself is stable (has a high degree of self-regulation), it makes sense to design an optimization system that finds the optimum of $L(\cdot)$ with respect to $\mathbf{u}$ as shown in Figure 2.16.3.

If $\mathbf{v}$ contains significant dynamic excitations, then it is natural to equip the process with ordinary feedback by means of controllers such as shown in Figure 2.16.4 applied to the process shown in Figure 2.16.1. In this case, feedback is obtained from measurements of x_1 and x_2 that are compared with the constant set points x_{10} and x_{20}. Since the control variables u_1 and u_2 are now generated on the basis of dynamic feedback, process disturbances will now lead to different values of the function $L(\cdot)$ compared with those obtained from the system shown in Figure 2.16.1. The independent variables will now be x_{10} and x_{20} instead of u_1 and u_2. Thus, we must consider that there exists a relationship between x_{10} and x_{20} as inputs and $L(\cdot)$ similar to that shown above.

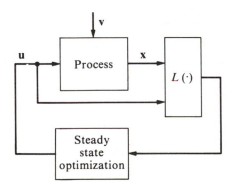

Figure 2.16.3. Multivariable process with system for steady-state optimization.

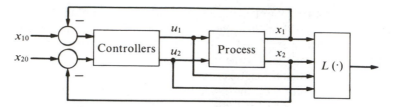

Figure 2.16.4. Multivariable, dynamic process with control and computation of objective function.

We can now provide the system shown in Figure 2.16.4 with a means of steady-state optimization as shown in Figure 2.16.6. This solution is called *set point optimization* because it causes a relatively slow adjustment of the set point of each individual controller in the system. This solution is widely used in industrial practice. The steady-state optimizing routine should actually include adjustment of certain parameters in the controller, the proportional gain setting for example, because these parameters determine the responses of **u**. This is seldom done in practice because the values of **u** are usually considered to be less important than the values of **x**.

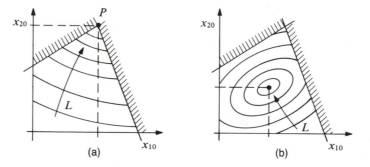

Figure 2.16.5. Steady-state objective function with constraints on state space.

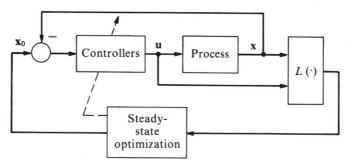

Figure 2.16.6. Steady-state optimization of set points in controlled, multivariable process.

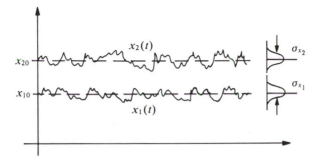

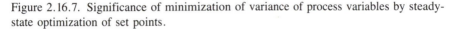

Figure 2.16.7. Significance of minimization of variance of process variables by steady-state optimization of set points.

In a typical industrial situation one would like to control a process so as to arrive at the "minimum variance" of $\mathbf{x}$ (and possibly $\mathbf{u}$ as well), as shown in Figure 2.16.7. The set point values x_{10} and x_{20} can then be moved closer to the boundaries, as indicated in Figure 2.16.5a. This can have great economic importance in cases where a process variable must not exceed a certain limit. An increase of the mean value of a process variable by as little as 1% or less can often lead to considerable economic gains in terms of such things as production, energy, or raw materials.

Steady-state set point optimization is especially common in large integrated process systems (oil refining, petrochemicals, pulp and paper, etc.) because small changes in process yield can seriously affect profits.

The usual way to set up a system for static set point optimization is to use a hierarchial structure in which the optimization calculations are at a level above the computations that provide the rapid process control functions. It often happens that installation of a high-level optimization system occurs only after successful solution of the control problem. Many times the control system is supplied by one vendor, while another may be responsible for the equipment and software for the optimization system.

A relatively recent technique (Balchen, 1984b), referred to as *quasi-dynamic optimal control,* addresses the problem of *dynamic optimization* of small, rapid disturbances around a *steady-state optimal solution* derived from a realistic profit criterion for the process. The control strategy is based on two different objective functions $L_1(\bar{\mathbf{x}}, \bar{\mathbf{u}})$ and $L_2(\Delta\mathbf{x}, \Delta\mathbf{u})$. The quantity $L_1(\cdot)$ is the realistic profit function operating on steady-state (or slowly varying) values $\bar{\mathbf{x}}$ and $\bar{\mathbf{u}}$ of $\mathbf{x}$ and $\mathbf{u}$. The quantity $L_2(\cdot)$ is a quadratic objective function operating on the deviations $\Delta\mathbf{x}$ and $\Delta\mathbf{u}$ around $\bar{\mathbf{x}}$ and $\bar{\mathbf{u}}$.

Figure 2.16.8 is a block diagram of the solution. It contains a nonlinear state estimator that produces estimates of the mean values of the disturbances ($\bar{\mathbf{v}} = \hat{\mathbf{v}}$), the mean values of the states ($\bar{\mathbf{x}}$), and the deviations of the states ($\Delta\mathbf{x}$). The quasi-dynamic optimal control vector $\mathbf{u}$ is made up of two components, $\Delta\mathbf{u}$

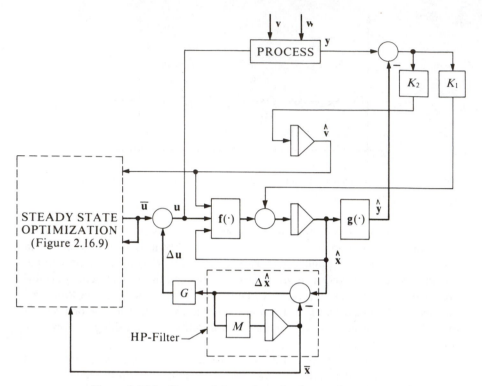

Figure 2.16.8. The quasi-dynamic optimal control strategy.

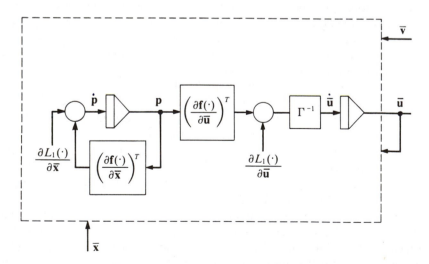

Figure 2.16.9. The steady-state optimizer.

generated by the LQG-strategy from $\Delta\mathbf{x}$, and $\bar{\mathbf{u}}$ generated by a steady-state optimization procedure from $\bar{\mathbf{x}}$ and $\bar{\mathbf{v}}$. The basic steady-state optimizer is shown in Figure 2.16.9. Because of stability considerations, a slight modification of the basic scheme is necessary (Balchen, 1984b).

2.17. PRACTICAL DESIGN OF PROCESS CONTROL SYSTEMS

In the previous sections, we discussed several basic control structures, but we have not said much about how they are to be realized in terms of actual equipment. In each case we started with the idea that the processes and the control functions were continuous, and we described their dynamic characteristics with differential equations and transfer functions. If we want to use digital systems (large or small programmable computers) to perform the control functions in actual applications, we must have computer programs to carry out the dynamic and algebraic functions needed for control. If the process sampling time is short compared with the dominant time constant of the process and the typical period of the disturbances, then it is quite simple to translate results from the continuous to the discrete case. However, if the sampling time is long compared with the dominant time constant of the system, then the effect of discretization will be more important. A very simple method for designing discrete control systems, which has the advantage of making it possible to use all of the techniques of frequency response analysis of continuous, linear, single-variable systems, is the w-transform technique (Balchen, 1967).

Use of the w-transform, which is described in Appendix A, can be summarized as follows:

1. Determine the discrete transfer function $h(w)$ with the help of the continuous transfer function $h(s)$ and the tables of Appendix A.
2. Design a discrete controller $h_c(w)$ in exactly the same manner one would use in setting up a continuous controller $h_c(s)$.
3. On the basis of $h_c(w)$, find $h_c(z^{-1})$ using:

$$q = \frac{2}{T} \cdot \frac{z-1}{z+1} = \frac{2}{T} \frac{1-z^{-1}}{1+z^{-1}}$$

4. The control algorithm in the time domain can then be written as:

$$\frac{u}{e}(z^{-1}) = h_c(z^{-1}) = \frac{g_0 + g_1\bar{z}^1 + g_2\bar{z}^2}{1 + f_1\bar{z}^1 + f_2\bar{z}^2}$$

which, after substitution, gives:

$$u(k) = -f_1 u(k-1) - f_2 u(k-2) + g_0 e(k) + g_1 e(k-1) + g_2 e(k-2)$$

for the case of a second-order controller.

Optimal control of multivariable systems, including state and parameter estimation, is usually described in discrete space form, as the discrete form is easier to derive and implement than the continuous form. A good example of this is the Kalman filter. There is now available a wide variety of computer software for discrete control, state estimation, and parameter estimation, which has been tested in industrial practice. These programs not only contain the basic control algorithms, but they also perform the important functions of reading and storing data, protecting the data, man–machine communication, output of data and results, and so on. The control algorithms are often just a small part of the total program package.

The description of the theory and principles of automatic control omits many practical details that can matter a great deal if they are neglected in practice. Two examples are:

1. Bumpless transfer of a controller from manual to automatic mode, or vice versa.
2. Automatic limiting of integral action when the control variable reaches saturation (prevention of reset windup).

Bumpless transfer has proved to be a problem in multivariable systems having multiple controllers with integral action. It is important that the integrators in the controllers are correctly charged (that is, they must give the proper values for the control variables) when the system is switched from manual to automatic control. There are various ways to do this, either with instrument hardware or computer programming, but how it is done will not be discussed here.

Reset windup is familiar to anyone who has watched the behavior of a system with integral control when the control variable reaches saturation. Since the control deviation will then be large for a long time, the integrator in the controller will integrate to a high value, usually its maximum. The integrator will not immediately follow the control variable as the variable begins to move away from the saturation limit. The result will be large oscillations due to the phase angle between the integrator and the variable.

There are various ways of preventing integrator (reset) windup. Some methods use the controller deviation $[e(t)]$ as the basis for a simple logical coupling to unload (unsaturate) the integrator. Others perform this function on the basis of the controller output to the control variable. Although we will not discuss any of these methods further, it is important to remember that the problem exists, and that practical solutions are available.

The practical implementation of large control structures can take many different forms. In the early days of process control, control functions were performed by analog equipment consisting of electric, pneumatic, or hydraulic components. Later (about 1960) digital equipment came into use. In 1986, well over 50% of all newly installed process control equipment used computers to perform the control algorithms. The availability of steadily less expensive computers having more computing power, more storage capacity, greater reliability, and better man–machine communication has made computer control very attractive. It does not matter, however, whether the required functions are performed by analog or digital equipment; the mathematics that describe the function are hardware-independent, in theory at least. Improvements in digital equipment and software will be responsible for the continuing replacement of analog equipment. There may be no ultimate limitations to what can be done with digital equipment, especially in terms of control structures and algorithms. However, analog equipment has several inherent advantages, and there is no doubt it will continue to have a place in process control.

There has been extensive development in the structural design of hardware and software for computer control of large industrial complexes since about 1960, when the first systems were put into operation. Contrary to what was (and is) the practice of associating one piece of apparatus with one function in analog control systems, the very first computer control systems were designed to use one central computer to perform all of the control functions for a large industrial process having many control loops. The object was to maximize the use of the expensive computer. This led to systems with high data-transmission costs and often quite poor reliability, in that the entire process operation depended upon the single computer. As computers became less expensive, it became practical to *decentralize* the control system by placing several smaller, less expensive computers throughout the industrial plant. The control computers in this *distributed control system* can communicate with a *supervisory* computer that coordinates the entire *network* (see Figure 2.17.1). Even if this structure does not provide the theoretical maximum possibility for communication be-

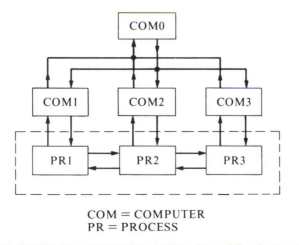

COM = COMPUTER
PR = PROCESS

Figure 2.17.1. Partitioning a process into sections for multilevel computer control.

tween the various process sections, in practice it meets most of the needs and has many practical advantages compared with the centralized computer structure. If one computer in the system should fail, the system can be manually controlled, or a spare computer can take over the function. This can be done either manually or automatically. There are many methods of accomplishing this backup, but they will not be discussed in detail here.

Since the early 1970s, we have seen the development of techniques and hardware for digital control systems using buses and networks for communication between different devices. These devices and communication protocols not only provide very high-speed communication, but they also allow for easy and almost random addition or removal of units from the system. The result has been decentralized, flexible control system networks. They have also made possible interactive operator stations with excellent man–machine communications through graphics as well as traditional forms of data display. Many vendors now supply complete systems of this type, but custom software is also very important. No industrially accepted standard for computer control has yet emerged. Because of the very rapid development in the field of data networking, the best source of information is the technical literature published by the equipment vendors (for example, Digital Equipment Corporation, 1982).

Chapter 3
PROCESS SYSTEMS AND PROCESS MODELS

The solution of a control problem generally requires some insight into the process that is to be controlled. A process can be anything from a control valve in a length of pipe (simple flow control) to an enormously comprehensive and complicated physical and/or chemical complex such as a system of distillation columns, or any complex process comprised of many process subsections. Many studies, calculations, and other considerations are involved in the planning, design, and construction of these subsystems, and in practice virtually all branches of the engineering profession become involved in putting together a large, complete process such as a pulp or paper mill, oil refinery, petrochemical plant, or electro-metallurgical plant.

3.1. GENERAL COMMENTS ON PROCESS DYNAMICS AND MODELS

It is undoubtedly a problem that the breadth of professional expertise necessary to solve both process and control problems is so wide that it is unreasonable to expect any one person to have complete competence in all areas. As a result, such projects are handled by groups of professionals, a situation that, in turn, creates an important need for good communications between engineers of different backgrounds. A chemical process engineer must, when designing a new process, know enough about the process control problem to communicate with the control engineer. On the other hand, the control engineer must know enough about the fundamentals of the process to be able to communicate with the chemical engineer and thus learn enough about the process to choose the best control solution.

Important means of communication between different professional disciplines include graphical means such as flowcharts, piping and instrument diagrams (P&IDs), and mathematical/numerical models of process units (transfer functions, frequency response representations, and state space equations).

Unfortunately, a systematic library of mathematical models of different process units has not yet been adequately developed. It would be very helpful to have good documentation, preferably in a standard format, and one or more mathematical models of each of the various types of subsections (*unit processes*) that are found in most industrial processes. These models could then be interconnected in the proper order to form a model of the total complex process. Because process equipment manufacturers normally do not supply this information, in this book we attempt to provide models to use as bases for the systematic formulation of control problems and their solutions.

We expect that within a few years it will be possible to apply CAD (computer-aided design) techniques to the development of models, with quantitative indications of process parameters for the most important unit operations, and to the combination of these models to form larger, complex process models. These techniques will enable the testing of different control strategies on the same process — a major step toward the general application of more advanced process control systems.

In this chapter we will present an overview of some of the most important industrial unit operations used as components of larger processes, and we will develop and/or present steady-state and dynamic models for some of them. This will provide a foundation for Chapters 4 and 5, where we will present control system solutions for elementary and more complex processes, and for Chapter 6, where we will discuss how all of this information fits together in the design and control of large industrial systems.

3.2. UNIT OPERATIONS

Unit operations play an important role in the development of chemical processes. They provide building blocks, or process segments, that can be analyzed and designed individually and then combined into larger processing systems. Unit operations include physical and chemical phenomena such as mass transport, mixing, distillation, separation processes, heat transfer, evaporation, mass transfer, gas absorption, combustion, extraction, crystallization, and drying. Most industrial plants include one or more of these elements. In addition, various chemical reactions with special characteristics can be carried out in specialized reactors or sequences of unit operations.

An accurate and detailed description of what happens in a unit operation must be based on the fundamentals of physics and chemistry. Therefore, it is important to use careful, unambiguous formulations and symbolic representations. However, not all of the mechanical design details are important to the control engineer, who considers the process from the standpoint of a mathematical/numerical model for use in solving the control problem.

The principles of dynamic modeling and automatic control can have a significant influence on the technology, design, and physical layout of modern production processes. As process sophistication increases, dynamic and automatic control considerations must become integral parts of process design. The technology and layout of a process that is to be automatically controlled can be influenced by considerations involved in good control system design. Such a process might be designed quite differently from one that must rely on a high degree of self-regulation. An example of this is a reactor that is theoretically unstable but can be made practical with stabilization by a control system.

Coordinated treatment of both the control problem and the process problem can produce changes in process design that will ultimately lead to a well-controlled efficient process. It should be noted that the demands on a process in terms of design and equipment from the control point of view can be quite different from those due to conventional process requirements. From the control standpoint, it is often necessary for one part of a process to respond very rapidly (have short time constants) if it is to influence other sections of the system. If this requirement is not recognized in the design of the process, the designer might get an unwelcome surprise when automatic control is attempted! Unfortunately, on some occasions the demands of the process technology are contrary to the demands of the control system, and it is not always easy to reach a compromise. It seems to have taken far too long for process designers to recognize this.

In this book, we usually simply present the static and dynamic characteristics of the various unit operations, but in some cases we derive them. The literature presenting the details of this important subject is extensive, and only a few major sources are listed here (Campbell, 1958; Buckley, 1964; Himmelblau and Bischoff, 1968; Douglas, 1972; Perry and Chilton, 1973; Foust et al., 1980; Geankoplis, 1983; McCabe, Smith, and Harriott, 1985).

No useful systematic overview of steady-state and dynamic characteristics of the various unit operations can be found in the literature; so in this chapter we will make a systematic presentation of information that has been published on the steady-state and dynamic behavior of unit operations.

3.3. STEADY-STATE AND DYNAMIC CHARACTERISTICS

Most chemical processing equipment now in use was sized according to expected steady-state operating conditions. Therefore, static models have been the customary basis for design. The purpose of most processes is to provide high production, high profit, good reliability, good product quality, low failure rates, and so on. Dynamic (nonstationary) considerations have too often been relegated to a very minor role in design considerations, at least until quite recently. Therefore, the dynamic characteristics have tended to be those that simply "came with the system" rather than the result of any particular intent. As we have said, this is sometimes quite acceptable, but in other cases it can lead to completely inappropriate dynamic behavior.

From the standpoint of process control, both static and dynamic characteristics are important. Obviously, if one has a complete dynamic model based on perfectly descriptive nonlinear partial differential equations, then the static characteristics can be treated as a special case of the dynamics. However, studies of such nonlinear models usually require extensive computation, and the proper presentation of the results becomes complicated and perhaps confusing. Therefore, it is not unusual to limit the model to *nonlinear static characteristics* and *dynamic characteristics* on the basis of a *linearization* of the behavior in response to small perturbations around nominal operating points. One can often use linearized models expressed in the form of transfer function matrices or frequency response matrices. If one can describe the process with reasonable accuracy by means of *lumped parameters,* which lead to ordinary differential equations, then a nonlinear model can be effectively expressed in the state space. From this it is quite easy to derive a transfer function matrix for the dynamics of the system linearized around a stationary operating point.

Assume we have a nonlinear state space model of a lumped parameter system given by:

$$\dot{\mathbf{x}} = \mathbf{f}(\mathbf{x}, \mathbf{u}, \mathbf{v}) \tag{3.3.1}$$

$$\mathbf{y} = \mathbf{g}(\mathbf{x}, \mathbf{u}, \mathbf{v}) \tag{3.3.2}$$

The steady-state (stationary) condition $\mathbf{x}^0$ (singular point) can be found by setting the right side of equation (3.3.1) equal to zero. Thus:

$$f(\mathbf{x}^0, \mathbf{u}^0, \mathbf{v}^0) = \mathbf{0}$$

So:

$$\mathbf{x}^0 = \mathbf{h}(\mathbf{u}^0, \mathbf{v}^0) \tag{3.3.3}$$

which expresses the static characteristics of the process states. Substituting the expression from equation (3.3.3) into equation (3.3.2) gives:

$$\mathbf{y}^0 = \tilde{\mathbf{g}}(\mathbf{u}^0, \mathbf{v}^0) \tag{3.3.4}$$

where $\tilde{\mathbf{g}}(\cdot)$ simply indicates that we have substituted one function into another. If we let the variables change slightly around the nominal values, we will have:

$$\delta\mathbf{x} = \mathbf{x} - \mathbf{x}^0, \qquad \delta\mathbf{u} = \mathbf{u} - \mathbf{u}^0, \qquad \text{and} \quad \delta\mathbf{v} = \mathbf{v} - \mathbf{v}^0$$

This can be used in the series:

$$\dot{\mathbf{x}} = \delta\dot{\mathbf{x}} = \mathbf{f}(\mathbf{x}, \mathbf{u}, \mathbf{v}) = \mathbf{f}(\mathbf{x}^0, \mathbf{u}^0, \mathbf{v}^0) + \left.\frac{\partial \mathbf{f}}{\partial \mathbf{x}}\right|_{\mathbf{x}^0} \delta\mathbf{x} + \left.\frac{\partial \mathbf{f}}{\partial \mathbf{u}}\right|_{\mathbf{u}^0} \delta\mathbf{u} + \left.\frac{\partial \mathbf{f}}{\partial \mathbf{u}}\right|_{\mathbf{v}^0} \delta\mathbf{v}$$

$$\tag{3.3.5}$$

which gives:

$$\delta\dot{\mathbf{x}} = A(\mathbf{x}^0, \mathbf{u}^0, \mathbf{v}^0)\delta\mathbf{x} + B(\mathbf{x}^0, \mathbf{u}^0, \mathbf{v}^0)\delta\mathbf{u} + C(\mathbf{x}^0, \mathbf{u}^0, \mathbf{v}^0)\delta\mathbf{v} \tag{3.3.6}$$

where $A(\cdot)$, $B(\cdot)$, and $C(\cdot)$ are matrices whose elements will be functions of the nominal values of the vectors.

If we look at the response of the measurement vector $\mathbf{y}$, we see the perturbations:

$$\delta \mathbf{y} = \mathbf{y} - \mathbf{y}^0 = \mathbf{g}(\mathbf{x}^0 + \delta\mathbf{x}, \mathbf{u}^0 + \delta\mathbf{u}, \mathbf{v}^0 + \delta\mathbf{v}) - \mathbf{g}(\mathbf{x}^0, \mathbf{u}^0, \mathbf{v}^0)$$

$$\cong \frac{\partial \mathbf{g}}{\partial \mathbf{x}}\bigg|_{x^0} \delta\mathbf{x} + \frac{\partial \mathbf{g}}{\partial \mathbf{u}}\bigg|_{u^0} \delta\mathbf{u} + \frac{\partial \mathbf{g}}{\partial \mathbf{v}}\bigg|_{v^0} \delta\mathbf{v}$$

$$= D(\mathbf{x}^0, \mathbf{u}^0, \mathbf{v}^0)\delta\mathbf{x} + E(\mathbf{x}^0, \mathbf{u}^0, \mathbf{v}^0)\delta\mathbf{u} + F(\mathbf{x}^0, \mathbf{u}^0, \mathbf{v}^0)\delta\mathbf{v} \qquad (3.3.7)$$

where $D(\cdot)$, $E(\cdot)$, and $F(\cdot)$ are also matrices whose elements are functions of the operating points.

We can use the Laplace transform of equation (3.3.6) to find the different transfer function matrices between the control variables and the disturbances to the system and between the process states and the output measurements. We then have:

$$H_{xu}(s) = (sI - A)^{-1}B \qquad (3.3.8)$$

$$H_{xv}(s) = (sI - A)^{-1}C \qquad (3.3.9)$$

$$H_{yu}(s) = D(sI - A)^{-1}B + E \qquad (3.3.10)$$

$$H_{yv}(s) = D(sI - A)^{-1}C + F \qquad (3.3.11)$$

It should be mentioned that this development also holds for discrete systems in which the z or w transformations are applied.

The relationships between the linearized dynamic characteristics given by equations (3.3.8) through (3.3.11) and the static characteristics given by equations (3.3.3) and (3.3.4) can be found by setting $s = 0$ in the transfer matrices. As an example:

$$H_{yu}(s = 0) = -DA^{-1}B + E \qquad (3.3.12)$$

It is interesting to see how the static characteristics, based on stationary conditions, and the dynamic characteristics, based on linearization of small perturbations around a fixed operating point, can be combined to give a satisfactory description of both static and dynamic behavior. One such combination is shown in Figure 3.3.1. In this situation, two external influences, $\mathbf{u}$ and $\mathbf{v}$, enter the system at the left and pass through two parallel process branches. The upper branch gives the stationary response, and the lower gives the dynamic response. In order to accomplish this, the blocks labeled LP in the upper branch must be diagonal (quadratic) transfer function matrices that act as "low pass filters" with cutoff frequencies about one decade below the lowest eigenvalue of the linearized process. This generates estimates of $\mathbf{u}^0$ and $\mathbf{v}^0$ that are needed in equation (3.3.3). The output of block $\mathbf{h}(\cdot)$ is the steady state $\mathbf{x}^0$. In the lower branch are "high pass filters" that will pass $\mathbf{u}$ and $\mathbf{v}$ at frequencies above the cutoff frequency described above. The perturbations $\delta\mathbf{u}$ and $\delta\mathbf{v}$ force the two transfer function matrices given in equations (3.3.8) and (3.3.9). These responses are added to $\mathbf{x}^0$ to give the total response $\mathbf{x}$. A corresponding diagram can be drawn for the case of the measured variables $\mathbf{y}$.

Sometimes it is of interest to develop an approximate dynamic model of a multivariable process for which there exist only a steady-state model $\mathbf{x} = \mathbf{h}(\mathbf{u})$ (which may be empirical) and some observations of dynamic responses (step responses, for example).

If the actual system is assumed to be described by

$$\dot{\mathbf{x}} = \mathbf{f}(\mathbf{x}, \mathbf{u}) \tag{3.3.13}$$

the steady-state model will be (when $\dot{\mathbf{x}} = 0$):

$$0 = \mathbf{f}(\mathbf{x}, \mathbf{u})$$

which leads to:

$$0 = -\mathbf{x} + \mathbf{h}(\mathbf{u}) \tag{3.3.14}$$

An indication of the procedure for making equation (3.3.14) dynamic is obtained by considering the linear case

$$\dot{\mathbf{x}} = A\mathbf{x} + B\mathbf{u} \tag{3.3.15}$$

which has the steady-state model

$$0 = -\mathbf{x} - A^{-1}B\mathbf{u} \tag{3.3.16}$$

where $-A^{-1}B\mathbf{u}$ plays the role of $\mathbf{h}(\mathbf{u})$ in equation (3.3.14). In equation (3.3.16) the 0 on the left side is replaced by the expression

$$-A^{-1}\dot{\mathbf{x}} \tag{3.3.17}$$

which becomes 0 in the steady state. Therefore, equation (3.3.14) is "made dynamic" by the expression

$$\left(\frac{\partial \mathbf{f}(\cdot)}{\partial \mathbf{x}}\right)^{-1} \dot{\mathbf{x}} = -\mathbf{x} + \mathbf{h}(\mathbf{u}) \tag{3.3.18}$$

where

$$\frac{\partial \mathbf{f}(\cdot)}{\partial \mathbf{x}} = A$$

must be estimated from the dynamic behavior. For the very simplest case, which is a linear diagonal process, we have:

$$A = \Lambda \tag{3.3.19}$$

where Λ is the eigenvalue matrix. Thus:

$$-A^{-1} = \begin{bmatrix} -\frac{1}{\lambda_1} & & & 0 \\ & -\frac{1}{\lambda_2} & & \\ & & \ddots & \\ 0 & & & -\frac{1}{\lambda_n} \end{bmatrix} = \begin{bmatrix} T_1 & & & 0 \\ & T_2 & & \\ & & \ddots & \\ 0 & & & T_n \end{bmatrix} = T$$

(3.3.20)

where T is a diagonal matrix of time constants. If it is not reasonable to assume a diagonal process, elements of T in addition to those on the diagonal must be estimated. Because

$$\frac{\partial \mathbf{f}(\cdot)}{\partial \mathbf{x}} = A(\mathbf{x}) \tag{3.3.21}$$

is a function of $\mathbf{x}$, the "time constants" of T could also be made functions of $\mathbf{x}$, but this concept would be applicable only to small perturbations around the steady state. Thus:

$$T(\mathbf{x})\dot{\mathbf{x}} = -x + \mathbf{h}(\mathbf{u}) \tag{3.3.22}$$

This model will give a correct description of system behavior in the steady state and an approximate description of behavior in the dynamic state.

A block diagram of the solution given by equation (3.3.22) is shown in Figure 3.3.2. As can be seen from the figure, this solution implements the dynamic effects in series with the steady-state model, whereas the two are in parallel in Figure 3.3.1. The solution shown in Figure 3.3.1 makes possible more elaborate dynamic descriptions.

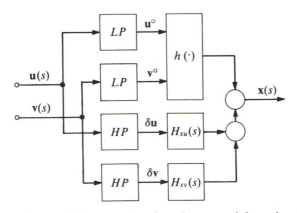

Figure 3.3.1. Parallel connection of steady-state and dynamic models.

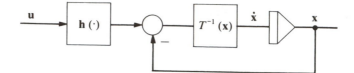

Figure 3.3.2. Solution of equation (3.3.22).

The transfer function matrices and the corresponding frequency response matrices give the most compact description of a dynamic process. The transfer function matrix is a parametric model that must include the gain factors, poles, zeroes, and transport delays. The frequency response can be expressed either analytically or graphically. The latter form has the advantage that one can immediately see the important characteristics of the system. From the point of view of process control, it is of relatively little interest to know the details of the frequency response at frequencies higher than that which gives a phase angle of $-180°$ $(-\pi$ radians); that is, higher than the point where $\omega > \omega_{180}$. This is easy to see on a plot of the frequency response (Bode plot), but it is quite difficult to determine from a transfer function. It is even more difficult to see if one has only the differential equations in the state space. The reason that one is not especially interested in the process phenomena that only respond at frequencies greater than ω_{180} is that that feedback control is not effective above that frequency. Therefore, we have a means of judging which simplifications can be used in dynamic modeling. This is also the reason that a simple model of the form

$$h(s) = \frac{Ke^{-\tau s}}{1 + Ts} \tag{3.3.23}$$

can be effective in many cases. By properly choosing τ in the term $e^{-\tau s}$, one can generate any desired phase angle.

When designing feedforward control (or any similar approach) based on dynamic models, one would like to achieve the best possible control at all frequencies. Therefore, as mentioned in section 2.3, it is very important to carefully tune the feedforward controller to get satisfactory action. Tuning errors in either the amplitude or the phase will produce about the same effect (see Figure 2.3.2), but it is unrealistic in most cases to expect feedforward to have much effect at frequencies above ω_{180}.

If the control system is to be based on state estimation, then it is probably best to use a model formulated in the state space, either continuous or discrete. Usually, it pays to include the most important nonlinearities in the model in order to get good estimation accuracy. Just as in the case of linear models, the gain matrices of the estimator can usually be constant. The method for improving the performance of a feedback control system for a process with dead time described in section 2.7 (see Figure 2.7.1), is a special case of the multivariable system described in section 2.13. Either state space models or transfer function matrix models can be used in such cases.

We will use transfer function matrices, frequency response matrices, and state space formulations to represent dynamic models of unit operations and complete process sections. We will also try to use consistent notation in order to simplify and systematize the material. Therefore, our notation will often differ from that used in the original sources in the literature.

3.4. TRANSPORT OF LIQUIDS, SOLIDS, AND GASES

One of the most basic unit operations is the movement of materials in different forms from one point to another between processes or within process segments. The method of transport depends on whether the material is a liquid, gas, or solid. Fluids are usually transported through pipes in which the driving force is a pressure drop created by a pump, compressor, or blower (fan). The description of a fluid transport system must include the operating characteristics of the pump, compressor, or blower expressed as a function of speed of the driving device, pressure drop, pipe dimensions, and so forth.

Transport of liquid usually occurs in pipes in which the pressure differential is created either by a pump or by a hydrostatic head. Here, it is important to know the pump characteristics as related to speed, input and output pressures, and the pipe dimensions and layout, in order to characterize the process. Physical characteristics of the liquid, such as viscosity and density, are also important. A major difference between liquids and gases is that for most practical purposes most liquids can be assumed to be incompressible. This makes liquid systems much simpler to analyze then gas or steam systems. Even though it is theoretically possible to analyze gases and liquids in the same manner and account for the differences as a function of the distance between individual molecules, this unified approach is virtually never used in industrial practice. As we will see, the equipment used for handling liquids is quite different from that used for gases.

Compression of Gases

Blowers (fans) and compressors are often used to provide the pressure differential needed to move gases through chemical processes. Blowers are effective up to pressures of about 0.3 bar, while compressors can develop up to about 4000 bar. A gas compressor is based upon the same principles as a pump that is used for liquids. For gases, however, the gas laws apply, and we see from the ideal gas law ($pV = nRT$) that the volume is reduced when the pressure is increased. During compression, the temperature of the gas also increases. Therefore, it is often necessary to cool gases between compression stages.

The flow characteristics of pumps and blowers are normally expressed in terms of either mass flow or volumetric flow converted to atmospheric pressure. The volume flow characteristic is usually specified in units of normal cubic meters (nm^3).

In a multistage turbo compressor consisting of several centrifugal blowers in series, the pressure increase in each stage is a function of turbine speed, mass flow, and absolute pressure. The absolute pressure and the mass flow have a dynamic relationship to the pressure increase in each step. As a consequence, a complete model of a compressor system can be very complicated.

In most compressors it is important that the pressure and temperature be such that the gas does not reach the critical point. At this point, condensation occurs, thus mixing liquid with the compressed gas. A compressed gas can be liquified by cooling, as shown in the P-V-T diagrams for water and carbon dioxide in Figure 3.4.1.

Depending on what happens to the energy in the gas during compression, the system can be isothermal, adiabatic, or polytropic. Under isothermal conditions, the temperature of the gas does not change because the heat from the gas is transferred to the environment. In adiabatic compression, the system is perfectly insulated from the surroundings so the energy of the system is constant. A real compressor will be a combination of these systems (polytropic), but compressors are often designed on the basis of adiabatic assumptions.

Under adiabatic conditions, the pressure, temperature, and volume of an ideal gas are given by:

$$\frac{p_2}{p_1} = \left(\frac{V_1}{V_2}\right)^k \tag{3.4.1}$$

$$\frac{T_2}{T_1} = \left(\frac{V_1}{V_2}\right)^{k-1} \tag{3.4.2}$$

$$\frac{p_2}{p_1} = \left(\frac{T_2}{T_1}\right)^{k/(k-1)} \tag{3.4.3}$$

where $k = c_p/c_v$ with c_p = specific heat at constant pressure and c_v = specific heat at constant volume.

Equation (3.4.1) gives the change in volume as a function of pressure. Equation (3.4.3) is the corresponding relation between pressure and temperature.

The power absorbed during compression can be calculated from the product of the absolute pressure and the mass flow of the gas. Under adiabatic conditions, we have:

$$P_{ad}(\text{kW}) = K_{01} \cdot \frac{k}{k-1} \cdot w \cdot R \cdot T_1 \left[\left(\frac{p_2}{p_1}\right)^{(k-1)/k} - 1\right] \tag{3.4.4}$$

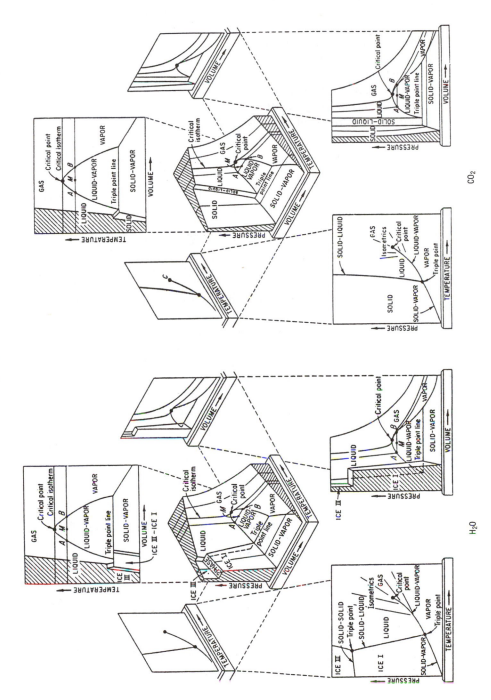

Figure 3.4.1. *P-V-T* diagrams for H_2O and CO_2.

where:

P_{ad} = power under adiabatic conditions
p_1 = absolute pressure at inlet (bar)
p_2 = absolute pressure at outlet (bar)
w = mass flowrate of the gas (kg/s)
R = gas constant, 8.314 (joules/g mole °K)
T_1 = absolute temperature of input gas (°K)
K_{01} = constant = .0019

Alternatively, we could write:

$$P_{ad}(\text{kW}) = K_{02} \frac{k}{k-1} Q_1 p_1 \left[\left(\frac{p_2}{p_1} \right)^{(k-1)/k} - 1 \right] \qquad (3.4.5)$$

where:

Q_1 = volumetric gas inlet flow (nm^3/sec)
K_{02} = dimensionless constant = 1.903×10^{-7}

Similar expressions for the multistage compressor can be found in Perry and Chilton (1973, pp. 6–16).

Although blowers and compressors operate on compressible gases, their principles of operation are the same as those of centrifugal or piston pumps operating on incompressible liquids.

One must carefully choose the proper control scheme and equipment for control of the states in the transport of gases and liquids. This choice must reflect the specific characteristics of each process unit. As we have said, the most important difference between a gas and a liquid is the fact that a gas is compressible. Therefore, the dynamic characteristics of a gas are very different from those of a liquid even though the static characteristics are similar. The pressure in a liquid system can build up very rapidly — so rapidly that the bandwidth of the pressure buildup is usually much higher than that of other components of the process. On the other hand, pressure buildup in a gas can be very slow, especially if the volume is large. The pressure effect in a long, small-diameter pipe is distributed along the pipe, thus requiring analysis by means of partial differential equations in a manner analogous to that used for electrical wave propagation along a line (Campbell, 1958).

Pumps used for liquids are usually easier to characterize than those used for gases. A typical static curve for a centrifugal pump is shown in Figure 3.4.2a, where the pressure increase due to the pump is expressed as a function of flowrate and pump speed. Many pumps have a region in this curve where the relation between pressure and pump speed is nearly proportional. In this region the pressure drop can also be assumed to be linear with increasing flow.

Figure 3.4.2b is a corresponding curve for a metering pump such as a piston pump, gear pump, or screw pump. We see that the pressure characteristic for a

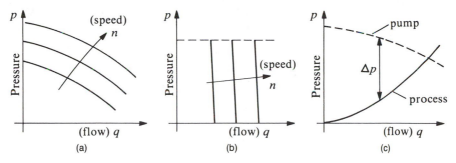

Figure 3.4.2. Various pump characteristics.

metering pump is nearly vertical. Thus, except for possible leakage from the pump, the flow will be proportional to the pump speed regardless of the pressure. The upper pressure limit is usually controlled by a safety valve that will release at the point shown by the dashed line. This pump can be described by a linear model in its normal operating range. Figure 3.4.2c is a typical pressure–flow curve for the process downstream of the pump. When the liquid flowrate increases, the pressure head opposing the pump will increase by approximately the square. The difference between the pump pressure (indicated by the dashed line) and the opposing pressure in the process is the driving force for the liquid in the system. When the two curves intersect, the difference becomes zero, and the flow from the pump ceases.

The most important question from the control standpoint is, how can pressure and flow be controlled in a pump, blower, or compressor?

Figure 3.4.3a shows a blower, or pump, that is driven by a motor whose speed can be used as a control variable. This has both advantages and disadvantages. One of the advantages is that only the power used to build up the pressure is consumed by the motor, assuming that the speed control of the motor itself is satisfactory. A disadvantage is that speed control of a large motor can be relatively expensive, and the bandwidth of the speed control of a large motor is usually quite low. For various reasons, many industries cannot use dc motors even though they are the easiest to control. Today, it is possible to achieve relatively good speed control of ac motors, and they are becoming the most common type used in the process industries. Large pumps and compressors are often driven by turbines, preferably gas turbines, which can be speed-controlled, but they usually have a relatively low bandwidth.

For liquids, the dynamic pressure and flow changes in the system shown in Figure 3.4.3a will be essentially determined by the dynamics of the speed changes of the pump, while the characteristics for a gas compressor can have extra dynamics due to the compressibility of the gas. This is discussed further in Appendix F.

Figure 3.4.3b shows another scheme for the control of pressure and flow from a blower, pump, or compressor. Here we see a control valve in a recircu-

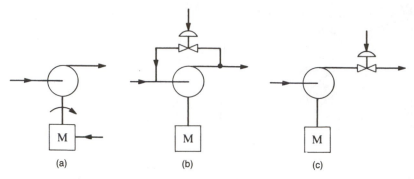

Figure 3.4.3. Methods of changing flow from a pump.

lating line around the pump. The amount of fluid recirculated is determined by the opening of the control valve. Figure 3.4.4 shows the pressure–flow characteristic of the pump (for a liquid) as a function of the opening of the recirculating valve. The points where the curves intersect indicate how the pressure of the pump will vary as the recirculating valve is opened when the pump is not loaded. Loading of the pump can be treated as a change in the characteristics of the recirculation valve. Figure 3.4.5 shows a typical relationship between pump loading (flow), pressure, and opening of the recirculating valve.

The arrangement shown in Figure 3.4.3b can be combined with either pressure or flow control on the secondary side of the pump (downstream side) to give a simple and reliable means of regulating the pump. In the case of liquids, the dynamics will usually be dominated by the dynamics of the control valve motor. There may also be a slight effect from a change in motor speed when the load on the pump increases because of an increase in flow. The pump motor speed may decrease unless the motor has very good speed control, but this

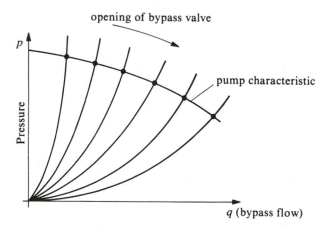

Figure 3.4.4. Pump pressure for system in Figure 3.4.3b as function of recirculation rate.

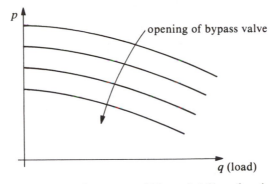

Figure 3.4.5. Pump pressure for system of Figure 3.4.3b as function of load (flow).

effect is usually very small. As in the previous case, compressibility can add a dynamic effect to gases (see Appendix F).

A disadvantage of the system shown in Figure 3.4.3b is that it is not energy-efficient. Although the total flow delivered by the pump may be large, the net flow can be low if the recirculation ratio is high. Therefore, the simplicity of the system is offset by the energy consumption, a factor that becomes increasingly important as the cost of energy rises.

If the pump is a metering pump with the characteristic shown in Figure 3.4.2b, and if motor speed is not used as the control variable, then the only reasonable solution for pressure control is a recirculating system, as shown in Figure 3.4.3b.

One also finds variable-capacity pumps in industry, but they tend to have limited, although extremely important, application. One quite common application of a variable capacity pump is use as the so-called fan pump that supplies pulp stock to a paper machine.

A third scheme for controlling a blower, pump, or compressor is shown in Figure 3.4.3c. Here we have a throttling valve on the output side of the pump. This solution can be used with centrifugal pumps and compressors and also with metering pumps if they are equipped with a recirculation valve for pressure control. A throttling valve controls the system about the same as the scheme shown in Figure 3.4.3b, except that now the pressure is greatest when the valve is wide open (see Figure 3.4.6). This method also has about the same advantages and disadvantages as the system that uses a recirculating valve.

In the operation of a gas compressor, at least one additional important difficulty must be kept in mind. There is a lower limit to the flow that a compressor can deliver without becoming internally unstable. At the point of instability, the gas tends to move back and forth inside the compressor, thus creating a strong oscillation and severe noise that, in the long run, damages the equipment. This lower limit, which is shown as the *surge line* in Figure 3.4.7, must not be violated in operation. Protection can be obtained by using a recirculating line as shown in Figure 3.4.3b, where the recirculating valve is opened when the pressure gets close to the surge limit (see section 5.9). There are other methods of

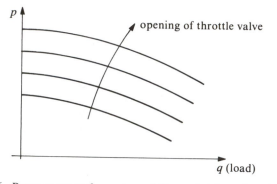

Figure 3.4.6. Pump pressure for system of Figure 3.4.3c as function of load.

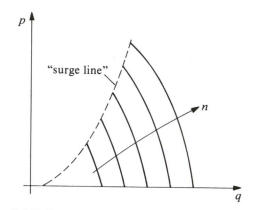

Figure 3.4.7. Pressure as function of flow in a gas compressor.

protecting against the surge limit, including direct programming of the surge function or feedback from measurement of the acoustic noise generated as the compressor approaches the surge limit.

Pressure and Flow Relationships in Pipes

Gases and liquids flowing in pipes undergo a pressure drop that depends upon physical characteristics such as pipe diameter, pipe length, the inner wall condition of the pipe, and the number and placement of pipe fittings (elbows, tees, etc.). An extensive amount of available theoretical and experimental information can be used to calculate the pressure drop of a flowing fluid.

The velocity profile of a fluid in a pipe depends upon whether the flow is laminar, turbulent, or in the transition region. In laminar flow there is no velocity component in the radial direction of flow, and the cross-sectional velocity profile is parabolic, as shown in Figure 3.4.8a. In turbulent flow, there is particle (element) motion in the radial direction, and the velocity profile over the

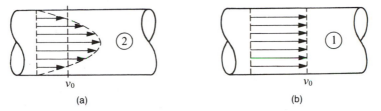

Figure 3.4.8. Velocity profiles for liquid flow in a pipe.

cross-sectional area of the pipe is uniform in the axial direction. This profile is shown in Figure 3.4.8b. Flow in the transition region occurs when the velocity in the axial direction is too high for laminar flow, but too low to develop full turbulent flow. One can calculate the Reynolds number to determine which region applies under any given flow conditions (Perry and Chilton, 1973).

In the case of laminar flow, the fluid elements in the center of the pipe have twice the average velocity of the elements in the pipe. This means that those elements in the center will move from one point to another in half the time of the average of the elements. Figure 3.4.9 shows the response to a step disturbance in some fluid property, perhaps temperature or concentration, for laminar and turbulent flow. For turbulent flow the change arrives at a downstream L at time $\tau = L/v_0$ wherre v_0 is the average velocity of the fluid. We see that the change arrives at L with exactly the same shape as the disturbance. For laminar flow, we see that the first elements arrive at L at time $\tau/2$, and the elements closest to the wall arrive considerably later. It should be mentioned that the response curve for laminar flow is *not* an exponential function. Appendix G contains a discussion of how the transfer function of a fluid flowing in a pipe is dependent upon the velocity profile.

We can conclude that the transfer function and frequency response in the transitional region will lie between the functions

$$h_1(s) = e^{-\frac{L}{v_0}s} \qquad (3.4.6)$$

and

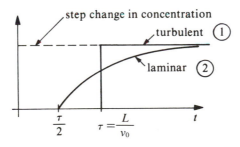

Figure 3.4.9. Response to step disturbance for laminar and turbulent flow.

$$h_2(s) = e^{-\frac{L}{2v_0}s} \cdot \frac{1}{1 + \frac{L}{2v_0}s} \tag{3.4.7}$$

which are the transfer functions for turbulent and laminar flow, respectively. The frequency responses of these two functions are shown in Figure 3.4.10, where the regions of possible amplitude and phase angles fall between the limits of the two functions.

Control Valves for Gases and Liquids

The automatic control valve is a very important element in a system for control of liquid and gas flow through pipes. The pneumatic motor, shown on the control valve in Figure 3.4.11a, is the most common type of valve motor used in the process industries. Figures 3.4.11b and 3.4.11c show a few of the many different types of valve bodies (Perry and Chilton, 1973, pp. 22–88, 22–89.)

The relation between the motion of the valve stem (the control variable) and the flow through the valve body, at constant pressure drop across the valve, is defined as the *valve characteristic*. Three types of valve characteristics are shown in Figure 3.4.12:

(a) "Above linear," which gives decreasing sensitivity as the valve opens (typical characteristic: square root).
(b) Linear, which has constant sensitivity.

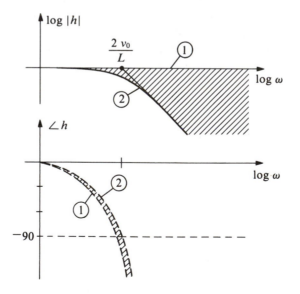

Figure 3.4.10. Frequency response in pipe for different velocity profiles.

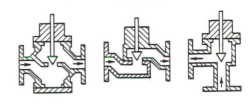

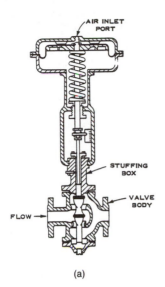

(a)

(b)

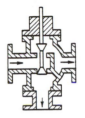

(c)

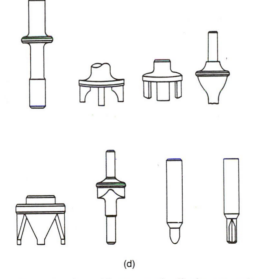

(d)

Figure 3.4.11. a. Control valve with pneumatic diaphragm motor. b. Various single-seated control valve bodies. c. Various double-seated control valve bodies. d. Various valve plug shapes.

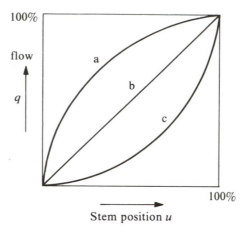

Figure 3.4.12. Valve flow characterisics.

(c) "Under linear," which has increasing sensitivity (typical characteristics: parabolic, equal percentage, logarithmic).

The actual shapes of curves (a) and (c) usually follow mathematical functions such as square root, logarithmic, equal percentage, and so on. Although linear valves may be the simplest to handle in the design of process control systems, other shapes are often required because of the flow characteristics needed by the process itself.

The valve characteristic shown in Figure 3.4.12 will usually change when the valve is installed in a piping network with a pump or compressor. The function of the valve is to vary the restriction in the pipe so that the flow will change while keeping the pressure drop constant. However, there is a relationship between the actual working valve characteristic and the pressure drop across the valve. As the pressure drop across the valve becomes a greater fraction of the total pressure drop in the system, the ability of the valve to perform its function increases. Therefore, it is very important to select a control valve that will assure sufficient pressure drop to retain the designed flow characteristic regardless of stem position. It is also necessary to choose a valve sized so as to operate in approximately the middle two-thirds of its range in order to enable it to operate according to its designed flow characteristic.

Figure 3.4.13 shows what will happen when valves with different characteristics, as shown in Figure 3.4.12, are put in service in typical situations where the downstream pressure increases as the square of the flow. We see that the net characteristic of a valve with characteristic (a) has become nearly an on-off action. On the other hand, the valve with characteristic (c) gives a nearly linear result. This is one of the reasons why valves with "under linear" characteristics are often chosen for control.

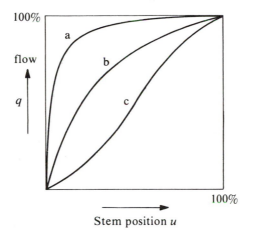

Figure 3.4.13. Characteristics of valves of Figure 3.4.12 installed in system with pressure drop.

Control valve manufacturers offer a wide array of theoretical and experimental information on which to base the choice and size of valves. Unfortunately, there is no standard notation used in this information. We will designate the functions shown in Figure 3.4.12 as $f(\cdot)$, and we will let $q = f(u)$ where q is the volumetric flow rate, and u is the valve stem position. We will assume that the pressure drop Δp is constant. For varying pressure we can write the relationship between flow and pressure drop as:

$$q = f(u)C_v \sqrt{\frac{\Delta p}{\rho}} \qquad (3.4.8)$$

where:
C_v = valve coefficient (supplied by vendor)
Δp = pressure drop across the valve
ρ = density of the fluid
Equation (3.4.8) applies only when the fluid is a liquid. For the case of a gas or vapor, the flow is given by (Driskell, 1974; Ferson, 1982):

$$w = f(u)C_{v_1}\left(1 - \frac{0.466}{k} \cdot \frac{\Delta p}{\Delta p_{rated}}\right)\sqrt{\Delta p \cdot \rho_1}$$

$$q = f(u)C_{v_2}\left(1 - \frac{0.466}{k} \cdot \frac{\Delta p}{\Delta p_{rated}}\right)\sqrt{\frac{\Delta p \cdot p_1}{\rho_1 T_1 z}} \qquad (3.4.9)$$

where:
w = mass flow (kg/s)
q = volume flow (nm^3/s)

C_{v_i} = valve coefficient
Δp = pressure drop (bar)
p_1 = inlet pressure (bar)
T_1 = absolute inlet temperature (°K)
ρ_1 = density at inlet (kg/m³)
z = compressibility factor ($z = 1$ for an ideal gas)

$k = \dfrac{c_p}{c_v}$ (the ratio of specific heats)

We will consider the question of control of mass and volumetric fluid flow again in section 4.3, where a mathematical model of a flow process is developed.

Transport of Solids

Many industrial processes must handle bulk solids, often in granular or powdered form. Various devices, such as conveyors, screw augers, vibrating platforms, pneumatic tubes, bucket elevators, and so on, can be used for transporting and metering these materials.

Automatic control requires that the transport device be controllable over a reasonable range by means of some type of control variable. The screw conveyor shown in Figure 3.4.14a varies the flow by varying the speed of the screw. Quite a different approach is shown in Figure 3.4.14b. Here, the amount of material allowed to drop from the silo onto the conveyor belt is controlled by the opening of the discharge chute from the silo. Even though the belt has

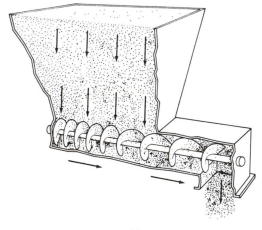

(a)

Figure 3.4.14. Different equipment for feeding solid materials. (Chemical Engineering, reprinted by permission of McGraw-Hill Book Co.)

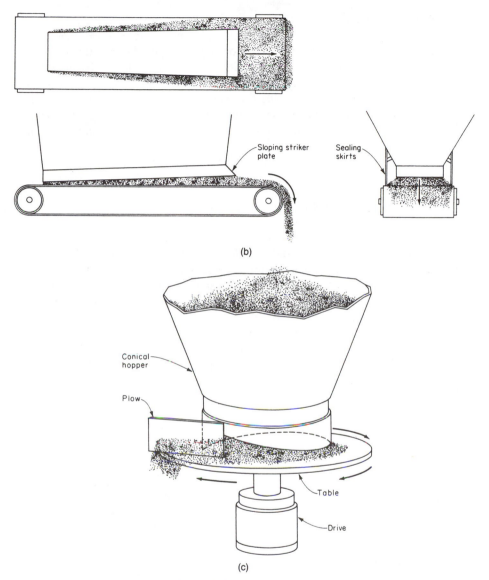

Figure 3.4.14. (continued)

constant speed, the total material delivered will be controlled. Flow from the hopper in Figure 3.4.14c is controlled either by controlling the rotation speed of the table or the position of the scraper (plow).

The system shown in Figure 2.7.3, section 2.7, provides precise mass flow metering of a solid material. Control can be achieved by applying this scheme to any of the devices shown in Figure 3.4.14.

If there is no measurement or estimation of the mass flow, as in Figure 2.7.3, one must expect the precision of metering a solid material to be lower than that for metering gases or liquids. The precision depends upon the particle size and shape, and in many cases the concentration of active material in the granules to be metered can be quite variable. The effect of nonuniform particle size is easily imagined in the case of coal, for example. The volumetric flowrate of solids, and perhaps the mass flowrate as well, must often be combined with other characteristics such as moisture content, porosity, and concentration in order to get a satisfactory description of what is actually moving in the stream. The dynamics of these systems are determined primarily by the dynamics of the motor. This problem will be seen again in several practical process examples.

3.5. METHODS FOR ADJUSTING PARTICLE SIZE OF BULK SOLIDS

Many industrial processes include operations for reducing the size of bulk raw materials or intermediate products through crushing, milling, or grinding. Other industries use unit processes for increasing particle size through agglomeration or pelletizing. Most of these processes must be automatically controlled to function satisfactorily.

An example of this type of process is a ball mill used to produce a controlled size of ore particles that may previously have been crushed to a reasonable size. In a ball mill (see Figure 3.5.1) the raw material enters the center of a large

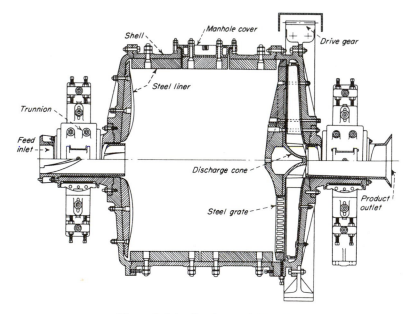

Figure 3.5.1. Continuous ball mill.

rotating cylinder lined with steel or rubber plates. The cylinder contains a large number of steel balls that, because of wear, usually have different diameters. The rotation of the cylinder causes the steel balls to rise and fall inside the cylinder, applying a grinding and crushing action on the material. The speed with which the size of the material is reduced depends on many characteristics of both the mill and the material. Some of the most important are:

- Level of material in the mill.
- Solid–liquid ratio, in the case of wet-milling processes.
- Rotation speed of the mill.
- Number of balls in the mill (which cannot be changed easily).

There have been many studies of the dynamics of ball mills (Olsen, 1972; Kelsall et al., 1967–70; Niemi, 1983). This research has led to the development of models that can be used to calculate grinding curves based on the various important mill properties listed above.

The dynamics of a mill depend primarily on whether the mill is operated as a closed-circuit system with a classifier and recycle of the rejects. The mill in Figure 3.5.2 is followed by a classifier that separates the oversized particles and sends them back to the mill. The process shown in Figure 3.5.2 is most typical of dry grinding. Figure 3.5.3 shows a two-step closed-circuit process used for wet grinding. In this system, a primary mill is followed by a rake classifier that returns the large particles ("rejects") to the primary feed, while the fine particles ("accepts") are sent to a bowl classifier. The rejects from this classifier are sent to another rake classifier that feeds the secondary mill. The accepts from the bowl classifier are sent to a thickening device, often a sedimentation tank.

Figure 3.5.2 can be used as the basis for a simplified analysis of the dynamic behavior of a ball mill in a closed-circuit configuration. (We would need a more detailed system diagram if we wanted to design an actual process.) We

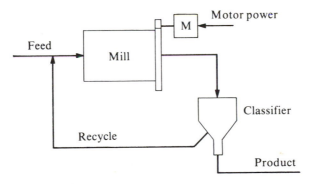

Figure 3.5.2. Continuous ball mill with classifier and recycle.

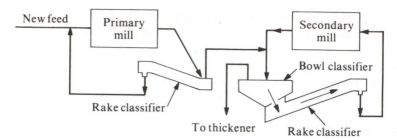

Figure 3.5.3. Two-stage closed-circuit system for wet grinding. (from Pit & Quarry, Feb./Mar. 1959.)

will use the following notation:

w_F: mass flow of the feed.
k_F: fraction of coarse particles in the feed.
$1 - k_F$: fraction of fine particles in the feed.
w_{1F}: mass flow of coarse particles in the feed.
w_{2F}: mass flow of fine particles in the feed.
W_1: mass of coarse particles in the mill.
W_2: mass of fine particles in the mill.
$W = W_1 + W_2$: mass of material in the mill.
w_{21}: grinding rate from large to small particles in the mill.
w_0: total mass flow out of the mill.
w_{10}: $w_0(W_1/W)$: mass flow of large particles out of the mill.
w_{20}: $w_0(W_2/W)$: mass flow of fine particles out of the mill.
w_R: mass flow of the recycle.
w_P: mass flow of the product.

Figure 3.5.4 is a simplified block diagram that describes the behavior of a closed-circuit mill process. The feed, entering the system at the left, is multiplied by the fractions of coarse and fine particles to give the quantities w_{1F} and w_{2F}. Because it is assumed that the recycle flow contains only large particles (rejects), the recycle flow w_R is added to w_{1F}. A mass balance on the large-particle component gives:

$$\dot{W}_1 = w_{1F} + w_R - w_{10} - w_{21} \qquad (3.5.1)$$

and a mass balance on the small particles is:

$$\dot{W}_2 = w_{2F} - w_{20} + w_{21} \qquad (3.5.2)$$

These two equations are represented in the block diagram.

At the right side of the diagram we see the connection between the mass and the volume of the material in the mill. This relationship depends on physical characteristics such as density of the material and the number of balls in the

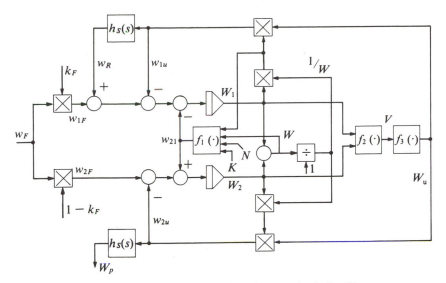

Figure 3.5.4. Elementary block diagram for ball mill.

mill. We also see how the volume affects the total mass flow (w_o) out of the mill. For simplicity, we assume here that the functions $f_2(\cdot)$ and $f_3(\cdot)$ are linear. This means that when the total mass flow in w_F changes, the response of the total mass flow through the mill will have the holdup time

$$T = W_n/w_{Fn}$$

where the subscript n refers to nominal conditions.

In the reduction of large particles, we are most interested in the grinding rate:

$$w_{21} = f_1[(W_1/W), W, N, B] \tag{3.5.3}$$

where N = rotation speed of the mill, and B = mass of balls in the mill. It has been shown experimentally that for a given mill speed and a given mass of balls in the mill, there is an optimum mass W with respect to w_{21}. This can be explained by the fact that for a given mass of balls in the mill, grinding can be very rapid if the absolute throughput of material is small. If the flowrate is increased, the effectiveness of the balls will be decreased. The grinding rate w_{21} also depends on the ratio of coarse particles to fine particles in the mill. When the amount of coarse material decreases, then w_{21} also decreases. There also exists a rotation speed that gives a maximum rate w_{21}. Finally, process operators are interested in the power used by the motor that drives the mill. This is closely connected to the grinding energy that is used to express the breakdown of the material. This subject will not be pursued here, however.

If the mill is optimally filled with raw material (W_{opt}), small changes in W will not give significant changes in w_{21}. However, the fractions of large and

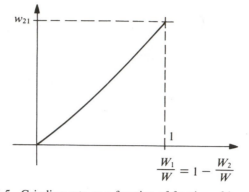

Figure 3.5.5. Grinding rate as a function of fraction of large particles.

small particles, W_1/W and $W_2/W = 1 - W_1/W$ respectively, will affect the grinding rate w_{21} approximately as shown in Figure 3.5.5. This means that we have negative feedback around the two integrators shown in Figure 3.5.4.

The second phenomenon of special interest is related to the two elements that contain the transfer function $h_S(s)$. This transfer function can be approximated as

$$h_S(s) = \frac{1}{1 + T_S \cdot s} \tag{3.5.4}$$

This transfer function represents a holdup of the material in the classifier screen. Therefore, the Laplace transform of the stream of large particles out of the mill [$w_{10}(s)$] must be multiplied by this transfer function to produce $w_R(s)$. If we let $T_S \approx 0$ as a first approximation, we see from Figure 3.5.4 that the flows $-w_{10}$ and $+w_R$ will cancel each other. The result is that we have an ideal positive feedback that will make the dynamics in the upper loop (around the upper integrator) different from those in the lower loop. Using the simplification $T_S = 0$, the dynamics in the upper loop can be characterized by the time constant:

$$T_1 = \frac{1}{\dfrac{\partial w_{21}}{\partial W_1}}$$

and the lower loop by:

$$T_2 = \frac{1}{\dfrac{\partial w_{21}}{\partial W_2} + \dfrac{\partial w_{20}}{\partial W_2}}$$

This means that a change in flow in the feed stream will lead to a slower change in the coarse particles (characterized by T_1) than the fine particles (characterized by T_2) in the mill. Since the total mass in the mill is the sum of all the particles, the response in total mass will be given by:

$$\frac{W}{w_F}(s) = K \frac{1 + T_0 s}{(1 + T_1 s)(1 + T_2 s)} \qquad (3.5.5)$$

where:

$$K = \text{constant}$$

$$T_0 = \frac{1}{\dfrac{\partial w_{21}}{\partial W_2} + k_F \dfrac{\partial w_{20}}{\partial W_2}}$$

To check this result, consider the special case where the feed consists of only coarse particles. In this case, $k_F = 1$, and equation (3.5.5) reduces to:

$$\frac{W}{w_F}(s) = K \frac{1}{1 + T_1 s} \qquad (3.5.6)$$

This tells us that the mass in the mill is now characterized by the largest time constant.

If the feed contains only fine particles, $k_F = 0$, and equation (3.5.5) becomes:

$$\frac{W}{w_F}(s) = K \cdot \frac{1}{1 + T_s s} \qquad (3.5.7)$$

which means that the mass in the mill will be characterized by the holdup time.

The most important control variable is certainly the feed rate w_F. We have seen that the reaction of the mass in the mill to changes in the feed is simple, and the phase angle is never larger than $-90°$. In a real system having a mill and classifier, the distribution of particle size is continuous (statistically distributed) rather than binary as in the ideal case we have considered. This simply means that real milling and screening are not perfect. It is possible to influence, and perhaps control, the size distribution by means of additional control variables. One such variable is the mill speed. This is expensive, however, and the decision on whether to use this control must include economic considerations such as capital and maintenance costs, energy consumption, and the process and product advantages (and disadvantages) of changing the size distribution. These questions are beyond the scope of our discussion.

Control systems for ball mills are relatively easy to design, and they are important elements of mineral dressing plants, which will be discussed further in section 5.1.

Particle Size Increase (Agglomeration) of Powdered and Granulated Materials

Many industrial processes require that the raw materials have a minimum particle size in order to assure proper processing or chemical reaction. This is the case, for example, in the reduction of iron ore where the various reactants must be in the form of pellets from 1 to 3 cm in diameter to react properly.

Perhaps the most common method of forming pellets (granules) is to roll dry powdered materials together with a little moisture. (The choice of the actual liquid depends upon the material and the use.) This action causes the powdered particles to stick together, forming granules of increasing size. Two such processes are shown in Figures 3.5.6a and 3.5.6b. The unit shown in Figure 3.5.6a is an inclined disc that rotates quite rapidly. This action produces an even distribution of granules up to the desired size. The larger granules will tend to move outward on the plate and eventually fall off the edge, thus undergoing a natural size selection process. The powder and liquid are added as shown. Each of the supply streams must be carefully controlled, and a possible control variable for granule size is the rotation speed of the disc. The angle of inclination of the disc could also be a control variable, but this is mechanically much more difficult to achieve. Therefore, it is seldom used for control. Most systems of the type shown in Figure 3.5.6a are not equipped with automatic control, but there have been several attempts to develop control in order to improve the quality of the granules (Perry and Chilton, 1973).

Figure 3.5.6b shows a drum process in which the powder and the liquid are mixed and rolled together. A classifier follows the drum, and the rejects are returned to the feed. In this case the rejects are the undersized particles, rather than the large particles which were the rejects in the case of the ball mill. It is extremely important that the classifier does not physically either break up the newly formed granules or cause them to stick together to form new larger par-

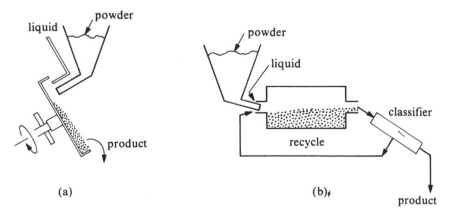

Figure 3.5.6. Two methods for agglomeration of fine powders.

ticles. Possible control variables for this system are the feed rate of powder from the silo, moisture–solid ratio, drum rotation speed, and angle of inclination of the drum. The dynamic characteristics of the process shown in Figure 3.5.6b are dominated by the holdup time of the drum. This is determined simply by dividing the drum volume by the net throughput rate.

Figure 3.5.7 shows the qualitative relationship between throughput, mean granule (pellet) size, and moisture content. Clearly, it is possible to get large granules when the moisture content is high.

Sintering is a process very often used to treat the product of a pelletizing process. In this process the pellets are heated to produce a final pellet that has good mechanical strength and high porosity. Strength provides mechanical stability, and porosity is important for good reactant penetration, which ultimately affects the reaction characteristics. This is important, for example, in reduction processes for metal ores.

Prilling is another widely used process for increasing the size of particles. In this process, drops of liquid are sprayed into the top of a high tower and allowed to fall against an upward stream of warm air. As they fall, the droplets coagulate and form nearly spherical particles, or prills, through a combination of drying and crystallization. This process is especially common in the fertilizer industry where ammonium nitrate, urea, and other materials attain their final form. When designing a prill tower, one must consider the material that is to be prilled in order to calculate the height of the tower and the number and size of the orifices that produce the liquid droplets.

The most reasonable control variables for control of a prilling tower are the liquid flow through the orifice, the vertical air flow, and the air temperature. Conventional control of prill towers consists of simply holding the process vari-

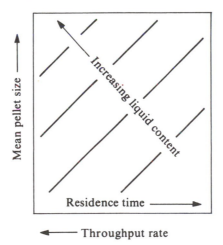

Figure 3.5.7. Approximate relationship between average pellet size and holdup time, with moisture as a parameter. (Reprinted by permission of Aufbreitungs-Technik)

ables constant, with possible long-term set point changes based on analysis of the product quality.

3.6. MIXING

There are many kinds of mixing operations needed in the process industries, and there are probably nearly as many variations of mixing devices and systems. The process used depends upon the job. Some liquids are soluble in each other while others are not, and the process and purpose of mixing can vary with varying solubility. Insoluble liquids can be mixed to promote a chemical reaction, to heat or cool one of the liquids, or to form permanent emulsions. Solids of various types, usually powders or granules, must also be mixed with each other for reasons similar to those for mixing liquids.

Good mixing—which is actually the best possible distribution of one material in the other(s)—requires creation of an internal mixing (or transport) velocity high enough to create a high probability of contact between elements of the different materials. A common mixer is a tank equipped with an agitator designed to create high turbulence in the materials to be mixed. High turbulence, in turn, gives a high probability of contact between the elements in the tank, and thus good mixing. To develop the concept of *mixing efficiency,* consider the following situation:

Consider a tank containing a single liquid. Now, add a tracer, or small amount of some material that has a high concentration of some easily measurable property. This could be a dye to change the color, a concentrated electrolyte to change liquid conductance, and so on. By observing the changing distribution of the tracer, one can determine the effectiveness of the tank agitator (mixer or propeller) and the uniformity of the mixture in the tank. This is shown in Figure 3.6.1. At time $t = t_1$ we see a high, narrow pulse caused by the addition of the tracer. Response curve a indicates perfect mixing, instantaneous and perfectly distributed. Because the volume of the tank is constant, perfect mixing will appear as a step change in the concentration of the mixture. Curve b is a

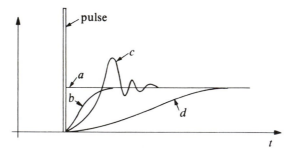

Figure 3.6.1. Typical response curves for concentration of a tracer added to a mixing tank.

more realistic response curve. The mixture reaches a constant concentration, but it does not do so instantaneously. Curve c shows a typical and often misleading result that can occur if the measuring element is placed close to the point where the tracer material is added and/or where agitation is poor and unable to produce turbulent flow. In these cases, one can get pockets (dead areas) or streaks of high concentration (bypassing) that can take a long time to dissipate. The response shown by curve c has nothing to do with the concept of the oscillation characteristic of an underdamped system. It is simply the result of placing the measuring element in a position where the concentration in the tank is not homogeneous. Curve d is the response obtained when one has uniform mixing but low internal velocity.

If we were to place several measuring elements throughout the tank, we would obtain an average output showing quick and uniform mixing, perhaps as shown by curve b, even if the mixing were poor. We can see that it would be easy to be fooled if we arranged the measurements in a manner that would even out the large variations in the process.

Another way to illustrate the importance of the internal transport velocity in a mixing process is shown in Figure 3.6.2. Assume a flow q_0 enters a tall tank and leaves at the bottom in such a way that there is no mixing in the tank; that is, there is a uniform velocity profile from top to bottom with no vertical exchange. A level controller (LC) keeps the volume of liquid in the tank constant. Part of the flow (q_1) from the outlet of the tank is pumped back to the top of the tank where it is mixed instantaneously and perfectly with the input stream q_0. This is a model of what happens in an ideal mixing process, in that a change

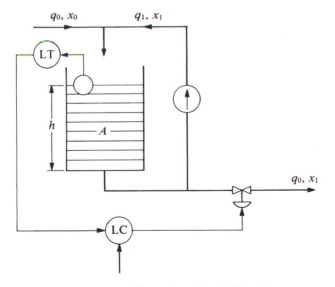

Figure 3.6.2. An illustration of a mixing process.

in the flow q_1 can change the makeup of the material flowing through the tank without changing the net flow through the system. If we increase q_1, the flow velocity in the tank will increase, thus decreasing the holdup time in the tank. If q_0 is constant, and the concentration x_0 changes, the situation in the tank will be that shown by the block diagram of Figure 3.6.3. The outlet concentration is indicated by x_1, and the concentration at the mixing point at the top of the tank is x_2. The transfer function between the input concentration $x_0(s)$ and the output $x_1(s)$ is:

$$\frac{x_1}{x_0}(s) = \frac{q_0}{q_1} \cdot \frac{\dfrac{q_1}{q_0 + q_1} \cdot e^{-\tau s}}{1 - \dfrac{q_1}{q_0 + q_1} \cdot e^{-\tau s}} \tag{3.6.1}$$

where the holdup time is given by:

$$\tau = \frac{h \cdot A}{q_0 + q_1} = \frac{V}{q}$$

and:

   h = height of liquid in the tank
   A = cross-sectional area of the tank
   V = volume of liquid in the tank
   q = total flow through the tank

Dividing and combining terms in equation (3.6.1), we have:

$$\frac{x_1}{x_0}(s) = \frac{q_0}{q} e^{-\tau s} \left(1 + \frac{q_1}{q} e^{-\tau s} + \left(\frac{q_1}{q} \right)^2 e^{-2\tau s} + \left(\frac{q_1}{q} \right)^3 e^{-3\tau s} + \cdots \right) \tag{3.6.2}$$

Let us define a factor β as the ratio of the recirculated flowrate (q_1) to the net flowrate (q_0):

$$\beta = q_1/q_0$$

where $0 < \beta < \infty$ when $0 < q_1 < \infty$. Inserting this ratio into equation (3.6.2)

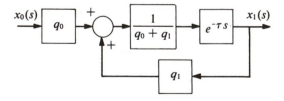

Figure 3.6.3. Elementary block diagram for description of a mixing process.

gives:

$$\frac{x_1}{x_0}(s) = \frac{1}{1+\beta} e^{-\tau s}\left(1 + \frac{\beta}{1+\beta} e^{-\tau s} + \left(\frac{\beta}{1+\beta}\right)^2 e^{-2\tau s} + \cdots\right)$$

$$(3.6.3)$$

It will be helpful to normalize the transport delay τ as:

$$\tau = \tau_0 \frac{1}{1+\beta}$$

where $\tau_0 = V/q_0$ is the holdup time in the tank.

Figure 3.6.4 shows the time response of concentration x_1 to a step change in input concentration x_0 for three different internal velocities (mixing intensities) that occur when $\beta = 0$, $\beta = 10$, and $\beta = 100$. For $\beta = 0$ (no mixing), we see the effect is a pure transport delay where $\tau = \tau_0$. For $\beta = 100$, $\tau \approx 0.01\tau_0$, and $x_1(t) \approx 1 - e^{(t/\tau_0)}$, which gives:

$$\frac{x_1}{x_0}(s) \approx \frac{1}{1+\tau_0 s}$$

From this we see that when $\beta = 100$ the process behaves approximately as a first-order process with a time constant τ_0.

According to Figure 3.6.4, the response when $\beta = 10$ is between the two extremes given above. This means we will get a transport delay of $\tau \approx 0.1\tau_0$ and a rising response approximating an exponential function. The time constant for this exponential function will be nearly τ_0.

Mixing is not so simple as one might think. There are many problems involved in designing mechanical equipment to obtain the required internal velocity characteristics. The physical form of the mixer also depends upon what is to be mixed. The basic phase possibilities are: gas–gas, gas–liquid, gas–solid, liquid–liquid, liquid–solid, and solid–solid.

Gas–gas mixing is probably the simplest problem, and the comments regarding liquid–liquid mixing also apply in principle to the mixing of gases.

Contact between *liquids and gases* is important in a long list of unit processes including distillation, absorption, adsorption, evaporation, humidification, dehumidification, and drying. We will not discuss all the various mechanisms used to develop good contact between gases and liquids — there are simply too many! In general, for a given mixer the holdup time and the internal velocity will determine the mixing efficiency. The control engineer would like to have a mixer with a very short holdup time and high turbulence. However, it should not be forgotten that a short holdup time requires very good control of the input streams, and such a mixer will be "nervous" in the sense that it will be very sensitive to small changes in the input streams. In many cases gas–liquid contact depends somewhat on mass transfer rates and/or chemical reaction rates,

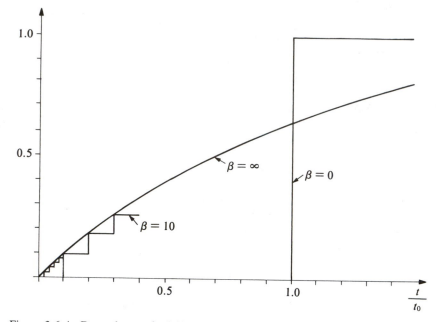

Figure 3.6.4. Dependence of mixing dynamics on internal velocity (mixing intensity).

so they may require some minimum holdup time. This in turn will be related to the gas pressure and temperature, and all will affect mixing efficiency. We will discuss these factors later when we discuss specific processes.

Mixing of *gases and solids* is especially important in drying and chemical reactions. One very common mixer for this combination is a rotating drum, mounted at a slight angle to the horizontal, in which the solid material is tumbled, preferably countercurrent to the flow of the gas. An increasingly popular and very efficient method is the fluidized bed where gas at high velocity actually "floats" the solid. In this case the solid circulates in the bed as the gas moves upward. This floating, or fluidization, provides excellent contact between the solid particles and the gas. A third method is the prilling tower, which was described earlier. These three methods are shown in Figures 3.6.5a, b, and c, respectively.

Holdup time is also important in these processes. In the system shown in Figure 3.6.5a, the usual control variable, the rotation speed of the drum, has an approximately inverse relationship to the holdup time of the solid, and the gas holdup time is determined by the gas velocity. Other control variables for this system are the temperature and the pressure of the gas. Flowrates of the solid and the gas are the important control variables for the fluidized bed. The height of the bed is usually fixed, which is why the holdup times of the materials are related only to the flows. Corresponding observations can be made with respect to the prilling tower.

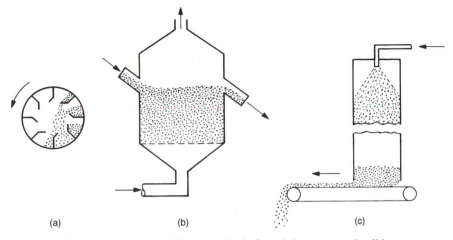

Figure 3.6.5. Three different methods for mixing gases and solids.

There are many types of equipment for *mixing liquids,* including agitators in tanks, injectors, centrifugal pumps with and without recycle, and so forth. The choice depends upon the materials to be mixed and the overall process requirements, and the mixing efficiency of most is usually closely related to the holdup time. However, as noted earlier, there are mixing devices that are very efficient in spite of very short holdup times, but they require careful and accurate control of the feed streams to assure the proper characteristics of the mixer output.

Many types of equipment can be used for *liquid–solid* mixing, but the object is always to obtain the best possible uniform mixture. The choice of equipment will depend on whether the liquid or the solid is the dominant material. For very viscous materials (dominated by the solid), a screw mixer, similar to the familiar kitchen meat grinder, is commonly used. Here again, the mixing efficiency is strongly dependent on holdup time and internal velocity. If a chemical reaction accompanies the mixing, then the holdup time must be sufficient for the reaction to reach the desired degree of completion. These relationships must usually be considered in the design of each individual case. Figure 3.6.6a shows a typical system for mixing liquids and solids. It consists of a silo fitted with a screw conveyor that has the dual function of mixing the materials and, in the lower section, removing the mixture from the silo. A conical screw mixer for mixing highly viscous materials is shown in Figure 3.6.6b.

Mixing of solids, usually powders and/or granules, has many industrial applications. This mixing problem is often difficult, not so much in the mixing, but in terms of natural separation that can occur during transportation or storage of the mixture. An example of such mixing occurs in the production and transportation of ground coke for the manufacture of carbon anodes in the electrometallurgical industry. In this case, the goal is to have a certain size distribution

(a)

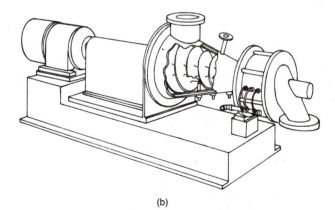

(b)

Figure 3.6.6. Two methods for mixing liquids and solids. [Photo (a) courtesy Day Mixing, Littleford Bros. Inc. (b) Baker Perkin, Inc.]

that is obtained by grinding large pieces of coke followed by classification. Unless continuous mixing action is provided, the sizes will segregate during storage and conveyor transport. The greater the differences in the size, shape, and specific weight of the particles, the greater will be the tendency to separate, with thinner, smaller, and rounder particles tending to sink through the mixture. Many times electrostatic charge will also play a role. As the charge builds up, it can tend to drive the particles apart (or, perhaps, hold them together) and upset the mechanical mixing effect. Usually, the charge builds as a function of time, thus making it desirable to find an optimum mixer holdup time that will minimize the buildup of the charge and yet maximize the mechanical mixing.

Many of the devices discussed earlier, such as the rotating drum and the unit pictured in Figure 3.6.6a, can also be used to mix solids.

Special care must be taken in mixing solids to minimize physical damage to the particles. For example, the mixing action can often reduce the size of the particles. This is sometimes a serious problem, and it can be a decisive factor in the choice of the particular mixing device best suited to the process. In this situation, a long mixing time conflicts directly with the desire to prevent damage to the particles. Here again we see the need for an optimum relation between mixing action and mixing time.

It is difficult to measure the degree of mixing obtained from any of the systems we have discussed. The major problem is that it is difficult to take samples that are representative in both time and space. Certainly, one cannot be sure that a small sample (short time and small volume) taken from a particular spot in the process is representative of all elements in time and space. Further, the very act of sampling can destroy the character of the sample (a well-known problem in consistency sampling of pulp slurries). Therefore, we must precisely define degree of mixing in terms of sample size, frequency, position, and method of sampling. The best way to proceed is to take frequent samples from many places in the process. Analysis of these data can identify concentration gradients caused by poor mixing or separation, and can also identify the dynamic behavior of the system. Because data obtained from poor measurements taken from poorly chosen sampling points in the process can lead to totally erroneous conclusions, it is vital that the sampling problem be taken very seriously.

3.7. SEPARATION PROCESSES

In section 3.6, we considered mixing of various materials. In this section, we will discuss means of separation of different materials and different phases.

The possible combinations that might have to be separated in industrial processes are: gas–gas, gas–liquid, gas–solid, liquid–liquid, liquid–solid, and solid–solid.

As expected, the choice of equipment for separation depends upon the materials to be separated. We can use screens or fiber filters for mechanical separation of solids from liquids, fiber or electrostatic filters for separating solids from gases, cyclones and centrifuges for separating solids from gases and liquids, magnetic separators for separating magnetic from nonmagnetic materials, flotation cells for separating materials with different surface properties, and so on.

We will not discuss *gas–gas* separation further at this point.

The problem of *gas–liquid* separation can take two forms: separation of a gas that is not soluble in the liquid, and separation by evaporation of the gas from the liquid. Separation of a gas from a quiescent liquid occurs in the form of gas bubbles that rise to the surface. The velocity with which the bubbles rise is given approximately by

$$v \approx 6 \cdot 10^{-5} d^2 \tag{3.7.1}$$

where v is bubble velocity (m/s) and d is bubble diameter (μm). This expression applies when the liquid is water at 20°C and the bubble diameter is $10 < d < 300$ μm. For liquids with different viscosity and density, the velocity difference from that of water is approximately proportional to density and inversely proportional to viscosity. For bubbles having diameters greater than 300 μm, the velocity will be somewhat greater than that predicted by equation (3.7.1). A change in pressure affects the bubble diameter according to

$$d_1 = \left(\frac{p_0}{p_1}\right)^{1/3} d_0 \tag{3.7.2}$$

This means that if the pressure is decreased by half, the velocity will increase by a factor of about 1.6.

Processes for driving gases out of liquids usually have two control variables: holdup time (determined by the average throughput velocity) and pressure.

Separation of gases from solids is required in many processes. One method is drying by means of either evaporation or sublimation. This very important process will be discussed further in section 3.11. Another gas/solids separation process is used for removing dust or dirt from gas. This is done by means of a cyclone separator (Figure 3.7.1), a bag filter (Figure 3.7.3), scrubbers (Figure 3.7.4), or an electrostatic filter (Figure 3.7.5).

The cyclone separator is probably the most widely used. In this device, gas that contains small particles of dirt or dust enters the upper part of the cyclone tangentially to the cyclone wall. Because they are heavier than the gas, the dirt particles are forced toward the wall. They then fall along the wall and eventually leave the cyclone at the bottom. The gas is forced to the center at the bottom, reverses direction, and flows upward and out the top of the cyclone. A cyclone has no direct control variables, but separation efficiency is a function of the inlet velocity of the gas, which can be controlled by manipulating the pressure drop across the cyclone. At any given velocity, the cyclone will re-

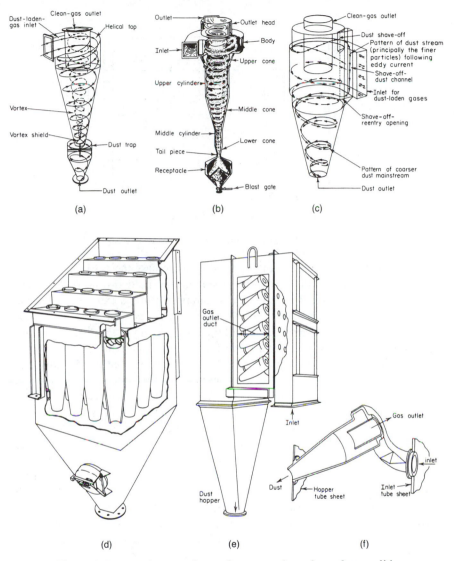

Figure 3.7.1. Various cyclones for separation of gas from solids.

move heavier particles better than light ones. This relationship is shown in Figure 3.7.2. If the particles do not influence one another in the cyclone, then it is possible, in theory at least, to determine the size distribution of the particles in the exit stream if the distribution in the feed is known.

In practice, if a single cyclone does not provide the required degree of separation, it can be followed by additional cyclone stages or other separation devices, or recirculation can be provided (much the same as that discussed

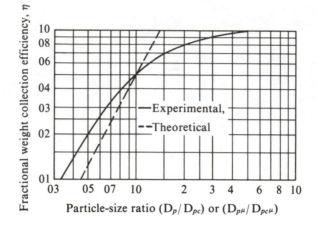

Figure 3.7.2. Separation characteristics of cyclones.

earlier with respect to the ball mill) in order to increase the net holdup time of the cyclone. As one should expect, increasing the holdup time increases the cyclone action and should improve the separation. But, as expected, recirculation decreases the net capacity of the cyclone. The dynamics of change between the solid content of the feed stream and the solid stream out of the cyclone can generally be treated as a first-order transfer function where the time constant is the ratio between the solid volume in the cyclone and the solids flow rate. However, relative to other units usually included in the system, the cyclone has a very wide bandwidth, and its dynamics can usually be neglected.

The bag filter for removing small particles from gas can take several forms in terms of the bag material and the method for collecting the dust (Figure 3.7.3). The bag can be made of relatively thin, fine-meshed material, or it can be made of quite thick filtering material. When thin materials are used, the mesh size is usually slightly larger than the size of the particles to be removed. The filtering action is provided by a filter cake, which is a buildup of filtered particles on the mesh material. The filter cake will continue to build up until the pressure drop across it gets so high that some of the cake must be removed. This is usually done either by periodically removing the filter from service and backflushing the bag with an air stream to blow the collected particles off the bag, shaking (vibrating) the bag to remove the cake, or agitating the bag surface with sound waves at about 200 Hz and 150 dB. Alternatively, high-pressure air counter-current to the normal gas flow can be applied during continuous operation.

Control of the bag filter is generally limited to monitoring the pressure drop across the filter cake and possibly measuring the particle content of the filtered gas. Such measurements can be used to determine when backflushing of the filter is required. Continuous control of bag filters is rather rare.

Cleaning, or washing, a gas in a scrubber (Figure 3.7.4) consists of literally washing the solid particles out of the gas by means of a countercurrent liquid

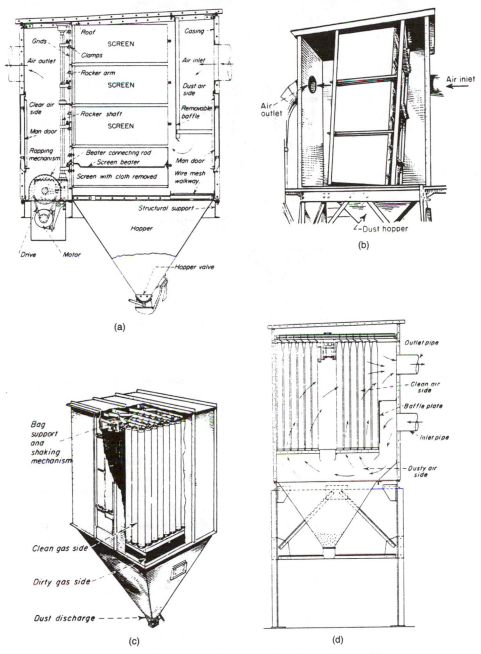

Figure 3.7.3. Bag filters for separation of gas from solids.

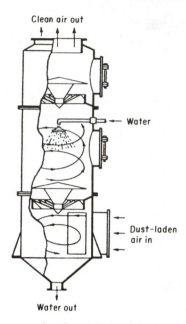

Clean air out

Water

Dust-laden
air in

Water out

Figure 3.7.4. Scrubber for removing fine solid particles in a liquid shower. (Courtesy of DUCON Company, Inc.)

shower, usually water. The actual physical mechanism involved in scrubbing is not clear, but apparently several phenomena are involved. The most important effect probably is the collision between a solid particle and a liquid droplet caused by gravitational momentum, as well as that caused by random motion and motion due to electrostatic attraction.

The actual layout of a scrubber system depends on the method used to atomize the liquid. The solids can be separated from the scrubbing liquid by a filtration process, and then either material can be used again, if desired, or discarded or destroyed if it has no value. A scrubber is usually designed to remove finer particles than either a cyclone or a bag filter removes, and therefore it is often installed in cascade following those devices if they cannot satisfactorily separate the solid particles from the gas.

The process dynamics of a scrubber can usually be characterized by a first-order transfer function in which, as usual, the volume and the throughput determine the time constant. If one is most interested in the gas phase, then the volume and gas flow will be used to determine the time constant of interest. The dynamics of the concentration of solids in the wash liquid must also be included in the dynamics of the solid phase. The dynamics of the solid phase will generally have a somewhat lower bandwidth than the dynamics of the gas phase. An important factor in the gas phase is the varying resistance to gas flow, which gives a varying pressure drop.

Electrostatic separators, often called electrofilters or electrostatic precipi-tators, usually consist of a large number of tubular metal electrodes that are connected in parallel and placed in a cylindrical column. A thin wire is placed in the center of each tube, and these wires are also connected in parallel. High voltage is applied to the wires (see Figure 3.7.5). This arrangement creates a field that ionizes the gas and charges the solid particles suspended in the gas. The particles then migrate to the wires, where the force of the electrostatic field is highest. The collected particles are removed from the wires by mechan-ical vibration.

An electrostatic filter presents virtually no restriction to the flow of gas, making it useful even when the pressure drop in the gas flow is very low. This is one reason why electrostatic precipitators are often used to clean industrial stack gases. The charge current in the filter can be measured and is a measure of the mass flow of the particles being removed. This measurement can also be used as a monitor and possible control variable for separation steps that might be in cascade with the electrostatic filter.

Liquid–liquid separation is used for the separation of a mixture of insoluble liquids such as the components of an emulsion. Settlers (gravity separators), cyclones, and centrifuges are used for this purpose. All operate on the principle that separation will occur because of differences in the specific weight of the component liquids.

If the components of a liquid mixture differ very little in specific weight, or if the viscosity of the mixture is high, it will take a long time for separation to occur in a gravity separator. Therefore, the gravity separator must have a rela-tively large volume and/or a low throughput. Here, too, the major time constant of the system will be the holdup time of the separator.

In a cyclone or centrifuge an apparent increase in the difference between the specific weights of the components to be separated will yield a corresponding decrease in the holdup time required for separation to occur. The most common form of centrifuge is shown in Figure 3.7.6. The mixture is added at the top center, but it actually enters the centrifuge at the bottom. The liquid rises through a series of conical metal plates that are separated by a distance of about 0.3 to 2 mm, depending on the material to be centrifuged. There are about one hun-dred of these plates, each of which has a number of holes. The heavier liquid is forced outward to the wall of the centrifuge, leaving the lighter liquid near the center, as shown in Figure 3.7.6.

A centrifuge rotates at very high speed, up to about 10,000 rpm, increasing the gravitational force by a factor of 10,000 to 15,000. This gives a high pro-duction capacity per unit volume. The holdup time in a centrifuge is determined by the ratio of the holdup volume to the throughput rate for each of the liquid components. Therefore, the only way to control the separation action in a cen-trifuge is to recycle one of the streams at a ratio that gives the desired separa-tion. Since a centrifuge can only separate "heavy" from "light" materials, a gradation of separation requires multiple centrifuges in series.

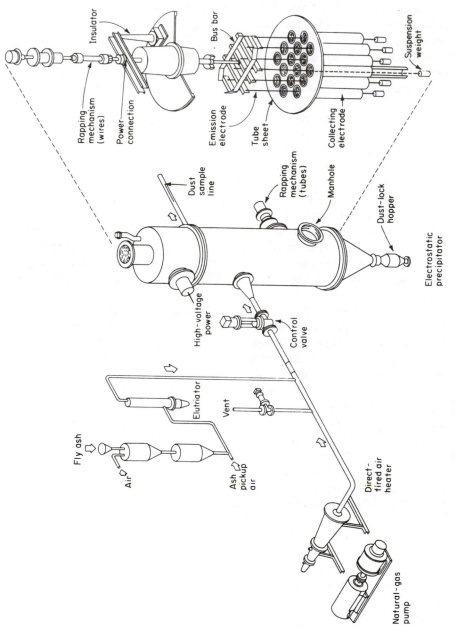

Figure 3.7.5. A common layout for an electrostatic filter.

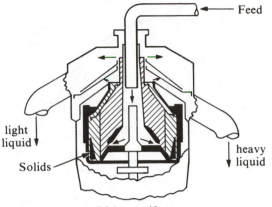

Disk centrifuge

Figure 3.7.6. Construction of a centrifuge for separation of different specific weights. (McCabe and Smith, reprinted by permission of McGraw-Hill Book Co.)

There are also many ways to separate *liquids from solids,* and vice versa. Common methods are:

- Evaporation.
- Crystallization based on evaporation or cooling.
- Leaching of a soluble fraction from an insoluble fraction.
- Filtering with a filter cloth, etc.
- Gravitational separation.
- Centrifuging.

We will discuss evaporators in section 3.9 and crystallizers in section 3.10. Centrifuges for separation of solids from liquids operate on the same principles as those described for liquid–liquid separation, so they will not be discussed further here. The main topics of this section are continuous filtration and gravity separators.

One common type of continuous filter is shown schematically in Figure 3.7.7 and in cutaway form in Figure 3.7.8. In the system shown in Figure 3.7.7, a rotating drum covered with some type of screen, either a wire mesh or perhaps a stronger construction with holes, revolves through a tank containing the suspension that is to be filtered. The inside of the drum contains a series of radial chambers coupled to a vacuum pump. By this means, about two-thirds of the drum surface operates under vacuum. This surface area starts at the point where the drum enters the tank and extends in the direction of rotation. The solid particles are held to the filter screen by the vacuum as the water is pulled through the developing mat (filter cake) and the screen. This mat of solid material then either is scraped off the drum by means of a scraper, often called a "doctor," or simply falls off, as it does in the unit shown in the cutaway view in Figure 3.7.8.

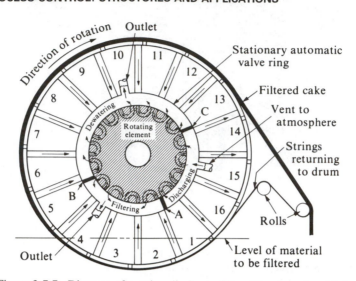

Figure 3.7.7. Diagram of a string-discharge filter. (Courtesy Ametek.)

The advantage of this arrangement is that it is not necessary to "doctor" the filter cake off the filter screen because it comes off by itself. In the third of the drum that is not under vacuum and where there is no filter cake, backwashing is usually provided as a means of cleaning the filter screen. The wash material can be water, solvent, or whatever is required by the specific situation.

The filter cake plays a significant role in the continuous filter described above. Filtering effectiveness, expressed as the amount of solid material in the final liquid, and filtering capacity, expressed as the amount of suspension that can be treated per unit time, depend upon the concentration of solids in the suspension. Several parameters can be varied to control filtration in a continuous rotary filter, the most important being the drum rotation speed, the vacuum, and the level of the suspension in the tank. If the level of the suspension is automatically controlled, then the filter capacity can be measured by the amount of suspension picked up by the drum. If the vacuum and rotation speed are not controlled, then the input stream must be controlled. The dynamics of the vacuum filter are determined by the holdup time in the tank and the transport time around the drum. For the liquid phase, the holdup of the tank will dominate, as the liquid passes quickly through the filter cake on the drum. However, for the solid phase, the transport time of the drum becomes an important addition to the total system dynamics.

Many other types of filters are used in industry, but from the point of view of automatic control, all have approximately the same characteristics as the vacuum filter described above.

Gravity separators (Figure 3.7.9) for the separation of solids from liquids are usually relatively shallow tanks equipped with slowly rotating rakes that remove the solids sludge. The suspension is added to the surface of the settler

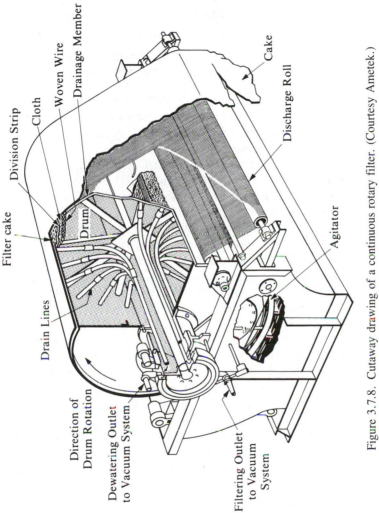

Figure 3.7.8. Cutaway drawing of a continuous rotary filter. (Courtesy Ametek.)

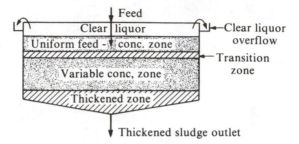

Figure 3.7.9. Principles of a gravity separator (clarifier). (Foust, Wentzel, Clump, Mause, and Andersen, Principles of Unit Operations, 1980. Reprinted by permission of John Wiley & Sons, Inc.)

at the center, and the clear liquid (supernate) leaves the tank by overflowing at the periphery. The rotating rake scrapes the filtrate (sludge) at the bottom of the tank toward the center, from which it is pumped out, either continuously or intermittently. This type of separator is often used for clarification in waste treatment plants. These devices are sometimes also known as "thickeners," but one should be careful with this term — in other contexts. For example, in the paper industry, a "thickener" can refer to a continuous vacuum drum filter!

The capacity of a gravity separator (clarifier) is expressed as the flow of clear liquid per unit time, and is essentially limited by the area of the tank. Because the liquor separates into distinct zones, as shown in Figure 3.7.9, the flow into the system must not be so large that the zones containing the settling solids rise high enough to allow solids to overflow the rim of the tank. This is precisely what happens when a sudden rainstorm overloads the capacity of a municipal sewage treatment plant. Automatic control is somewhat difficult because the relationship between volumetric flow and the amount of suspended solids can change quickly and without any cause-and-effect relationship. When the flow into the system cannot be controlled, the plant must be designed on the basis of the worst possible case (maximum expected input). The sludge removal rate must not exceed the settling rate, and this can be controlled on the basis of the height of the thickened zone (Figure 3.7.9).

Many processes require separation of different fractions of solids from each other, which can be accomplished in many ways. Separation by size (classification) can be done with screens (filters), vibration separators, or cyclones. Other characteristics on which separation systems can be based include density difference magnetic characteristics, electrostatic characteristics, or surface characteristics.

A typical classifier for granulated material is the mechanical system shown in Figure 3.7.10. The material is placed on the first of two screens, or plates, at the left end. (Other systems may have several screens.) The first screen has a mesh or hole size determined by the largest granules that are to pass through the screen. Different sizes are used on succeeding screens. To aid screening,

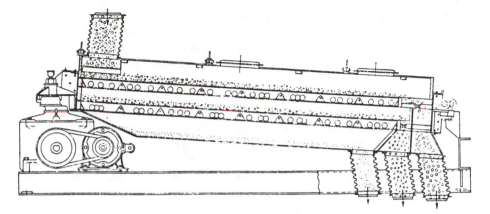

Figure 3.7.10. Continuous screen for classification of granular material — Rotex (R) screener.

the screen plates usually oscillate in the longitudinal and transverse directions, driven by an eccentric that in turn is driven by a constant-speed motor. Since the screens are at a slight angle to the horizontal, the vibration of the screen will easily pass the granules from one end to the other. Figure 3.7.10 shows that the two-screen system produces a stream of granules classified into three fractions, the finest having passed through both screens, the middle fraction having passed through one screen, and the coarsest having been rejected by the first screen. Solid particles are prevented from sticking in the screen holes by a number of balls that bounce up against the bottom of the screen plates, knocking the particles back out of the holes. The motion of these balls is caused by action of the same eccentric used to vibrate the screens.

Whether a screen separator is equipped with automatic control depends on its role as a unit operation in a larger process. In some applications, it is operated without any control, and the classification is simply a result of the distribution of the sizes in the feed stream. If the function of the system is to produce a particular size distribution of the solid, then the relative distribution of the flows of the various fractions must be determined on the basis of the needs of the product. These fractions must satisfy the material balance of the production system, and an imbalance will lead to a surplus of material in one or more of the fractions. One can choose between the case in which the size distribution can be controlled by prior processing, and the case in which the material is simply taken as it comes.

If the classification unit follows an earlier process in which the particle size was reduced, at least two situations can occur:

1. There is an excess of particles in the large fractions and a shortage in the fine fractions.
2. There is an excess of fine particles and a shortage of large ones.

The first case presents no problem because one or more of the large-particle fractions can be recirculated to the upstream size reduction process for reprocessing. The second case is far more serious unless one has available a process for increasing the size of the particles (agglomeration by sintering, granulation, etc.).

If the output of the screening unit appears to have too many large particles on some occasions and too many small ones on others, one should consider the possibility that there has been segregation in storage and/or transportation in the processing upstream of the screen classifier. It is well known, for example, that material drawn from the bottom of a silo is likely to contain a high fraction of small particles compared with the silo contents taken as a whole. When this happens, it may be profitable simply to store the small particles and use them when there is a shortage of small particles in the system output. This can be more economical than reprocessing (milling or grinding) large particles in order to maintain the desired size distribution.

The best solution to the problem described above lies in maintaining an economic and technical balance with respect to the storage of large quantities of material and the cost of installation of additional processing equipment.

The dynamic behavior of the classifier shown in Figure 3.7.10 is dominated by a first-order transfer function with a time constant given by the ratio of the mass volume on each screen (plate) to the mass flow of each fraction. There is also a transport delay that can be approximated as about half the travel time of the material across each screen plate.

Separation of solid particles on the basis of density or size can also sometimes be done by dispersing the solids in a liquid. This process is also called classification, and it is based on the following concepts:

1. Large particles will sink faster than fine particles with the same specific density.
2. Particles with high specific density sink somewhat faster than particles having the same size but lower specific density.
3. The sinking velocity of solid particles decreases with increasing viscosity and/or density of the liquid.

When this approach is appropriate, it must be followed by a separator such as a cyclone or a centrifuge.

Separation of solids by magnetic means is widely used in the mining and mineral refining industries. Figure 3.7.11 is a schematic diagram of a counterrotation magnetic separator for a suspension of small iron ore particles in water. The rotating drum, made of a nonmagnetic material, is partially submerged in the suspension. A number of fixed permanent magnets mounted inside the drum create a magnetic field that holds the magnetic material, such as magnetite (Fe_3O_4), on the surface of the drum. The nonmagnetic material remains in suspension and leaves the system. The magnetic particles remain on the

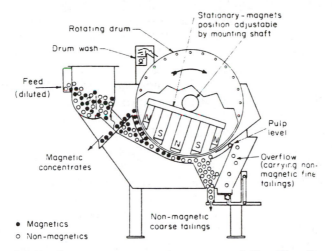

Figure 3.7.11. Countercurrent magnetic separator—Jeffrey Div., Dresser Co.

drum until they leave the field of the last magnet where they will drop and/or
be scraped off the drum. This principle can also be applied to magnetic separation of dry materials.

The efficiency, or separation quality, of a magnetic separator is characterized
by the amount of magnetic material remaining in the suspension relative to the
amount of nonmagnetic material. It depends on many factors, including the size
of the magnetic particles relative to the size of the nonmagnetic particles, the
magnetic field strength, the drum rotation speed, and the rate of flow through
the separator. In order to improve separation, it is often necessary to increase
the effective holdup time of the separator. This is usually done by recirculating
a portion of one of the streams. Just as in the case of the closed-circuit mill
described in section 3.5, recirculation will increase the time constant of the
system but decrease the production capacity. Another problem is that a strong
magnetic field will attract large nonmagnetic particles that contain small magnetic particles, thus holding an amount of nonmagnetic material in the separated, or accepts, stream. These will not separate unless they are first returned
to a milling stage for particle size reduction. In order to manipulate one or more
of the above variables to control the separator, one would like to measure the
amount of magnetic material in the tailings (the nonmagnetic stream) from
the separator.

The principle of electrostatic separation is similar to that of magnetic separation except that the particles are given an electric charge. In the separator, the
charged particles react to an electrode of opposite polarity, whereas the un-
charged particles do not respond. Thus a selective separation occurs.

Flotation is another separation process widely used for mineral dressing and
in the chemical process industries. This process is used for separating metal
ores and pyrites that have been reduced to particles of less than about 0.3 mm.

The solids are suspended in water at a consistency of 15 to 35% solids. The principle of flotation is based on the tendency of the suspended particles to fasten themselves to small, rising air bubbles in the flotation chamber. This tendency is promoted by the addition of a collector, or promoter agent. The choice of agent for this function depends upon which materials are to be separated. For flotation of metal sulfides (pyrites), xanthates are used as the collector. Flotation can also be assisted by the use of frothers such as various oils, polypropylene glycol, or methylamyl alcohol.

A typical flotation cell is shown in Figure 3.7.12. Usually these cells are connected in series, where the tailings, not the accepts, are used as the feed for the next cell. The product (concentrate), supported by the foaming agent, rises to the top of the cell and simply flows over the rim to a collecting tray.

There are many configurations possible for flotations cells and systems. Often, an impeller is placed in the bottom of the cell to agitate the suspension, and air may be added through a vertical tube around the impeller shaft. The feed, perhaps from another cell, enters at the left (Figure 3.7.12) through a controllable opening.

Figure 3.7.13 is a flowchart showing the major components of a complete flotation system. At the top, the ore and/or pyrites are crushed in a multistage

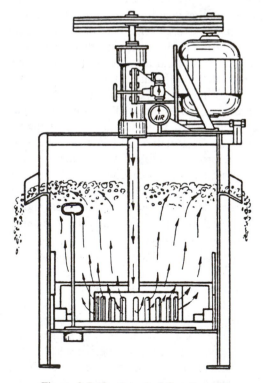

Figure 3.7.12. A typical flotation cell.

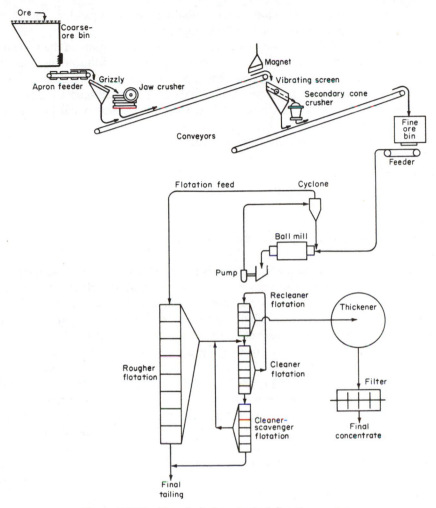

Figure 3.7.13. Flowchart for a typical flotation system.

unit, and then sent to a mill operating in a closed circuit with a cyclone separator. The flotation section is also multistage. First is a rougher flotation section consisting of eight cells in series, in which the tailings from one cell are the feed to the next. The collected concentrate from the rougher flotation unit passes through three different flotation units. The first is a cleaner stage having six cells. The concentrate from this unit goes on to a four-cell recleaner unit. Concentrate from this stage is the final product. The accepts from the scavenger stage are mixed with the flow from the rougher stage and returned to the cleaning stage, while the tailings from the scavenger stage are mixed with the tailings from the rougher stage and discharged from the system. The accepts from the recleaner unit are sent to a thickener (usually a gravity separator), and then to a filter that produces the final product concentrate.

Control of the flotation process can be applied at many points in the system, and can be achieved by several different means. Investigation of the control characteristics of flotation processes has led to a good understanding of how this process should be manipulated to obtain the best possible separation quality and capacity (Niemi, 1966; Olsen, 1975; LeGuea, 1975; Koivo and Cojocariu, 1977; Kaggerud, 1983).

The most important control variables are:

- Airflow to the cell.
- Flow of tailings from the cell.
- Flow of collector.
- Flow of frother agents.
- Water/solid ratio in the suspension (pulp).

As we can see in Figure 3.7.13, there is positive feedback in that the concentrate from the cleaning cell passes through the recleaning cell and returns as feed to the cleaning cell. A second positive feedback loop has the concentrate from the scavenger cell recirculating through the cleaning cell and appearing again as feed to the scavenger cell. As we saw in the case of closed-circuit milling (section 3.5), positive feedback leads to slow dynamics. As a result, there is a real need for automatic control. In spite of this need, application of automatic control is not widespread because of difficulties with measurement instrumentation. There has been little application of automatic control to individual flotation cells, but this situation is improving. The most common practice is to carefully control the frother and collector flows. The flotation process is becoming increasingly important as the world supply of high-grade mineral ores is depleted. An even better understanding of the fundamental principles of flotation units, as well as a greater ability to automatically control them, will be required in order to operate them most efficiently.

We will discuss some practical control solutions for this process in section 5.1.

3.8. HEAT GENERATION AND HEAT EXCHANGE

The generation, transfer, and exchange of heat are very important unit processes in industry and are the focus of many process control systems. Control of heat, as well as all energy, has become especially important as energy conservation has become a major concern. The economic survival of many industries demands precise control of energy costs.

The most important heat energy sources are:

- Combustion of organic materials such as coal, oil, and gas.

- Conversion of electricity to heat by resistance heating, inductive heating, dielectric heating, microwave heating, and infrared heating.
- Nuclear fission (atomic energy).

Other sources of energy, including solar energy, geothermal energy, wind power, and so on, are, at least for now, relatively insignificant. Electric power is usually produced by rotating generators driven by steam- or water-powered turbines. Most steam turbines are driven by steam generated by the combustion of organic fuels or by nuclear power.

We will discuss only the most common industrial method of heat generation, the burning of organic fuel. We will not discuss the production of electrical energy by turbines and generators or the control problems associated with the conversion of electricity to heat. These subjects are so broad that they require separate treatment, but many of the fundamental problems that apply to fossil fuel heat generation also apply to these systems.

Combustion Processes

Combustion of gases, liquids (oil), and solids (coal, wood, etc.) is a major source of thermal energy. Combustion can be a very difficult process to describe, especially if the fuel is complex (wood chips, for example). Although the combustion of wood is perhaps one of the oldest sources of thermal energy, an accurate description of the process has been of little interest until recently.

The combustion process is fundamentally the oxidation of carbon, hydrogen, and sulfur to produce carbon dioxide, water vapor, and sulfur dioxide. Oxygen is supplied by air, and the general stoichiometry is given by:

$$C_m H_n + \left(\frac{4m + n}{4}\right) O_2 = m CO_2 + \left(\frac{n}{2}\right) \cdot H_2O \qquad (3.8.1)$$

where m and n represent the contents of carbon and hydrogen, respectively in the actual hydrocarbon. A simple, common hydrocarbon is methane (CH_4). From equation (3.8.1) we find that one mole of methane, for which $m = 1$ and $n = 4$, requires $[(4 \times 1) + 4]/4] = 2$ moles of oxygen for complete combustion. Because air is not pure oxygen, each mole of oxygen supplied by air will carry with it 3.8 moles of nitrogen. Nitrogen is inert in combustion, but the process equipment must account for its volume.

If the relative weight distribution of carbon, hydrogen, oxygen, and sulfur in a fuel is known, an estimate of the amount of oxygen needed for total combustion can be made using the expression

$$22.41\left(\frac{C}{12} + \frac{H_2}{4} - \frac{O_2}{32} + \frac{S}{32}\right) \qquad (3.8.2)$$

where C, H_2, O_2, and S are the decimal fractions of each of the elements in the fuel. The amount of oxygen is given in normal cubic meters (nm^3), which is the volume at 0°C and atmospheric pressure (1 bar). If the oxygen is to be supplied by air, the factor 22.41 in equation (3.8.2) must be replaced with the factor 106.74.

Table 3.8.1 is a list of some important constants for a number of basic fuels. Some more complex industrial fuels are also included in the table.

In order to control the combustion process, it is important that the process description focus on those parameters that most influence (and are influenced by) the control manipulations.

One important parameter is the heat content of the fuel; that is, the amount of heat that can be generated per unit weight of fuel. This quantity, which can often be found in tables (or determined by laboratory analysis), determines the fuel mass flow required for a given heat requirement. Therefore, control of the fuel mass flow is a possible means of controlling the heat generated by the combustion process.

Another important parameter is the amount of air theoretically required for complete combustion of the fuel. This too can often be found in tables such as Table 3.8.1. Because the fuel component ratios are never ideal or precisely known, it is always necessary to provide an amount of *excess air* relative to the theoretical requirement in order to minimize the unburned hydrocarbons and carbon monoxide (CO) in the exhaust (flue) gases.

A third parameter of interest is the relationship between oxygen in the exhaust gas and the amount of excess air added. Figure 3.8.1 contains curves showing that oil, coal, and natural gas require about 10% excess air in order to have about 2% oxygen in the flue gas. Basic oxygen process gas requires about twice as much excess air to provide the same exhaust oxygen content as the other fuels.

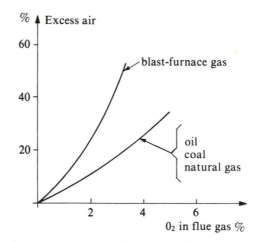

Figure 3.8.1. Relation between excess air and oxygen in exhaust gas for different fuels.

The final major parameter of interest to the control engineer is residence time in the combustion chamber, which is approximately the volume divided by the flue gas flowrate. Because the combustion process causes a very fast temperature rise, temperature must be considered in the calculation of the product gas volumetric flow rate. The masses of the three most important products of combustion, CO_2, H_2O, and N_2, are given at the far right side of Table 3.8.1. For temperatures that usually occur in actual combustion processes, the volumetric flow rate of the product gas (m³/s) is approximately:

$$q = (2 \cdot 10^{-3} \cdot w_{CO_2} + 5 \cdot 10^{-3} \cdot w_{H_2O} + 2.9 \cdot 10^{-3} \cdot w_{N_2})\theta\left(\frac{m^3}{s}\right)$$

$$(3.8.3)$$

where the w_i are the mass flows (kg/s) of carbon dioxide, water vapor, and nitrogen, and θ is the absolute temperature (°K). As an example, consider the combustion of methane. From the right side of Table 3.8.1, we can find the mass flow of the products per kilogram of methane burned. If we assume that the product gas temperature is 200°C, which is about 473°K, we find that the gas volume is 26 m³/kg methane. Since the heat content of methane is about 52×10^6 joules/kg, we find the gas flow to be about 0.5×10^{-3}(m³/s-kW). Therefore, a methane furnace that produces 500 kW and has a chamber volume of 10 m³ will have a holdup time of

$$\frac{V}{q} = \frac{10}{0.5 \cdot 10^{-3} \cdot 500} = 40(s)$$

The process dynamics between changes in fuel or air feed rates and changes in the exhaust gas will generally be characterized by a first-order transfer function:

$$h(s) = \frac{K}{1 + Ts} \qquad (3.8.4)$$

where $T = V/q$, and K depends on which component of the product gas and which input variable are under consideration. The dynamics of the combustion process are, then, relatively simple, and the most commonly used control loops usually have no problems with instability. We will discuss this further in section 4.7.

The means of mixing the fuel and air required for combustion depends upon which fuel is used and its form. Certainly, the equipment needed for oil is different from that used for coal, wood chips, or gas. There are many configurations for mixing fuel and air, and Figure 3.8.2 shows several different forms of oil burners. Similarly, there are many types of burners and mixers for solids and gases.

Table 3.8.1.

NO.	MATERIAL	FORMULA	MOLECULAR WEIGHT	kg/m³ 15°C 1 BAR	m³/kg 15°C 1 BAR	DENSITY REL. AIR	HEAT OF COMBUSTION kJ/m³		kJ/kg	
							GROSS	NET	GROSS	NET
1	Carbon	C	12.01	...	...	...	...	...	32809	32809
2	Hydrogen	H_2	2.016	0.0849	11.7786	0.0696	12116	10252	142241	120178
3	Oxygen	O_2	32.000	1.3553	0.7378	1.1053				
4	Nitrogen	N_2	28.016	1.1919	0.8390	0.9718				
5	Carbon monoxide	CO	28.01	1.1855	0.8435	0.9672	12004	12004	10120	10120
6	Carbon dioxide	CO_2	44.01	1.8743	0.5335	1.5282				
Paraffins										
7	Methane	CH_4	16.041	0.6792	1.4723	0.5543	37764	34037	55590	50099
8	Ethane	C_2H_6	30.067	1.2864	0.7785	1.0488	66806	61176	51961	47566
9	Propane	C_3H_8	44.092	1.9160	0.5219	1.5617	96555	88913	50427	46430
10	n-butane	C_4H_{10}	58.118	2.5344	0.3946	2.0665	125634	116053	49605	45815
11	Isobutane	C_4H_{10}	58.118	2.5344	0.3946	2.0665	125373	115754	49486	45696
12	n-pentane	C_5H_{12}	72.144	3.0902	0.3236	2.4872	149717	138272	49100	45436
13	Isopentane	C_5H_{12}	72.144	3.0902	0.3236	2.4872	149418	138533	49009	45345
14	Neopentane	C_5H_{12}	72.144	3.0902	0.3236	2.4872	148859	137675	48818	45154
15	n-hexane	C_6H_{14}	86.169	3.6429	0.2745	2.9704	177527	164479	48748	45170
Olefins										
16	Ethylene	C_2H_4	28.051	1.1951	0.8368	0.9740	60170	56405	50387	47247
17	Propylene	C_3H_6	42.077	1.7782	0.5624	1.4504	87086	81494	48983	45841
18	n-butene	C_4H_8	56.102	2.3710	0.4218	1.9336	114972	107553	48516	45387
19	Isobutene	C_4H_8	56.102	2.3710	0.4218	1.9336	114375	106956	48259	45121
20	n-Pentene	C_5H_{10}	70.128	2.9669	0.3371	2.4190	143006	137414	48218	45077
Aromatics										
21	Benzene	C_6H_6	78.107	3.3001	0.3030	2.6920	139837	134245	42393	40693
22	Toluene	C_7H_8	92.132	3.8945	0.2568	3.1760	167164	159708	42928	41019
23	Xylene	C_8H_{10}	106.158	4.4904	0.2227	3.6618	194974	185654	43417	41345
Misc. Gases										
24	Acetylene	C_2H_6	26.036	1.1166	0.8956	0.9107	55883	53981	50052	48367
25	Napthalene	$C_{10}H_8$	128.162	5.4212	0.1845	4.4208	218237	210781	40270	38896
26	Methanol	CH_3OH	32.041	1.3553	0.7378	1.1052	32359	28631	23883	21136
27	Ethanol	C_2H_5OH	46.067	2.0442	0.4892	1.5890	59648	54093	30639	27770
28	Ammonia	NH_3	17.031	0.7305	1.3689	0.5961	16440	13607	22507	18626
29	Sulphur	S	32.06			...			9272	9272
30	Hydrogen sulphide	H_2S	34.076	1.4594	0.6852	1.898	24120	22219	16529	15237
31	Sulphur dioxide	SO_2	64.06	2.7763	0.3602	2.264				
Liquid Fuels										
32	Gasoline			750	0.0013				45000	42000
33	Diesel Oil			835	0.0012				45900	43000
34	Fuel oil EL			840	0.0012				45500	42700
35	Fuel oil L			880	0.0011				44800	42000
36	Fuel oil M			920	0.0011				43300	40700
37	Fuel oil S			970	0.0011				42700	40200
Solid Fuels										
38	Anthracite coal								~33000	
39	Birchwood (green)								~11000	
40	Birchwood (air dry)								~13000	
41	Peat								~10000	
Various										
42	Steam	H_2O	18.016	0.7626	1.3114	0.6215				
43	Air	...	28.9	1.2271	0.8149	1.000				

mole/mole OR m³/m³ COMBUSTIBLE						kg/kg COMBUSTIBLE					
REQ'D FOR COMBUSTION			FLUE PRODUCTS			REQ'D FOR COMBUSTION			FLUE PRODUCTS		
O_2	N_2	AIR	CO_2	H_2O	N_2	O_2	N_2	AIR	CO_2	H_2O	N_2
1.0	3.76	4.76	1.0	...	3.76	2.66	8.86	11.53	3.66	...	8.86
0.5	1.88	2.38	...	1.0	1.88	7.94	26.41	34.34	...	8.94	26.41
0.5	1.88	2.38	1.0	...	1.88	0.57	1.90	2.47	1.57	...	1.90
2.0	7.53	9.53	1.0	2.0	7.53	3.99	13.28	17.27	2.74	2.25	13.28
3.5	13.18	16.68	2.0	3.0	13.18	3.73	12.39	16.12	2.93	1.80	12.39
5.0	18.82	23.82	3.0	4.0	18.82	3.63	12.07	15.70	2.99	1.63	12.07
6.5	24.47	30.97	4.0	5.0	24.47	3.58	11.91	15.49	3.03	1.55	11.91
6.5	24.47	30.97	4.0	5.0	24.47	3.58	11.91	15.49	3.03	1.55	11.91
8.0	30.11	38.11	5.0	6.0	30.11	3.55	11.81	15.35	3.05	1.50	11.81
8.0	30.11	38.11	5.0	6.0	30.11	3.55	11.81	15.35	3.05	1.50	11.81
8.0	30.11	38.11	5.0	6.0	30.11	3.55	11.81	15.35	3.05	1.50	11.81
9.5	35.76	45.26	6.0	7.0	35.76	3.53	11.74	15.27	3.06	1.46	11.74
3.0	11.29	14.29	2.0	2.0	11.29	3.42	11.39	14.81	3.14	1.29	11.39
4.5	16.94	21.44	3.0	3.0	16.94	3.42	11.39	14.81	3.14	1.29	11.39
6.0	22.59	28.59	4.0	4.0	22.59	3.42	11.39	14.81	3.14	1.29	11.39
6.0	22.59	28.59	4.0	4.0	22.59	3.42	11.39	14.81	3.14	1.29	11.39
7.5	28.23	35.73	5.0	6.0	28.23	3.42	11.39	14.81	3.14	1.29	11.39
7.5	28.23	35.73	6.0	3.0	28.23	3.07	10.22	13.30	3.38	0.69	10.22
9.0	33.88	42.88	7.0	4.0	33.88	3.13	10.40	13.53	3.34	0.78	10.40
10.5	39.52	50.02	8.0	5.0	39.52	3.17	10.53	13.70	3.32	0.85	10.53
2.5	9.41	11.91	2.0	1.0	9.41	3.07	10.22	13.30	3.38	0.69	10.22
12.0	45.17	57.17	10.0	4.0	45.17	3.00	9.97	12.96	3.43	0.56	9.97
1.5	5.65	7.15	1.0	2.0	5.65	1.50	4.98	6.48	1.37	1.13	4.98
3.0	11.29	14.29	2.0	3.0	11.29	2.08	6.91	9.02	1.92	1.17	6.93
0.75	2.82	3.57	...	1.5	3.32	1.41	4.69	6.10	...	1.59	5.51
			SO						SO		
1.0	3.76	4.76	1.0	...	3.76	1.00	3.29	4.29	2.00	...	3.29
1.5	5.65	7.15	1.0	1.0	5.65	1.41	4.69	6.10	1.88	0.53	4.69

13.5
14.8
14.8
14.5
14.1
13.8
~9.4
~4.0

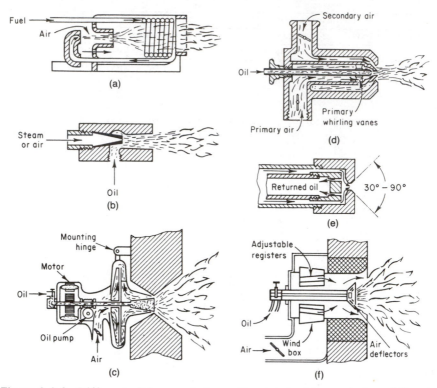

Figure 3.8.2. Different oil burner designs. (Reprinted with permission of Hauck Mfg. Co.)

The most important control aspect of the combustion process is control of the fuel and air flows at rates that produce the desired heat and at a ratio to each other that provides optimal combustion. In some cases, it is best to control the air flow relative to the fuel flow; in others, the opposite may be true. One important practical argument favoring the use of the air flow as the primary variable (proportioning the fuel flow to the air flow) is that if the air flow should stop, or fail, the fuel flow would automatically stop. This eliminates the danger of explosion.

Even if the fuel and air are properly and automatically proportioned, it is still necessary to measure the actual combustion. A number of different measuring systems are commercially available for the measurement of free O_2 or the CO/CO_2 ratio in the flue gas. The control structures for combustion control will be examined in section 4.7.

Heat Exchangers

The conversion of thermal energy from one medium to another, generally referred to as heat transfer, or heat exchange, plays a major role in many industrial production processes. Heat exchangers are most commonly used for the transfer of energy between two fluids. Exchangers can be operated in many configurations, and the fluid can be gas or liquid on either the primary or secondary side of the exchanger. We will see many industrial applications of heat exchangers in the systems described in this book. Therefore, in this section we will discuss several different heat exchanger configurations.

The efficiency of a heat exchanger depends very much on its being designed to meet the capacity required by the particular process in which it is to be installed. The control characteristics are usually a matter of secondary importance, but if the exchanger is to be operated in a dynamic mode, the overall efficiency and performance will also be determined by the control characteristics.

The most common heat exchanger consists of a cylindrical tank filled with a number of parallel tubes aligned in the axial direction, as shown in Figures 3.8.3a, b, c, and d. If we let the primary medium (the hot medium) flow through the tank, and the secondary medium (the material to be heated) flow through the tubes, we will have the two situations shown in Figures 3.8.3a and b. These configurations are known respectively as co-current and countercurrent heat exchangers because the two media flow, respectively, in either the same direction or the opposite direction. The basic design is usually referred to as a shell and tube heat exchanger. In Figure 3.8.3c, we see a multi-pass heat exchanger in which the tubes double back inside the shell, and the primary medium is channeled as desired by a series of baffles. A somewhat specialized design is shown in Figure 3.8.3d. Here, the primary medium flows through tubes that are submerged in a large chamber containing a liquid secondary medium. The secondary tank is divided into two parts separated by an overflow dam. The secondary liquid is partially vaporized, and vaporization is aided by the large surface area of the chamber. The vapor leaves the chamber at the top, and the remaining secondary liquid leaves at the bottom of the small (lower) chamber. This design is frequently used in the reboiler section of a distillation column (see section 3.12).

Another type of heat exchanger, which has many different configurations, is a condenser. The purpose of this unit is to condense a gas, or vapor, to the liquid phase. This is done through cooling by means of a second medium in the exchanger. One of the many possible configurations is shown in Figure 3.8.4. In this case the tubes are vertical, and the vapor flows into the shell. The cooling medium flows through the tubes from the bottom. By throttling the outlet flow from the shell, the level of condensate can be held at any desired level, as shown by the dashed line. This is an effective means of controlling the amount of condensation and/or heat exchange, as only the tube area exposed to the vapor can cause condensation.

The heat transfer in an exchanger in which a liquid film is formed by partial condensation (the film moves downward along the tubes in Figure 3.8.4), or where there is evaporation (giving a partial liquid and partial vapor contact), is very complex and difficult to model. Designers of heat exchangers must be

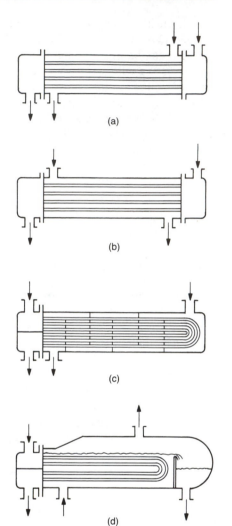

Figure 3.8.3. Common heat exchanger configurations.

very careful to provide suitable heat transfer characteristics over the entire load range for which the unit is intended. Fortunately, there is a great deal of theoretical and experimental information available regarding the design of heat exchangers (Perry and Chilton, 1973).

The design of heat exchangers is an important and very broad subject. There are so many types and configurations that it is impossible to include here detailed mathematical descriptions of them all. Therefore, we will limit our discussion to those points that are most essential to modeling of heat exchangers from the point of view of automatic control (Campbell, 1958; Thal-Larsen, 1960; Hempel, 1961; Buckley, 1964).

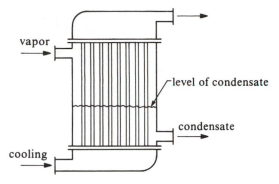

vapor

level of condensate

condensate

cooling

Figure 3.8.4. Vertical condenser with variable condensate level.

The basis for a steady-state model of a heat exchanger is the law of conservation of energy: the energy supplied to the exchanger must equal the energy removed. If the heat exchanger is well insulated from the environment, then all the energy that is lost by the primary medium must leave the exchanger with the secondary medium. Consider the case in which both media are liquids (Figure 3.8.5). Figure 3.8.5a shows the temperature profile along the tubes of a countercurrent heat exchanger operating at steady state. In this case the temperature difference between the two media is relatively constant over the length of the exchanger. On the other hand, the temperature difference for a co-current exchanger (Figure 3.8.5b) changes a great deal. It is not obvious which system should be used; there are many processes in which a large temperature difference cannot be tolerated. Therefore, both types of heat exchangers are very common in industry.

The first problem in the calculation of the steady-state performance of a heat exchanger is to estimate the heat transfer coefficients at various points along the tubes of the exchanger. As a first approximation, the curves shown in Figure 3.8.5 may be assumed to be linear. It is then possible, using a simple energy balance, to calculate the temperatures of the primary and secondary media leaving the unit when the input temperatures are known. A review of the steady-state energy balance equations is given in Appendix H.

The situation becomes more complicated when one or both of the media undergo either evaporation or condensation within the heat exchanger. A common but very complicated situation occurs when the temperature of the tube is high enough to cause vaporization at some point along the tube. When this occurs, vapor bubbles form a varying boundary layer along the tubes, and thus radically change the heat transfer coefficients at those points along the tube. Obviously, this vaporization will change the performance of the entire heat exchanger. The problem can usually be reduced by maintaining turbulent flow and good circulation of the secondary medium in the shell of the exchanger, thus constantly removing any vapor bubbles from the tube surface. One way to accomplish this is to use a circulating pump to maintain a high velocity in the

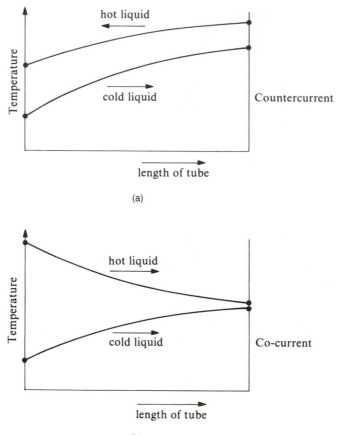

Figure 3.8.5. Stationary temperature profile in countercurrent and co-current heat exchangers.

secondary medium (Figure 3.8.6). With good circulation, the temperature distribution profile of the secondary medium will be smoothed considerably. In fact, as a first approximation it is often reasonable to consider the temperature of the secondary medium to be uniform.

In formulating a dynamic model of a heat exchanger, it is important to be very clear as to which variables might be used as control variables and which variables are to be controlled. Some important heat exchanger configurations are:

• Heating the secondary medium (liquid–liquid exchanger, boiler) by means of a suitable primary medium prepared for this purpose (hot water, steam, etc.). The heat content (enthalpy) or temperature of the liquid, or pressure or temperature of the saturated steam, will then be states that one might want to control.

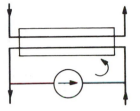

Figure 3.8.6. Heat exchanger with forced circulation on the secondary side.

- Heating the secondary medium with a primary medium that comes from another function in the process whose heat one wishes to use by cooling or condensation. This situation is especially likely to occur as part of the energy conservation system in large integrated processes. In these cases there will be constraints on the way in which the primary medium can be manipulated because it belongs to another system.
- Cooling the primary medium (perhaps producing condensation) by means of a secondary medium used only for that purpose (cold water, for example). The variables one would want to use here are the temperatures of the primary medium and condensate stream, and the vapor pressure on the primary side ahead of the condenser.

Control actions for the various heat exchanger configurations include the following:

- Restrict the flow of the primary medium into the exchanger by means of a control valve. This requires that the outlet of the primary have a restriction. This option can be used for both liquid and gaseous media (see Figure 3.8.7a).
- Throttle the primary flow of the outlet side of the exchanger. With this approach the pressure of the primary on the inlet side will be nearly constant. This configuration also makes it possible to control the heat transfer rate by controlling the level of condensate on the primary side, as was shown in Figure 3.8.4.
- Throttle the secondary flow at either the inlet (Figure 3.8.7c) or the outlet (Figure 3.8.7d). These two solutions are most applicable to cases where the desired result is cooling the primary medium.
- Use combinations of the three above configurations when both the primary and secondary media can be manipulated.
- Since the heat content of a fluid is a function of both the mass flow and the temperature, use the temperature of the medium as a control variable. Further, if the medium is saturated steam, then the pressure is a measure of the temperature. Pressure control of the steam on the primary side of a heat exchanger provides a rapidly responding change for the energy input.
- Vary the flow of the secondary medium. The arrangement shown in Figure 3.8.7e is especially interesting for the case where the secondary

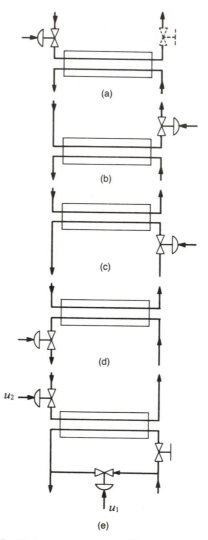

Figure 3.8.7. Various control possibilities for a heat exchanger.

medium is a liquid that must be held at constant temperature in an exchanger where the entering primary medium undergoes relatively rapid temperature changes. The valve u_1 makes it possible to change the ratio between the amount of the secondary that flows through the exchanger and the amount that bypasses it. This varies the mixture of heated and unheated liquid. The manual valve in the secondary branch of the exchanger sets the balance between the two streams. Control variable u_2 is used to vary the energy input such that valve u_1 always works in its mid-range.

The dynamics of the various types of heat exchangers depend on the design of the exchanger and the media used (see Appendix H). However, certain phenomena are common to most configurations. Usually, the dynamics will be determined as follows:

1. The ratio between the thermal mass $(\rho_i c_i V_i)$ and the total heat transfer coefficient $(A_i h_i)$ will give the thermal time constant for an arbitrary thermal unit. In an actual heat exchanger we have the thermal time constants for the metal tubes and for the primary and secondary media. The jacket, or tank, containing the tube pack also has thermal mass that absorbs and emits energy, thereby contributing to the thermal sluggishness of the total system. These properties give the system thermal mass, which is responsible for the sluggishness of the heat transfer in the system.
2. The holdup time of the medium in the heat exchanger is responsible for a dynamic phenomenon that can be characterized by a transport delay in the system.

To illustrate the phenomenon indicated in point 2 above, consider the simple system shown in Figure 3.8.8. A straight, thin tube is surrounded by a shell that contains saturated steam. Liquid flows through the tube. The temperature of the saturated steam is assumed to change only and instantaneously with the pressure, which can be controlled. We assumed that the liquid in the tube has perfect mixing and exchange of energy in the radial direction, but there is no

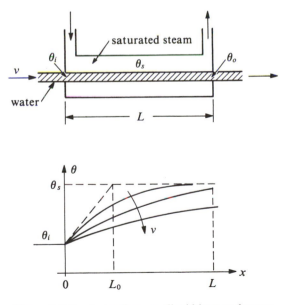

Figure 3.8.8. A simple vapor/liquid heat exchanger.

mixing or heat transfer in the axial direction. We further assume that the tube has negligible mass and therefore no thermal capacity. The heat transfer from the steam to the liquid will then be determined by the net heat transfer coefficient, which is a combination of the coefficients for the transfer from the steam to the tube and from the tube to the liquid, and by the temperature difference between the steam and the liquid. The steam, being saturated, has a constant temperature in the exchanger, so the liquid temperature will increase as the liquid passes through the tube. Consider now the rise in temperature when the steam temperature is constant and the velocity of the water flowing through the tube is constant. This situation is illustrated by the graph in Figure 3.8.8. The figure gives three curves for the liquid temperature. The top curve applies to the case where the velocity is low, and the bottom curve applies to the case where the velocity is high. As usual, the holdup time is given by $\tau = L/v$ where L is the length of tube, and v is the velocity of flow. These curves represent exponential functions with the length of the tube as the independent variable and with a "time constant" L_0, shown in the figure as $L_0 = vT$ where:

$$T = \frac{\rho c 2\pi r^2 dL}{2\pi r dL \cdot h} = \frac{\rho c r}{h}$$

and T is the thermal time constant for heat transfer to the liquid in the tube. The temperature at various locations is given by:

$$\theta(x) = \theta_i + (\theta_s - \theta_i)(1 - e^{-x/L_0}) \tag{3.8.5}$$

and the temperature at the outlet is:

$$\theta_o = \theta(L) = \theta_i + (\theta_s - \theta_i)(1 - e^{-L/L_0}) \tag{3.8.6}$$

where:

$$\theta_i = \text{inlet temperature of the liquid}$$

$$\theta_o = \text{outlet temperature of the liquid}$$

$$\theta_s = \text{steam temperature}$$

The last factor in equation (3.8.6) can also be written as:

$$(1 - e^{-\tau/T})$$

where τ/T is the ratio of the liquid holdup time in the heat exchanger to the thermal time constant of the heat transfer.

The above applies to steady-state conditions. If we let the inlet temperature (θ_i) change slightly, the change in outlet temperature (θ_o) will be given by the transfer function

$$\frac{\Delta\theta_o}{\Delta\theta_i}(s) = e^{-\tau/T} \cdot e^{-\tau s} \tag{3.8.7}$$

which is derived in Appendix H. The first factor in equation (3.8.7), which is the gain, decreases as the holdup time increases relative to the thermal time constant. This means that the outlet temperature becomes less sensitive to variations in the inlet temperature. The second term in equation (3.8.7) is a pure transport delay.

If we keep the input temperature of the liquid constant but vary the steam temperature around some operating point, we will get a change in the outlet temperature (around a steady-state operating point) according to the transfer function:

$$\frac{\Delta\theta_o}{\Delta\theta_s}(s) = \frac{1}{1 + Ts}(1 - e^{-\tau/T} \cdot e^{-\tau s}) \tag{3.8.8}$$

The first factor in this expression is characterized by the time constant T, which is independent of the length of the heat exchanger. The factor in parentheses is a characteristic seen in many processes of this type. It can be interpreted by means of the frequency response, which is shown in Figure 3.8.9, where we see the amplitude and phase plotted as functions of frequency for various values of the factor τ/T. In this figure, the abscissa is normalized.

For purposes of illustration, let $\tau/T = 1$. We then get the result shown in Figure 3.8.10. The term in parentheses in equation (3.8.8) shifts the amplitude away from the 0 dB line, and the amplitude curve has the same general shape as that of a first-order transfer function with time constant T. The phase curve is also close to that of a first-order system, but with a slightly less negative phase angle at low frequencies. In any case, since the phase angle is close to $-90°$, it will not matter much if we neglect the transport delay given by the term in parentheses of equation (3.8.8).

If the transport delay is significantly larger than the time constant ($\tau/T \gg 1$), then the transfer function given by equation (3.8.8) will be dominated by the first factor. If the transport delay is relatively small ($\tau/T \ll 1$), we can use the approximation:

$$e^{-\tau/T} \approx \frac{1 - \dfrac{\tau}{2T}}{1 + \dfrac{\tau}{2T}} \quad \text{and} \quad e^{-\tau s} \approx \frac{1 - \dfrac{\tau}{2}s}{1 + \dfrac{\tau}{2}s}$$

and we can write equation (3.8.8), for the case of low frequencies, as

$$\frac{\Delta\theta_o}{\Delta\theta_s}(s) \approx \frac{\dfrac{\tau}{T}}{1 + \dfrac{1}{2}\tau \cdot s} \tag{3.8.9}$$

This expression shows that the effect of the thermal time constant has become

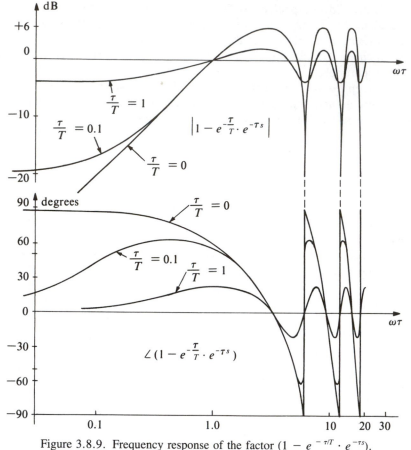

Figure 3.8.9. Frequency response of the factor $(1 - e^{-\tau/T} \cdot e^{-\tau s})$.

very small, and that the transport delay has become dominant. The frequency response for this case is shown in Figure 3.8.11, where we see that the amplitude characteristic has low gain at low frequencies [as expected because the numerator in equation (3.8.9) is small when τ/T is small], and it is nearly flat up to the frequency $\omega = 1/\tau$. At higher frequencies, the change in amplitude is determined by the thermal time constant of the heat exchanger. The lower graph of Figure 3.8.11 is the phase curve, and we see that the phase angle reaches its greatest negative value, given by:

$$\angle \frac{\theta_o}{\theta_s}(j\omega)\Big|_{max} \approx -154°$$

at the frequency $\omega \approx 5.8/\tau$.

If we look at the response of the temperature of the liquid in the heat exchanger in Figure 3.8.8 when both the steam temperature and the inlet water

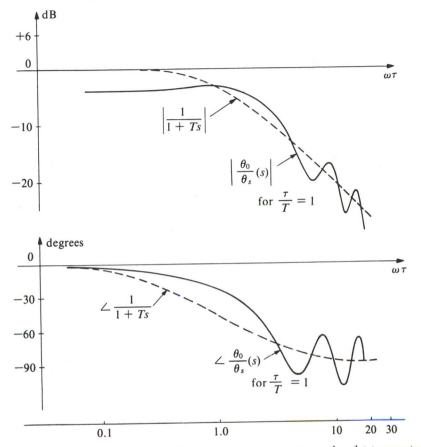

Figure 3.8.10. Frequency response between steam temperature and outlet temperature of the secondary medium $\Delta\theta_0/\Delta\theta_s$ for the case $\tau/T = 1$.

temperature are held constant but the velocity is allowed to change, we find that (Appendix H):

$$\frac{\Delta\theta_o}{\Delta v}(s) \approx \frac{k}{s}(1 - e^{-\tau s}) \qquad (3.8.10)$$

The frequency response of this expression is shown in Figure 3.8.12. The amplitude characteristic at low frequencies behaves as if the system has a time constant of about $\tau/2$, while the phase appears as if the system can be characterized by a transport delay of $\tau/2$.

The observations above refer to a simplified heat exchanger, but they are useful because they illustrate phenomena that are not so clear and easy to discover in the context of a more complicated description. Two factors dominate the dynamic behavior: the thermal time constant (or the sum of several thermal

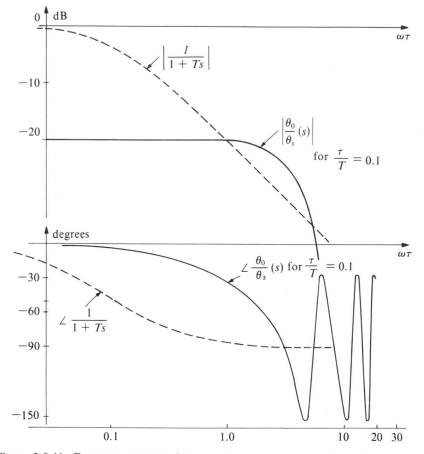

Figure 3.8.11. Frequency response between steam temperature and outlet temperature of secondary medium for the case $\tau/T = 0.1$.

time constants), which is determined by the thermal capacity divided by the heat transfer coefficient, and the holdup time (transport delay) in the exchanger.

Since the holdup time is inversely proportional to the throughput velocity, it makes little sense to use the velocity as a control variable if the velocity is low.

Appendix H contains a summary of mathematical models of different types of heat exchangers based on acceptable, practical simplifications. These models are suitable for choosing an appropriate control structure for systems that include these exchangers. In the simplest models, there is no change of physical phase of either medium in the exchanger. Phase changes such as evaporation or condensation are important unit operations in themselves, for which one needs to know the dynamic characteristics, and which we will examine later. Using the detailed steps found in Appendix H, one can find a reasonably precise mathematical model that can be used once the design of the exchanger has been determined.

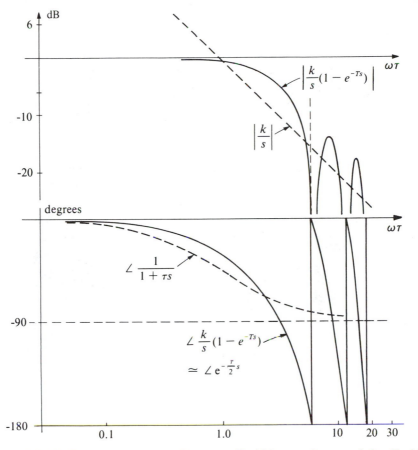

Figure 3.8.12. Frequency response of a steam–liquid heat exchanger relating liquid flow velocity and liquid outlet temperature.

Boilers

The heat energy liberated by combustion usually must be transferred to some other medium. A common example is water, which can absorb heat energy as either hot water or steam. Steam boilers are widely used in the process industries either for the production of "process steam" or for driving steam turbines.

Figure 3.8.13 shows a typical boiler and its associated equipment. The system consists of a combustion chamber containing a number of burners (in this case, for liquid or gas fuels). A furnace for burning solid fuels would contain a grate of some sort on which primary combustion would take place. In the boiler shown in Figure 3.8.13, the burning gases pass around a group of vertical tubes. These tubes, called risers, carry a mixture of water and steam. At the top of the risers is a drum, which is a horizontal cylinder kept about half full of water. The upper part of the drum contains steam. The tubes that leave the

Figure 3.8.13. A typical boiler with associated equipment. (Reprinted with permission of Sunrod International AB.)

bottom of the drum (the downcomers) are insulated from the combustion chamber and carry the water down to a somewhat smaller drum where the mud is separated from the water. The water is then fed to the risers where vaporization takes place. The steam that is produced passes from the drum to the superheaters, which are also located in the combustion chamber. The superheated product steam is then sent to the process where the energy is removed. The liquid condensate is returned to the drum where it begins the cycle again. The entire system is closed, and in theory no water would have to be added. However, most steam systems have water and/or steam leakage losses, so it is necessary to maintain the water in the system by adding feedwater to the drum. It should be noted that it is usually advantageous to preheat the boiler feedwater.

Boilers of this type have been subject to a great deal of study. Mathematical models have been developed that provide very detailed descriptions of the dynamic phenomena in the various sections of the system (Schulz, 1973; Bell et al., 1977; Whalley, 1978; Tyssø, 1981).

Although a steam boiler must be described by a nonlinear mathematical model if the model is to apply over the entire operating range of the boiler, it is possible to use linearized transfer function matrices to describe the dynamic behavior around any fixed operating point.

The most important variables for control of a steam boiler are:

- Flows of fuel and air.
- Flow of feedwater to the drum.
- Water injection after the superheater for steam temperature control.
- Steam flow.

The most important state variables in the system are:

- Pressure in the drum (saturated steam).
- Water level in the drum.
- Temperature of the steam after the superheaters.
- Pressure of the superheated steam.

A state space model of a boiler is developed in Appendix I.

Typical transfer functions for various control variables and states are given in section 5.2. Of particular interest is the transfer function between the opening of the main steam valve (controlling the steam flow out of the drum) and the water level in the drum. This transfer function contains a zero in the right half-plane that is the result of the following physical situation:

If the steam flow suddenly increases, the pressure in the drum will decrease. This causes an increase in the size of the steam bubbles in the drum and risers. These bubbles require more volume in the steam–water system, and will force the water level in the drum to rise. This is known as the *swelling* effect. After a while, this transient phenomenon will disappear. The drum level will then drop because the increased steam flow will have taken more water for steam production.

The phenomenon is sketched in Figure 3.8.14. Control of the feedwater supply on the basis of drum level alone will lead to problems of a special type, which we will discuss in section 5.2. In some boilers it is important to prevent large changes of the water level in the drum. This is especially true when the drum is small as in ship boilers, where it is essential that the system respond rapidly to changes in steam demand during maneuvering of the ship. If the level were to rise too far, water from the drum could get into the superheaters and eventually into the turbine itself. Therefore, it is very important to avoid large level changes in the drum.

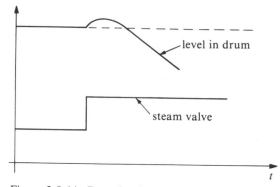

Figure 3.8.14. Drum level "shrink and swell" effect.

3.9. EVAPORATORS

Evaporation is such an important heat transfer process that it is treated as a separate unit operation. The purpose of an evaporator is to concentrate a solution by evaporating the more volatile portion of the solution. In many, but not all, cases this component is water. The material remaining after evaporation of the more volatile component of a solution is called the concentrate, or in some industrial jargon, the "thick liquor." Evaporation differs from drying in the sense that even though both processes drive off a volatile, the concentrate, or product of drying, is a solid material, while for evaporation it is a liquid. The difference between evaporation and distillation is that distillation may yield several fractions, while the vapor produced by evaporation is usually a single material. Even if this is not the case, no attempt is made to fractionate the vapor, as is done in distillation. Evaporation differs from crystallization in that one tries to operate an evaporator in such a way as to obtain a concentrate that does not crystallize. This is not always easy to achieve, as sometimes the concentrate (liquor) does, in fact, contain crystals.

Evaporators are used in many industries such as the pulp and paper industry, the fertilizer industry, mineral processing, food processing, and others.

Because the amount of volatile liquid that can be evaporated in a single evaporator stage is small, many industrial processes use several stages in cascade. Such a system is usually known as a multi-effect evaporator. Because many materials cannot tolerate high temperatures, evaporators are often operated at reduced pressure in order to reduce the boiling temperature of the material to be evaporated. As a simple example, at atmospheric pressure water boils at 100°C, but in order to boil water at 30°C, the pressure must be reduced to 0.06 bar.

Depending on the particular situation, the product of an evaporator can be either the vapor or the concentrate. Most often, the product is the concentrate, but an example of the opposite situation is the purification of water by separating it from pollutants such as minerals and salts. Although this process reminds one of distillation, it is much simpler because the vapor is not fractionated.

In a single-effect evaporator, the vapor that is produced is condensed and removed from the system. In a multi-effect evaporator, the vapor produced by one stage is used to heat the following stage, and so on through the entire series of stages. This is very significant in terms of the energy requirements of the system, but, as one might expect, it creates many problems in terms of operation and process control.

Examples of some typical evaporators are shown in Figure 3.9.1. These differ from each other in two major respects: the lengths of the evaporator tubes and the alignment of the tubes (one is horizontal and the other vertical). In some designs the evaporation stage is outside the vessel in which the thick liquor is produced and where the vapor is driven off. In other designs, the evaporator is inside the same vessel. This is clearly seen in Figure 3.9.1.

The system shown in Figure 3.9.2a is a triple-effect evaporator consisting of three identical effects in cascade. In order to use the steam from one effect to heat the next, the pressure in the upstream effect must be higher than that in the following effect. The pressure difference, which also gives the temperature difference because the steam is saturated, is determined by the pressure drop across the surface of the heat exchanger. The solution that is to be evaporated enters at the left side of the diagram through control valve F_1, and its level is controlled just high enough to cover the heat exchanger surface. The concentrate leaves the bottom of the first effect through another control valve, F_2, and flows to the second effect. The pressure difference between these two effects

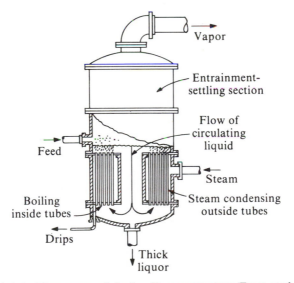

Figure 3.9.1. Three types of single-effect evaporators (Foust et al, 1980).

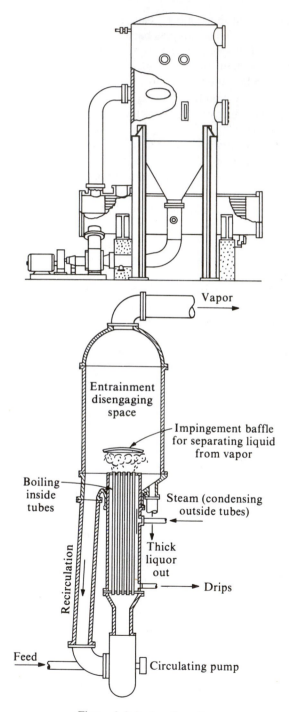

Figure 3.9.1. (continued)

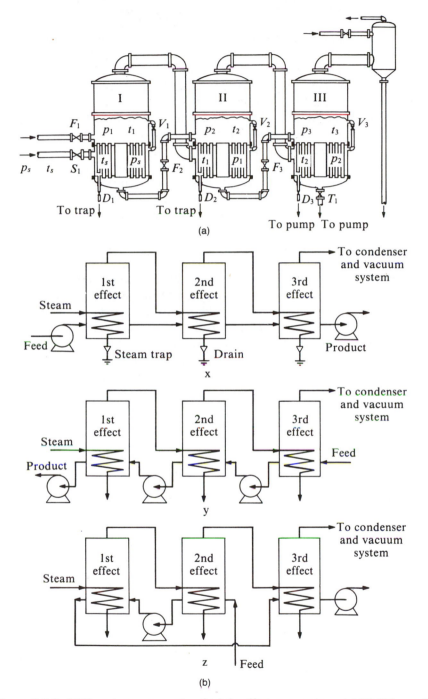

Figure 3.9.2. Different arrangements of triple-effect evaporators. (a) McCabe and Smith, reprinted by permission of McGraw-Hill Book Co. b) Foust et al, 1980.)

causes the liquid to flow. An ejector pump combined with the condenser creates a slight vacuum in the third effect. Therefore, the thick liquor must be pumped out of this effect. The primary steam that heats the first effect is condensed outside the heat exchanger. The condensate can either be discarded or used to preheat the feed to the system. The condensate from the secondary effect is also collected outside the heat exchanger. If it is the same material as the primary condensate (water, for example), it can be mixed with the primary condensate and used to heat the feed stream. Figure 3.9.2b shows several arrangements of flows in triple-effect evaporators.

Another reason why a slight vacuum is used in the last effect is that the boiling point of the liquid after concentration is higher relative to that of the solvent medium (water, for example) at the same pressure.

Design of evaporators, from the point of view of the chemical engineer, is largely concerned with energy conservation. Since the heat transfer surfaces in an evaporator usually become fouled, it is important to have some idea of the way in which the heat transfer coefficients change during operation. One way this can be estimated is to relate the temperature difference across the exchanger to the heat flow (which is the steam demand per product unit).

Mathematical models have been developed for single- and multi-effect evaporators (Newell and Fisher, 1972). For the case of a double-effect unit, one can choose a state space model having five state variables: x_1, the volume of liquid in the first effect; x_2, concentration of the liquid in the first effect; x_3, enthalpy in the first effect; x_4, liquid volume in the second effect; and x_5, concentration in the second effect.

There are three important control variables for this system: u_1, the primary steam flow; u_2, flow of liquor out of the first effect; and u_3, flow of liquor from the second effect.

Finally, there are three dominant disturbances: v_1, the feed stream flow; v_2, the feed concentration; and v_3, the enthalpy of the feed.

A mathematical model of a multi-effect evaporator is developed in Appendix J. Experience with this model shows that:

- The many small dynamic phenomena that are neglected must be replaced with an equivalent transport delay in each of the transfer functions appearing in the transfer function matrix developed in Appendix J. These transport delays will account for 10 to 100% of the dominant time constant, depending upon which transfer function is considered. As a result, the bandwidth that can be obtained from any single control loop will be lower than one would expect from the theoretical model.
- There is significant cross-coupling in the process, which means that one should expect strong interactions between individual control loops.

The question of control strategy will be discussed in section 5.3, where different solutions will be compared.

3.10. CRYSTALLIZATION

Crystals are solids of very high purity that can be produced from solutions containing significant pollutants. Many products are marketed in crystalline form, making crystallization an important industrial process. The crystallization process requires very little energy compared with distillation and other methods of separation.

Not all materials form crystals, so crystallization is limited to certain types of materials. The most common, but not the only, solution medium in which crystallization takes place is water. The theoretical basis of crystallization is very extensive and will not be discussed here. We will provide a qualitative process description sufficient for the development of satisfactory control systems. A detailed mathematical model is developed in Appendix K.

The concentration at which crystallization will occur, in a solution consisting of a solid such as a salt dissolved in a liquid, depends on temperature. The phase change from liquid to solid can be characterized, under steady-state conditions, by a phase diagram. A phase diagram for magnesium sulfate dissolved in water is given in Figure 3.10.1, where we see that the amount of magnesium that can be dissolved in water increases with temperature.

The conditions for the formation and growth of crystals are: (1) there must be small particles that can form the nuclei for growth, and (2) the solution must be supersaturated. At supersaturation the concentration of the material in solution is higher than that which appears on the phase diagram at any temperature (or pressure).

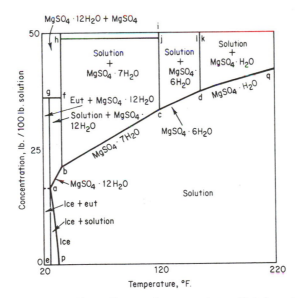

Figure 3.10.1. Phase diagram for magnesium sulfate in water.

Although there are many types of crystallizers, the type shown in Figure 3.10.2 may be the most widely used. A cylindrical tank is mounted integrally with a conical section at the bottom. The solution is circulated to the bottom, out of the bottom and into a single-effect evaporator (often heated with steam), and back to the cylindrical tank. The steam (vapor) is driven out of the top, and when the solution reaches supersaturation, crystals form. These crystals can be removed from the system by means of a vacuum pump. A variation on this principle is the so-called Oslo-crystallizer, which has an extra tank in which the crystals form, and which can also provide cooling to promote better supersaturation. This type of crystallizer is shown in Figure 3.10.3.

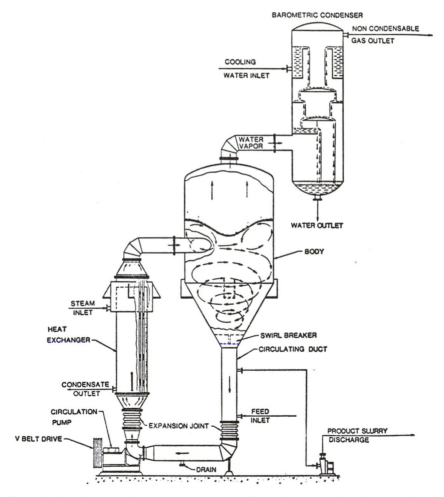

Figure 3.10.2. Swenson forced circulation crystallizer. (Courtesy Signal Swenson Div.).

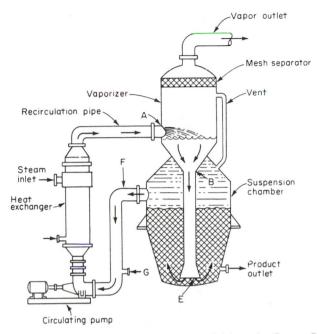

Figure 3.10.3. Oslo-crystallizer (Unitech Division, the Graver Co.).

The growth of crystals in solution is mainly determined by the degree of supersaturation and the temperature. A solution that has a certain super-saturation at one temperature can be cooled to produce a higher degree of supersaturation (for most solutions) and thereby provide faster crystal growth. Theoretically, the growth rate of the crystals is proportional to the degree of supersaturation, so the process should be characterized by a pure integration. However, there is some breakdown of crystals from large to small particles due to wear during mixing and transport. This does not significantly affect the amount of crystallized material, but it does affect the crystal size distribution.

We can easily identify the process variables by examining the crystallizer in Figure 3.10.3. The most important control variables are the feed of the solution that is to be crystallized, the flow rate of crystals out, the steam supply flow to the evaporator, and the vapor flow out of the tank.

Another control variable that has been proposed is nucleus seeding in order to promote the growth of crystals, and still another control variable is recirculation.

The most important process variables are: the amount of crystallized material, the concentration of the mother liquor (the feed solution), the temperature of the mother liquor, and the pressure of the vapor above the mother liquor.

A mathematical model of a typical crystallizer is given in Appendix K.

The dynamics of the crystallization process include the following characteristics:

1. There is a strong coupling between the heat balance and the mass balance, but this coupling is one-sided and does not cause any problems during normal operation.
2. The process is of low order.
3. The production rate of crystals is determined by the degree of supersaturation, which in turn is determined by the concentration and the temperature. It can generally be expected to be proportional to the degree of supersaturation and gives a process dynamic that is a pure integration.
4. Since the energy of the steam input is the most important control variable for the crystallization process, it is satisfactory to let the transfer function between the steam flow and the crystal production be characterized by an integration and a single time constant.

Control systems for crystallization processes will be discussed in section 5.4.

3.11. DRYING

Drying is another widely used unit process in industrial systems. The purpose of this unit operation is to separate liquids from solids by evaporation. Mechanical processes for removing liquid from solids are not considered to be drying operations, but they often precede drying processes in large operations. An obvious example is mechanical pressing prior to steam drying in a paper machine.

There are many commercial drying processes, a few of which are illustrated in Figure 3.11.1. Figure 3.11.1a is a rotary dryer, which is a long cylinder with its axis tilted slightly with respect to the horizontal. Cylinders of this type often contain baffles, or plates, on the inner surface, which carry the solid material around part of the inner periphery of the cylinder in order to increase exposure of the material to the drying medium and to agitate the solid material. The combination of the lifting by the baffles and the tilt of the axis of the cylinder carries the material through the dryer in a direction opposite to that of the flow of the drying gas. Sometimes in this type of operation there is a chemical reaction as well as drying. An example of this is a rotary kiln used for the production of cement. In this process the first part of the oven can be considered to be a rotary dryer.

A fluidized bed is shown in Figure 3.11.1b. Here, the solid material is suspended in fluidized form by a strong gas stream. This provides very effective contact between the material and the drying gas, which leads to evaporation of the liquid in the solid material.

Figure 3.11.1c shows how moist particles can be dried by simply dropping them through a countercurrent gas stream.

Figure 3.11.1d illustrates drying of a solid on a perforated conveyor. The drying gas passes through the material and the conveyor from both the top and the bottom.

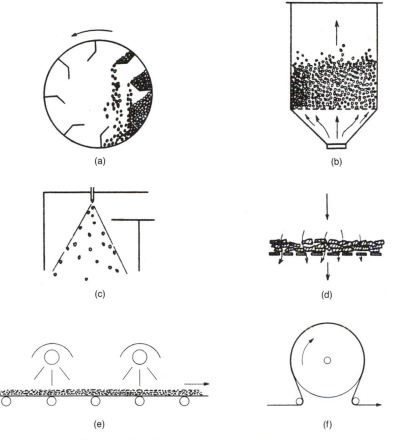

Figure 3.11.1. Types of continuous drying processes.

In Figure 3.11.1e we see a drying system that uses infrared radiation to dry a solid material on a conveyor.

A cylinder dryer for continuous drying of web materials such as paper and textiles is shown in Figure 3.11.1f. The steel-surfaced cylinder is heated internally by steam, and the continuous web of material passing over the cylinder is dried by evaporation. In a paper machine, there is a series of cylinder dryers that enable the drying to occur over a long, temperature-controlled distance through the machine.

Finally, in Figure 3.11.1g we see a simplified drawing of a flash drying system. In this system, the moist solid material is suddenly subjected to a reduced pressure. This causes the moisture to "flash," or vaporize. Systems of this type are common in the food processing industry and are becoming popular for drying ground wood pulp in the paper industry.

The design of equipment for drying must take many factors into account, including the nature of the material to be dried. In the following, we will consider primarily those basic principles that are pertinent to the question of how drying processes can be controlled.

Drying processes are distinguished from pure evaporation processes by the fact that, in drying, the liquid that is to be evaporated is distributed and held within a solid material. This leads to concern for a series of elementary processes such as heat transfer into the material, evaporation, and transport of the vapor out of the material.

The drying rate depends on, among other things, how effectively energy can be transferred to the liquid to cause evaporation. If the solid can tolerate a high temperature, one can use an inert gas, such as air, at high temperature to transfer energy and cause vaporization of the liquid phase. The higher the temperature of the inert gas, the higher the drying rate will be. However, if the solid cannot tolerate high temperatures, one must either use an inert gas at low temperature, but with a longer drying time, or reduce the pressure so that evaporation can occur more quickly at a low temperature (an advantage of flash drying).

In drying processes that do not involve agitation of the solid material, such as those shown in Figures 3.11.1d and 3.11.1e, the transport of vapor from the material will take much longer. An example of very good internal agitation is the fluidized bed shown in Figure 3.11.1b. The vapor pressure around the particles also affects the rate at which the vapor is driven out of the material.

The possible variations of the process dynamics of the different drying systems are too numerous to be described here. Therefore, as an example we will concentrate on one of the most widely used systems, the rotary dryer.

A typical drum dryer with direct heating by the countercurrent flow of an inert gas (air) is shown in Figure 3.11.2. The moist solid enters the drum at the left and is transported through the drum by the drum rotation. It moves in an axial direction opposite that of the flow of the hot gas, which enters the system at the right end, where the dry solid also leaves the drum. Figure 3.11.3 shows how the gas temperature, solid temperature, and moisture in the solid change along the drum between the inlet and the outlet. If we consider the drying action in more detail, we can define about three moisture stages, which are shown in Figure 3.11.4. Figure 3.11.4a shows the moisture content of the material as a function of time (about the same function that one would get by plotting moisture versus drum length), the drying rate as a function of moisture content is shown in Figure 3.11.4b, and the drying rate as a function of time is shown in Figure 3.11.4c. From Figure 3.11.4b, we see that when the moisture content is high, the drying rate will be constant. This means that drying is limited by evaporation at the surface of the material, and that there is good transfer of moisture from within each particle to the surface. We then have equilibrium between the liquid and the vapor, and the temperature of the liquid will be constant, as determined by the vapor pressure. When the moisture falls below a

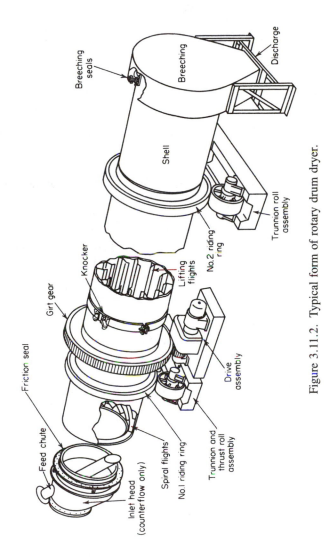

Figure 3.11.2. Typical form of rotary drum dryer.

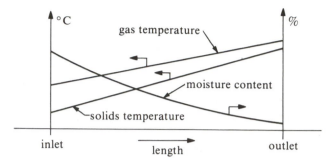

Figure 3.11.3. Profiles of gas temperature, solids temperature, and moisture in a countercurrent drum dryer.

certain limit, depicted by point C, the drying rate will decrease, roughly linearly down to E. In this moisture range, the drying rate is limited by the rate at which the internal moisture can move to the surface, as either liquid or vapor.

In the moisture range where the drying rate is constant, the mass transfer rate is in balance with the heat transfer rate. Therefore, this region is easy to model, as the drying rate is directly proportional to the temperature drop in the vapor.

Calculation of the drying rate in the lowest moisture region is complicated, but there are many theoretical results that, for purposes of automatic control, can give a satisfactory description of the process. It is important to recognize that in the moisture region where the drying rate decreases rapidly, the transport of moisture from the interior to the surface is controlled by mechanisms such as diffusion, capillary action, pressure gradients due to shrinking, and so on. Because this transport can be slow, it is necessary that a sufficient holdup time in the dryer by provided. If not, the material will leave the dryer with a dry surface and moist interior. After a certain period, the material will regain moisture equilibrium and again be moist throughout.

A mathematical model for a drum dryer is presented in Appendix L. This model is based on realistic simplifications of the process. The following comments apply to the use of a dynamic model of a drying process for the purpose of process control:

- If it is possible to measure the moisture in the material at the outlet of the dryer together with the temperature of the gas phase, it is possible to realize a relatively simple and effective control system.
- If it is not possible to measure the moisture content of the material, one must measure the temperature of the gas phase and compute the moisture content of the material, preferably on the basis of the model of the process dynamics. A method that uses measurements to update the model will give the best results.

It is often necessary to use a so-called psychometric diagram in the study of drying processes. Figure 3.11.5 is such a diagram. The abscissa is the dry bulb

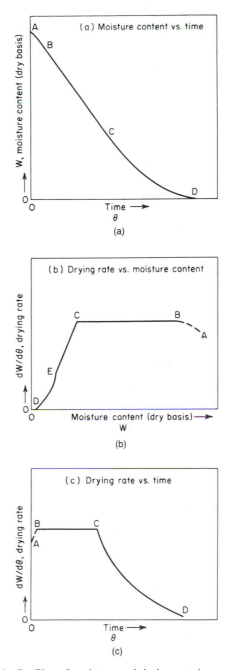

Figure 3.11.4. Profiles of moisture and drying rate in granular material.

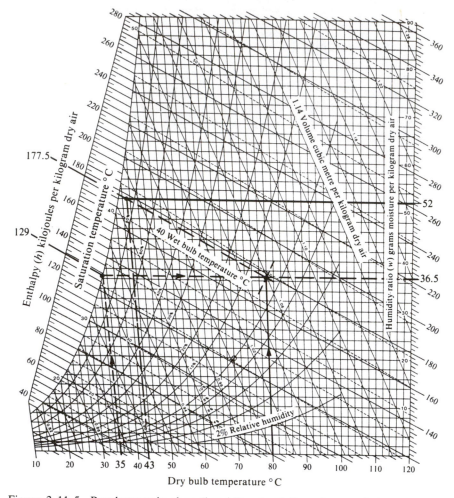

Figure 3.11.5. Psychrometric chart (humidity chart) for mixture of air and water vapor. [Used by permission of the American Society of Heating, Refrigerating, and Air-Conditioning Engineers, Inc. (ASHRAE), Atlanta.]

temperature of an air–water vapor mixture. The relative content of moisture in dry air (kg/kg) is given on the ordinate. Using the saturation curve, which is the curve angling up toward the right, we can determine how much moisture the air will hold at different temperatures. The data on this diagram apply at atmospheric pressure.

The dew point temperature is given along the saturation curve. This is the common set of values for moisture and temperature at saturation. If the temperature rises above that which gives saturation at any moisture, then we will have a relative humidity that can be given in terms of the percent of what one would have at saturation. Relative humidity can be read from the curves that are "parallel" to the saturation line.

The wet bulb temperature is the temperature measured by a thermometer that has its bulb covered with a moistened cloth sleeve. This provides a moisture layer that is in equilibrium with the surrounding atmosphere. In the figure, the "constant wet bulb" lines are nearly parallel to the "constant enthalpy" lines and intersect the saturation curve at the dew point temperatures. The wet bulb temperatures will always be lower than the dry bulb temperatures. (Dry bulb temperatures are obtained by measurement without the moist sleeve on the thermometer bulb.)

The absolute moisture content (kg water/kg dry air) can be determined from the partial pressure, which is defined as the absolute pressure multiplied by the mole fraction of water in the mixture. Since the molecular weight ratio of air to water is 29 to 18, the absolute moisture will be:

$$H = \frac{18\bar{p}_v}{29(1 - \bar{p}_v)}$$

where $\bar{p}_v$ is the partial pressure of the water vapor (bar).

The relative humidity (H_r) is defined as the ratio between the partial pressure of the water vapor and the vapor pressure of the water at the ambient air temperature. H_r is expressed in percent where 100% relative humidity means that the air is saturated with water, and 0% relative humidity means there is no water vapor in the air (i.e., dry air). Relative humidity curves are shown in Figure 3.11.5.

Different materials will have different moisture equilibrium values, as shown in Figure 3.11.6. Here the relative moisture is a free variable, and the water content of the material (kg water/kg dry weight) is the dependent variable. Even if it takes a long time to reach equilibrium, it is clear that one must control the relative humidity in order to control the absolute moisture in the material.

To illustrate the use of the diagram in Figure 3.11.5, consider the following example. Assume that a granular solid containing 93% water is to be dried in a dryer with ideal contact between the air and the material at a rate that produces 100 kg of dry material per hour. Atmospheric air enters through a heat exchanger and is heated to 80°C, which is the maximum temperature the material can tolerate. The air removes moisture from the material and leaves the dryer 90% saturated. If we know the atmospheric conditions, for example, an air temperature of 35°C and 100% relative humidity (about the worst possible situation), we can calculate how much air is required to produce 100 kg/hour of dry material.

From Figure 3.11.5 we see that the incoming air has an absolute humidity of 0.0365 kg water/kg air. When the air goes through the heat exchanger, the absolute humidity does not change, but the relative humidity is reduced. Since

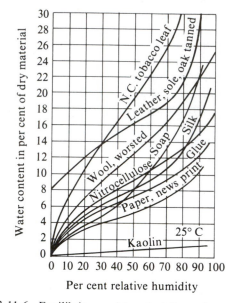

Figure 3.11.6. Equilibrium moisture (w_e) for various materials.

we know that the air temperature is raised to 80°C, we can use Figure 3.11.5 to find the corresponding relative humidity to be 12% and the wet bulb temperature to be 42°C. If we assume adiabatic conditions, there is no energy transfer except that between the air and the moist material and therefore no change in enthalpy. Therefore, the wet bulb temperature will change very little during drying until the relative humidity reaches 90%. We can see this by following the constant enthalpy line in Figure 3.11.5 to a relative humidity of 90%, where we find the air temperature to be about 43°C. The absolute moisture in the air will then be about 0.052 kg water/kg dry air.

Each kg of air added to the dryer will remove 0.052 − 0.0365 = 0.0155 kg water. The total amount of water removed is determined by the fact that the material added has 93% water (i.e., 7% dry solids). Therefore, the total amount of water removed will be (100/0.07) − 100 = 1329 kg. The amount of air required to produce 100 kg dry material per hour is, then, 1329/0.0155 ≈ 85,700 kg/hour.

The amount of heat required to evaporate a given amount of water is also found from Figure 3.11.5. The enthalpy change of air warmed from 35°C to 80°C is:

$$177.5 \times 10^3 - 129 \times 10^3 = 48.5 \times 10^3 \text{ joule/kg dry air}$$

Therefore, the total heat requirement is:

$$48.5 \times 10^3 \times 85,700 = 4.156 \times 10^9 \text{ joule/hour} = 1155 \text{ kW}$$

We see that a great deal of heat is involved, and it is apparent that one can achieve considerable energy savings by using the heat that is in the air leaving the dryer to preheat the fresh air that is added to the heat exchanger.

As shown in Figure 3.11.4b, the drying rate will be nearly constant when the moisture in the material is above a critical value, at which drying becomes pure evaporation. We can use the following model for the drying rate in this region:

$$\dot{w} = -\frac{h_t a}{\rho_s \lambda}(\theta - \theta_s) \qquad (3.11.1)$$

where:

h_t = total heat transfer coefficient (joule/s m^2 °C)
a = m^2 heat transfer surface/m^3 material (1/m)
ρ_s = density of the dry material (kg/m^3)
λ = heat of vaporization of water (joule/kg)
θ = air temperature (°C)
θ_s = surface temperature of the material (°C)

The value of a can often be estimated from the degree of packing of the solid $(0 < f < 1)$ according to:

$$a \approx \frac{6(1 - f)}{d}\left(\frac{1}{m}\right)$$

where d is the mean diameter of solid particles (m). If we assume that the air in the dryer is well mixed so that there are no temperature or moisture gradients in the gas phase, the temperature θ will be uniform through the entire dryer. The temperature θ_s when the surface of the solid is saturated with water will be the boiling point temperature (100°C at atmospheric pressure).

When the water content of the material becomes less than the critical value w_c, the model for drying becomes:

$$\dot{w} = -\frac{h_t a}{\rho_s \lambda}(\theta - \theta_s) \cdot \frac{w - w_e}{w_c - w_e} \qquad (3.11.2)$$

where:

w_c = critical moisture (kg/kg)
w_e = equilibrium moisture (kg/kg)

We see that at critical moisture w_c, equations (3.11.1) and (3.11.2) are the same. We also see that equation (3.11.2) is a first-order differential equation if we assume w_c and w_e to be constant.

Table 3.11.1 is a list of the values of critical moisture for several different materials.

Table 3.11.1. Approximate Critical Moisture Content w_c Obtained on the Air Drying of Various Materials, Expressed as Percentage Water on the Dry Basis.*

MATERIAL	THICKNESS, cm	CRITICAL MOISTURE, % WATER, DRY BASIS
Barium nitrate crystals, on trays	2.5	7
Beaverboard	0.4	above 120
Brick clay	1.6	14
Carbon pigment	2.5	40
Celotex	1.1	160
Chrome leather	0.1	125
Copper carbonate, on trays	2.5–3.8	60
English china clay	2.5	16
Flint clay refractory brick	5.0	13
Gelatin, initially 400% water	0.25–0.5 (wet)	300
Iron blue pigment (on trays)	0.6–1.9	110
Kaolin	. . .	14
Lithol red	2.5	50
Lithopone press cake, in trays	0.6	6.4
	1.3	8.0
	1.9	12.0
	2.5	16.0
Niter cake fines, on trays	. . .	above 16
Paper, white eggshell	0.02	41
fine book	0.01	33
coated	0.01	34
newsprint	. . .	60–70
Plastic clay brick mix	5.1	19
Poplar wood	0.4	120
Prussian blue	. . .	40
Pulp lead, initially 140% water	. . .	below 15
Rock salt, in trays	2.5	7
Sand, 50–150 mesh	5.1	5
Sand, 200–325 mesh	5.1	10
Sand, through 325 mesh	5.1	21
Sea sand, on trays	0.6	3
	1.3	4.7
	1.9	5.5
	2.5	5.9
	5.1	6.0
Silica brick mix	5.1	8
Sole leather	0.6	above 90
Stannic tetrachloride sludge	2.5	180
Subsoil, clay fraction 55.4%	. . .	21
Subsoil, much more clay	. . .	35
Sulfite pulp	0.6–1.9	60–80
Sulfite pulp (pulp lap)	0.1	110
White lead	. . .	11
Whiting	0.6–3.8	6.9
Wool fabric, worsted	. . .	31
Wool, undyed serge	. . .	8

*Based on Perry

The temperatures θ and θ_s can be interpreted as the dry and wet bulb temperatures, respectively. Both equations (3.11.1) and (3.11.2) show that the drying rate is proportional to the difference between the wet and dry bulb temperatures. These temperatures can often be measured directly in the dryer, making it possible to achieve direct control of the drying rate.

If w_e is very small ($w_e \approx 0$), equation (3.11.2) can be simplified. If we assume that the transport rate of the solid through the dryer is constant (on the average, a reasonable assumption because this is a function of rotation speed and the angle of inclination), the drying time will be determined simply as a function of the length and speed of rotation of the dryer. Finally, as a first approximation, assume that the temperature difference $(\theta - \theta_s)$ is constant. The drying profile will then be that shown in Figure 3.11.7, where we see a straight-line segment from the feed moisture w_1 to time t_1 where the moisture reaches the critical value w_c. Time t_1 is given by:

$$t_1 = \frac{\rho_s \lambda}{h_t a} \cdot \frac{w_1 - w_c}{\theta - \theta_s} \tag{3.11.3}$$

The location of this point in the dryer can be described in the z direction as $z_1 = vt_1$, where v is the transport velocity of the material (solid) in the dryer.

After this point, the drying profile will be exponential:

$$w(t) = (w_c - w_e)e^{-(t-t_1)/T} + w_e \tag{3.11.4}$$

where:

$$T = \frac{\rho_s \lambda}{h_t a} \cdot \frac{w_c - w_e}{\theta - \theta_s}$$

At the discharge of the dryer, $t = t_2 = L/v$, where L is the length of the dryer. We also find:

$$w(t_2) = (w_c - w_e)f_1(\theta - \theta_s)f_2(w_1) + w_e \tag{3.11.5}$$

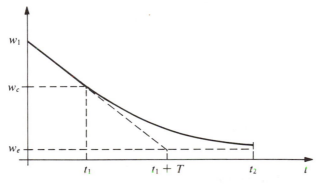

Figure 3.11.7. Drying profile in a rotary drum dryer.

where:

$$f_1(\cdot) = e^{-\frac{L}{v} \cdot \frac{h_t a}{\rho_s \lambda} \cdot \frac{\theta - \theta_s}{w_c - w_e}}$$

and:

$$f_2(\cdot) = e^{\frac{w_1 - w_c}{w_c - w_e}}$$

From equation (3.11.5) we see that the final moisture in the solid will decrease when the air temperature increases, and will increase when the input moisture content increases.

In order to design a dryer, one should know the value of the required holdup time and the specified maximum product moisture content. Equation (3.11.5) can then be solved with respect to the total time required:

$$t_2 = \frac{L}{v} = \frac{\rho_s \lambda (w_c - w_e)}{h_t a (\theta - \theta_s)} \left[\ln\left(\frac{w_c - w_e}{w(t_2) - w_e}\right) + \frac{w_1 - w_c}{w_c - w_e} \right]$$

$$(3.11.6)$$

The logarithmic term gives the time the material remains in the constant drying rate zone. It should also be recognized that equation (3.11.6) does not include the time it might take to raise the liquid in the material to the boiling point. This must be added to t_2.

A satisfactory dynamic model for this process can be derived on the basis of reasonable simplifications (see Appendix L). A detailed model is very complicated, and the physical phenomena are not completely understood.

There is a vast amount of literature on drying, but some basic, useful information can be found in Myklestad (1963a, b), Keey (1972), and Shinskey (1978).

3.12. DISTILLATION

General Principles

Distillation is a very important separation process that presents a big and difficult control problem. Distillation makes use of the fact that two or more materials can be separated on the basis of their different boiling points. In the following discussion we will discuss binary distillation, which is the separation of two components. The same principles can be extended to the case of multicomponent distillation.

A schematic diagram of a distillation column is shown in Figure 3.12.1. This column contains N trays, or plates, shown in Figure 3.12.2, and shown schematically in Figure 3.12.3. Liquid and vapor are in contact with each other at essentially the same temperature and pressure at each tray. Each tray contains a number of holes, each with a cover (cup), as shown in Figure 3.12.3. Vapor produced by the reboiler enters at the bottom and rises through the column,

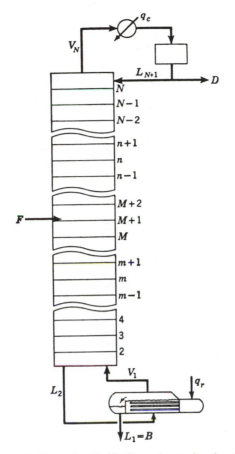

Figure 3.12.1. Schematic view of a distillation column showing feed, condenser, and reboiler. The tray numbering follows the most common convention.

passing through each tray where it bubbles through the liquid. The liquid moves down through the column, through a space that acts as a liquid seal. The liquid leaves each tray by passing over a barrier that, combined with the hydrodynamic conditions in the seal chamber, determines the liquid level on the tray. The holes in the trays are provided to assure good contact between the vapor and the liquid. The resistance to upward flow of the vapor gives the pressure gradient in the column, which is such that the absolute pressure is highest at the bottom and lowest at the top of the column. Likewise, there will be a temperature gradient over the column, with the highest temperature at the bottom and the lowest at the top.

In binary distillation, the liquid is assumed to have two components. It is fed into the system at some tray, designated $M + 1$, somewhere between the top and the bottom of the column. Because of its weight, the liquid will flow

Figure 3.12.2. Distillation column trays. (Foust et al, 1980.)

downward from tray to tray to the bottom of the column, where part of it is evaporated by the reboiler. The remainder of the liquid is removed from the system as bottoms product.

The vapor that leaves the column at the top is condensed and collected in an accumulator. Some of the liquid from the accumulator is recycled to the column (reflux), and the remainder is drawn off as top product (distillate).

The countercurrent flow of the rising vapor and the falling liquid provides very good contact between them; and when this good contact exists, the temperature, pressure, and component concentration at each tray will be in (or close to) equilibrium.

The more volatile component of the binary system will vaporize at a lower temperature than the other component, and it will tend to have the higher concentration in the vapor phase. The heavier component, which vaporizes at a higher temperature, will tend to have the higher concentration in the liquid phase. The result is that the vapor phase will become richer and richer in the light component as the vapor moves up the column, and the liquid phase will become richer in the heavier component as it moves downward.

How well the two components can be separated into top and bottom products depends on the relative volatility of the components, the number of trays in the column, the point of entry of the feed into the column, and the ratio between the liquid and vapor flowrates.

The column is divided into two sections, the upper being the rectification section and the lower known as the stripping section. The division between the two sections is the point at which the feed enters the column.

Plan view of top plate

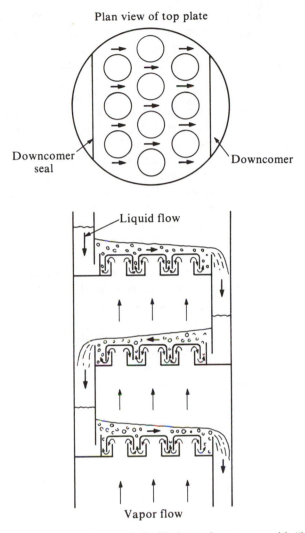

Figure 3.12.3. Schematic diagram of distillation column trays with "bubble caps." (Foust et al, 1980.)

Distillation columns are used in a very broad range of applications. Perhaps the best known is the distillation of a mixture of water and different alcohols produced by the fermentation of sugar. The object of this process is to separate the valuable alcohols, C_2H_5OH for example, from the water and the less desirable alcohols.

Pure oxygen, used for medical purposes and in steel production, is produced by the distillation of air that has been liquefied by compression and cooling. This process is also very important in the ammonia industry, where the raw material is nitrogen.

Crude oil can be separated into many fractions in large, multicomponent distillation columns. These fractions include such products as gas oil, naphtha, kerosene, gasoline, diesel oil, lubricating oils, heavy fuel oils, and asphalt. Each of these fractions can be distilled further to yield products of different qualities.

The gas that is usually found along with oil in the oil fields of the world contains many different hydrocarbon fractions, which can be separated by distillation. One of these fractions is ethane (C_2H_6). It, in turn, is a raw material from which lighter hydrocarbons can be produced by the process known as *cracking*, or pyrolysis. These products can also be further separated by low-temperature distillation. Common products of this process are ethylene (C_2H_4), methane (CH_4), propylene (C_3H_6), propane (C_3H_8), and butane (C_4H_{10}).

Large integrated separation process can contain many cascaded distillation stages. It is critical that the values of temperature and pressure at various locations in the process be controlled with respect to each other throughout the system. A stage that follows another will have a lower pressure than the upstream stage unless a compressor has been placed between the stages.

Evaporation in the reboiler at the bottom of the distillation column requires energy, but condensation at the top releases energy. Because energy is expensive, engineers try to use the energy released by the condenser for evaporation elsewhere in the process. The various parts of the total process thus are interconnected in a manner that makes the dynamics very complex, usually leading to positive feedback and sometimes to problems with stability. Therefore, automatic control is absolutely necessary.

Distillation Theory

The details of mass and energy transfer in a distillation column are very complicated, and any attempt to describe them in detail will lead to a very comprehensive and possibly unmanageable model. In order to simplify the model, it is common to assume that each tray is in equilibrium. This means that the liquid and vapor that leave the tray will be in complete equilibrium with each other, and thermodynamic relationships can be used to describe the concentration of the two components in the stream. In practice, the assumption of equilibrium does not hold. Therefore, when designing a column, one must consider a parameter known as *tray efficiency*. The assumption regarding equilibrium not only simplifies the model of the steady-state behavior of the column, but it also makes the dynamic description somewhat too simple. A real column will be "more dynamic" (that is, show greater negative phase shift, longer time constants, longer transport delays, etc.) than the idealized column.

Although the assumption regarding binary distillation will not be correct for stepwise multicomponent separation in a cascade system of columns, it will lead to a reasonable approximation if the boiling points for the various other components are reasonably different from the two components being considered.

The literature concerning mathematical modeling of distillation columns is extensive. A few important basic references are McCabe and Smith (1956); Campbell (1958); Buckley (1964); Perry and Chilton (1973); Shinskey (1977/84); Henley and Seader (1981); Deshpande (1985); and Buckley, Luyben, and Shunta (1985).

One can approach the design and/or study of distillation columns in two different ways: graphically and analytically. The graphical method has significant advantages because it makes it easy to observe the various important phenomena, but the method also has serious limitations, especially with respect to dynamic characteristics of the process. On the other hand, the analytical method is well suited to computer simulation and gives the best accuracy, but it is not well suited to providing a quick overview of the process.

Two different tray numbering systems are used in the literature. One starts numbering from the bottom and the other from the top of the column. Both methods have advantages and disadvantages, but we will use the notation that has the lowest tray number at the bottom and the highest at the top.

Since distillation of a binary mixture is based on the difference of boiling points, let us look at the basic characteristics of a binary mixture. A boiling point diagram for a binary mixture of components A and B is given in Figure 3.12.4. Concentration, preferably in terms of mole fraction (ratio of the number of moles of a component to the total number of moles in the mixture) is given along the abscissa. Component A is the more volatile and has the lower boiling point, θ_A. Temperature is given along the ordinate. The two curves meet

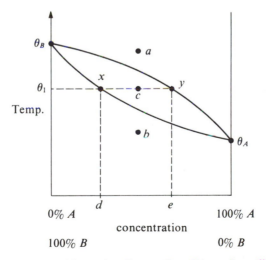

Figure 3.12.4. Boiling point diagram for mixture of two liquids.

at the end points where the concentration of A is either 0% or 100%. A point on the upper curve, y for example, represents vapor that is just about to condense at temperature θ_1. The concentration of A at point y is given by point e on the abscissa. The concentration of the first drop of condensate is found from the lower curve and is read from point d on the abscissa. The upper curve, called the dewpoint curve, represents the temperature–concentration relation at which condensation just begins. Similarly, the lower curve, known as the bubble point curve, represents the temperature–concentration relation at which evaporation just begins. At any point above the dewpoint curve, point a for example, both components of the mixture will be in the vapor phase. Below the bubble point curve, at b for example, both components will be in the liquid phase. At any point c between the two curves, the mixture will be two phases, liquid and vapor.

Because temperature and vapor pressure are uniquely known for different materials, at any pressure one can get a unique picture of the relationship between concentration (mole fraction) in the vapor phase and concentration (mole fraction) in the liquid phase, as shown in Figure 3.12.5. Since this diagram applies to the more volatile component, A, the higher concentration is in the vapor phase.

Azeotropic mixtures behave differently from the mixtures of Figures 3.12.4 and 3.12.5. Equilibrium curves for these mixtures can have the characteristics shown in Figure 3.12.6, where we see for curve a that the concentration in the vapor phase can be less than that in the liquid phase when the latter is high. Curve b shows a mixture where the concentration in the vapor phase is lower than that in the liquid phase when the latter is low. Distillation of azeotropic

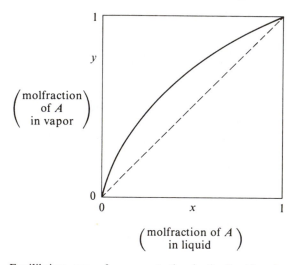

Figure 3.12.5. Equilibrium curve for concentration in the liquid and vapor phases for the more volatile component of a binary mixture.

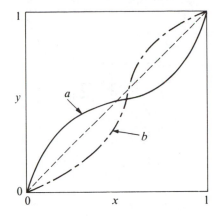

Figure 3.12.6. Equilibrium curves for azeotropic mixtures.

mixtures is more difficult than distillation of the mixtures described by the curve in Figure 3.12.5, and will not be discussed further here.

One approach used to describe a binary mixture is the *relative volatility* of components *A* and *B* as given by:

$$\alpha_{AB} = \frac{\dfrac{y_A}{x_A}}{\dfrac{y_B}{x_B}} = \frac{\dfrac{y_A}{x_A}}{\dfrac{1 - y_A}{1 - x_A}} = \frac{y_A(1 - x_A)}{x_A(1 - y_A)} \qquad (3.12.1)$$

which gives:

$$y_A = \frac{\alpha_{AB} \cdot x_A}{1 + (\alpha_{AB} - 1)x_A} \qquad (3.12.2)$$

In many distillation processes the relative volatility α_{AB} can be assumed to be approximately constant over the entire column. In this case it is easy to sketch the equilibrium curve shown in Figure 3.12.7 and based on Figure 3.12.2. It is easy to distill a binary system that has a high relative volatility. Even if the relative volatility is not constant over the column, and varies, for example, as a function of concentration along the column, one can use a mean relative volatility that can be calculated as either the arithmetic mean or the geometric mean of the relative volatilities at the bottom ($\alpha_{AB}^{(1)}$) and at the top ($\alpha_{AB}^{(N)}$), respectively. That is, either

$$\bar{\alpha}_{AB} = \frac{\alpha_{AB}^{(1)} + \alpha_{AB}^{(N)}}{2} \qquad (3.12.3)$$

or

$$\tilde{\alpha}_{AB} = \sqrt{\alpha_{AB}^{(1)} \cdot \alpha_{AB}^{(N)}} \qquad (3.12.4)$$

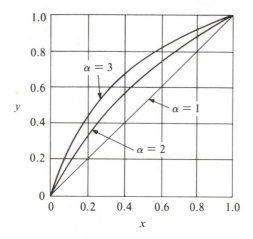

Figure 3.12.7. Equilibrium curves for materials with different relative volatility.

The McCabe-Thiele method of analysis (McCabe and Smith, 1956) is a general graphical method that can be used to understand the design of a distillation column. It actually provides a graphical representation of the material balance equations of the column under steady-state conditions.

Using the notation presented in Figure 3.12.8, let us write the material balance equation for the more volatile component (here assumed to be component A) of the binary mixture. We will do this for tray j, and designate the liquid mass on that tray as m_j. Let L_j be the liquid mass flow leaving tray j, and V_j be the vapor mass flow leaving the tray. The concentration (expressed as the mole fraction) of A in the vapor is y_j and in the liquid it is x_j. The material balance for the jth tray written in terms of the A component is:

$$\frac{d}{dt}(m_j x_j) = L_{j+1} \cdot x_{j+1} - L_j \cdot x_j + V_{j-1} \cdot y_{j-1} - V_j \cdot y_j \qquad (3.12.5)$$

Under stationary (steady-state) conditions, the derivative can be set to zero.

From equation (3.12.2) we find that the concentration of A in the vapor phase is:

$$y_j = \frac{\overline{\alpha} x_j}{1 + (\overline{\alpha} - 1)x_j} \qquad (3.12.6)$$

where $\overline{\alpha}$ is the mean relative volatility.

The total mass balance on tray j is given by:

$$\frac{dm_j}{dt} = L_{j+1} - L_j + V_{j-1} - V_j \qquad (3.12.7)$$

where the liquid leaving the tray is a function of the liquid level on the tray.

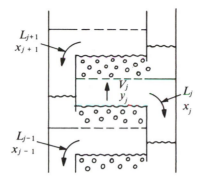

Figure 3.12.8. Definition of dynamic variables in a distillation column.

This relationship can reasonably be assumed to be linear. Thus:

$$\Delta L_j \approx \frac{\partial L_j}{\partial m_j} \cdot \Delta m_j \tag{3.12.8}$$

Equations (3.12.7) and (3.12.8) indicate that the mass dynamics on each plate are characterized by a first-order differential equation. A similar analysis applies to equations (3.12.5) and (3.12.6).

Let us assume steady-state conditions and examine just one of the sections of the column, either above or below the feed tray. With reasonable precision we can then assume that the liquid and vapor flows are the same, from tray to tray, but with different values above or below the feed tray. If we call the liquid and vapor streams in the rectification section (above the feed tray) L and V, and the same flows in the stripping section (below the feed ray) L' and V' respectively, under stationary conditions we will have the following simplified relation for the rectification section:

$$y_j - y_{j-1} = \frac{L}{V}(x_{j+1} - x_j) \tag{3.12.9}$$

The same equation applies to the stripping section, but with L' and V' instead of L and V. Equation (3.12.9) is the equation of a straight line and is the starting point for construction of the McCabe-Thiele diagram, which is shown in Figure 3.12.9. The abscissa is the concentration of the more volatile component in the liquid phase, and the ordinate is the concentration of that same component in the vapor phase. The equilibrium line is the bow-shaped curve. If the distillate (product) has a concentration x_D that is known (as an example, let $x_D = 0.95$), we can construct the line according to equation (3.12.9). The details of this construction can be seen in Figure 3.12.10 where the various values are clearly indicated. A corresponding straight line can be constructed using the concentration of the bottoms product, x_B (for example, $x_B = 0.03$) as the starting point and using L'/V' in equation (3.12.9).

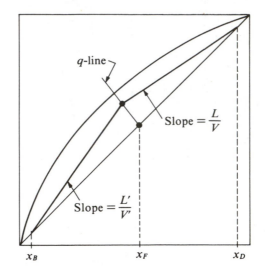

Figure 3.12.9. McCabe-Thiele diagram for binary distillation.

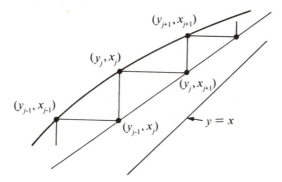

Figure 3.12.10. Graphical determination of concentration in the liquid and vapor phases by means of the McCabe-Thiele diagram.

The ratio L/V that gives the slope of the upper part of the straight line in Figure 3.12.9 is called the internal reflux ratio. This can be determined if one knows the *external reflux ratio R*, which is the ratio between the liquid returned to the top tray from the accumulator (L_{N+1}) and the flow of condensed distillate out of the system D. Therefore:

$$R = \frac{L_{N+1}}{D} \qquad (3.12.10)$$

Since we have assumed that:

$$\frac{L}{V} = \frac{L_{N+1}}{V_N} \qquad (3.12.11)$$

and since:

$$V_N = D + R \cdot D \tag{3.12.12}$$

we can write:

$$\frac{L}{V} = \frac{R}{1 + R} \tag{3.12.13}$$

Therefore, we can change the slope of the line by changing the external reflux ratio R.

The relationships between L and L' and V and V' can be expressed in terms of the feed flow F. We find that:

$$L' = L + qF \tag{3.12.14}$$

where q is a measure of the thermal state of the feed and is the number of moles of saturated liquid in the feed per mole of feed. We also have:

$$V = V' + (1 - q)F$$

where the value of q is, for various situations:

feed of supercooled liquid: $q > 1$
feed of saturated liquid: $q = 1$
feed of mixture of liquid and vapor: $0 < q < 1$
feed of saturated vapor: $q = 0$
feed of superheated vapor: $q < 0$

The two straight-line segments of differing slopes in Figure 3.12.9 will intersect at a point determined by the value of q. This point lies on a straight line, called the q-line, that is given by:

$$y = \frac{q}{q - 1}x - \frac{1}{q - 1}x_F \tag{3.12.15}$$

where x_F is the concentration of the more volatile component in the feed. If the feed is a liquid at its boiling temperature ($q = 1$), the q-line will be vertical. Figure 3.12.9 shows the case for $0 < q < 1$. The q-line is drawn between the point of intersection of the two straight-line segments and the point on the diagonal line, $x = y$, that is given by x_F. When we know the concentration of the distillate (the desired value), the external reflux ratio R, the feed concentration x_F, and the factor q, then the point of intersection of the two line segments and the slope of the lower segment will be known. As shown by the figure, the concentration of the bottom product, x_B, can then be determined. If the thermal state of the feed is changed, q will change but not x_F. This in turn changes the position of the point of intersection, and the equilibrium conditions of the column will be changed.

The technique described above can be used to determine the minimum number of trays required to separate a binary mixture when the final desired concentrations of the distillate x_D and bottom product x_B are specified. This is illustrated by Figure 3.12.11 for the case of a column with eight trays.

If the number of trays is given, the technique can be used to determine the concentrations obtainable at each tray as well as in the distillate and bottom products.

From equation (3.12.12) we see that as the external reflux ratio R approaches infinity, which is the case as the amount of distillate D withdrawn from the system approaches zero, the slope of the upper line segment (L/V) will approach 1, and the line segments will coincide with the diagonal. This represents a limiting case that gives the minimum number of trays that can be used to separate the feed into two specified streams. However, at this limit there will be no product, as $D = 0$ and $B = 0$. In order to obtain product from the column, the number of trays must be greater than the theoretical minimum. The limiting case is shown in Figure 3.12.12.

If we reduce the external reflux ratio R, the slope of the line segment (L/V) will decrease, and the point of intersection between the two segments will

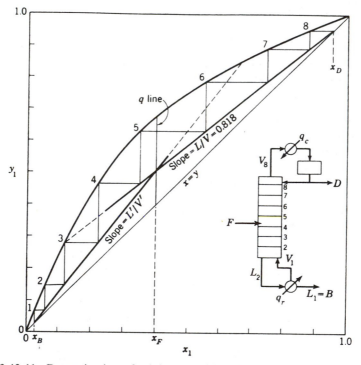

Figure 3.12.11. Determination of minimum number of trays needed in a distillation column.

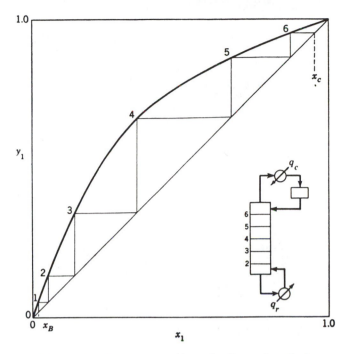

Figure 3.12.12. Column with total reflux and no feed.

move along the q-line toward the equilibrium curve. This is shown in Figure 3.12.13 for the case where the concentrations of the distillate x_D and the bottoms product x_B are held constant. We can now determine the number of trays necessary to obtain distillation. On the other hand, if we hold the number of trays constant, we can determine the product concentrations as a function of reflux ratio. This is illustrated by Figure 3.12.14.

According to Figure 3.12.7, the distance between the equilibrium curve and the diagonal will increase when the relative volatility $\bar{\alpha}$ increases. For a given relative volatility, we can easily find the point on the equilibrium curve that lies farthest from the diagonal by finding the point where the tangent to the equilibrium curve will have a slope of 1. Then we can determine the distance between the equilibrium curve and the diagonal δ from

$$\delta = \frac{\sqrt{2}}{2} \frac{(\sqrt{\bar{\alpha}} - 1)^2}{(\bar{\alpha} - 1)} \qquad (3.12.16)$$

The coordinates for the point on the equilibrium curve that lie farthest from the diagonal are:

$$y_0 = \sqrt{\bar{\alpha}} \cdot \frac{\sqrt{\bar{\alpha}} - 1}{\bar{\alpha} - 1} \qquad (3.12.17)$$

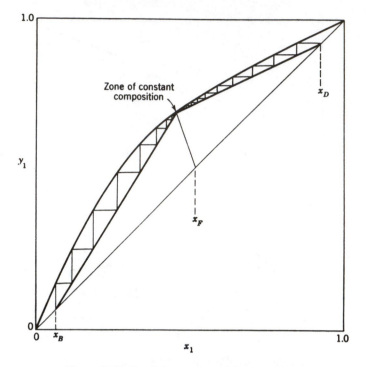

Figure 3.12.13. Column with minimum reflux.

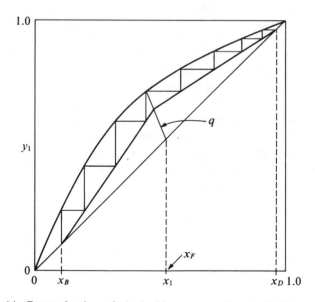

Figure 3.12.14. Determination of obtainable concentration in distillate and bottoms product in column with given number of trays and reflux ratio.

$$x_0 = \frac{\sqrt{\overline{\alpha}} - 1}{\overline{\alpha} - 1} \tag{3.12.18}$$

It is now possible to determine the operating curve that defines the lowest possible external reflux ratio R when the feed concentration and thermal state q are known, and when the desired concentration x_D of the distillate is specified. This is shown in Figure 3.12.15. For the extreme case, let $x_D = 1$, and we find:

$$R_{min} = \frac{1}{\sqrt{\overline{\alpha}} - 1} \tag{3.12.19}$$

The optimal external reflux ratio will be higher than the minimum, and for economic reasons the operating external reflux ratio is usually 25 to 30% higher than the minimum R_{min}. Equation (3.12.19) is shown graphically in Figure 3.12.16.

There are a number of approximate methods for calculation of the important parameters of distillation columns, given the initial specifications (Jafarey, Douglas, and McAvoy, 1979). An example is an equation for predicting the required number of trays in a column:

$$N = \frac{\ln\left(\frac{x_D}{1 - x_D} \cdot \frac{1 - x_B}{x_B}\right)}{\ln\left(\overline{\alpha}\left(1 - \frac{R + q}{(R + 1)(Rx_F + q)}\right)^{1/2}\right)} \tag{3.12.20}$$

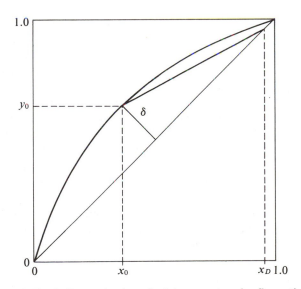

Figure 3.12.15. Determination of minimum external reflux ratio.

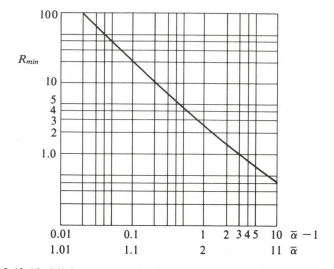

Figure 3.12.16. Minimum external reflux ratio as a function of relative volatility.

The numerator of equation (3.12.20) contains the separation factor, which is given by:

$$S = \frac{x_D}{1 - x_D} \cdot \frac{1 - x_B}{x_B} \tag{3.12.21}$$

This is a good measure of the demand on separation quality.

Distillation is very closely related to the boiling point of the materials at different pressures (or the vapor pressure at various temperatures). Figure 3.12.17 shows the relationship between vapor pressure and temperature for several different materials. Figure 3.12.17a lists several common hydrocarbons, and Figure 3.12.17b is a list of various alcohols and water. The curves are valid only for the pure substances. If there are only two components in a mixture, we can get an idea of the difficulty of separating them by looking at the difference in the boiling points at the same pressure. From Figure 3.12.17a, we see that at a pressure of 100 bar, methane boils at about −50°C, while ethylene boils at about +50°C. Therefore, these two components would be very easy to separate. In the region between the two curves of a boiling point diagram such as that shown in Figure 3.12.4, methane will be in the vapor phase, and ethylene will be a liquid.

For a mixture of propylene and propane at a pressure of 10 bar, the situation is quite different. In this case, propylene will boil at about 16°C and propane at about 28°C. This indicates that the relative volatility for these components in a binary mixture is small.

In a multicomponent mixture the relative volatility is defined as:

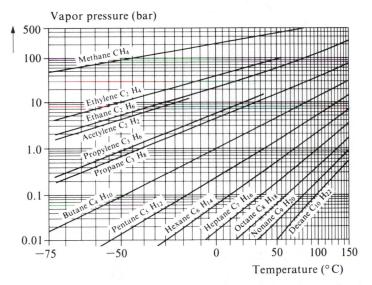

Figure 3.12.17a. Vapor saturation pressure as a function of temperature for the most important hydrocarbons.

$$\alpha_{ij} = \frac{\dfrac{y_i}{x_i}}{\dfrac{y_j}{x_j}} = \frac{K_i}{K_j} \qquad (3.12.22A)$$

If the components satisfy Raoult's law, which states that the partial pressure of one component in a solution is the product of the vapor pressure of the component and the mole fraction of the component, and Dalton's law, which says that the partial pressure of an ideal gas in a mixture of gases is proportional to the mole fraction of the gas in the mixture, then the relative volatility given by equation (3.12.22A) can be written as:

$$\alpha_{ij} = \frac{p_i^0}{p_j^0} \qquad (3.12.22B)$$

where p_i^0, p_j^0 are the vapor pressures of pure liquids i and j.

The relative volatility for different materials can be read from the distance between the curves in the pressure direction in Figure 3.12.17a. Since the ordinate of this plot is logarithmic, the relative volatility is essentially constant over quite a wide temperature range for the curves that lie close to each other (these curves are nearly parallel). This is the basis for the concept of mean relative volatility, which we encountered earlier in equations (3.12.3) and (3.12.4). As an example, we see that the relative volatility of propylene and propane is about 1.23 at 20°C. Figure 3.12.18 shows corresponding curves for many of the

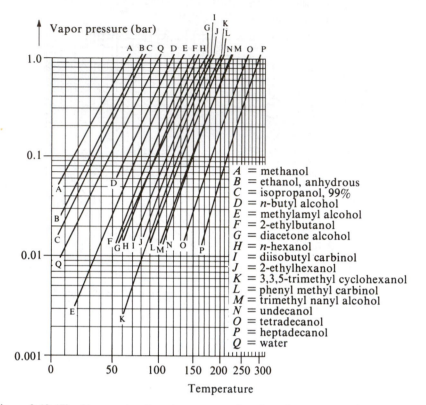

Figure 3.12.17b. Vapor saturation pressure as a function of temperature for some alcohols and water.

most important hydrocarbons based on the K_i ratio of equation (3.12.22A) for the various materials.

From Figure 3.12.17b we see that distillation of a mixture of water, methanol, ethanol, and methylamyl alcohol must begin with a separation of a mixture of methanol and ethanol from water and methylamyl alcohol. This occurs at a temperature of about 100°C, the boiling point for water, if the distillation is to take place at atmospheric pressure. The methanol and the ethanol can be separated after they have been removed, as a mixture, from the top of the column. Extrapolation on Figure 3.12.17b indicates that the relative volatility of these two alcohols is about 1.8 at about 80°C.

Dynamic Models of Distillation

The simplified steady-state description given above is necessary in order to develop a general picture of how the different factors influence the distillation process during steady-state operation and how the various components

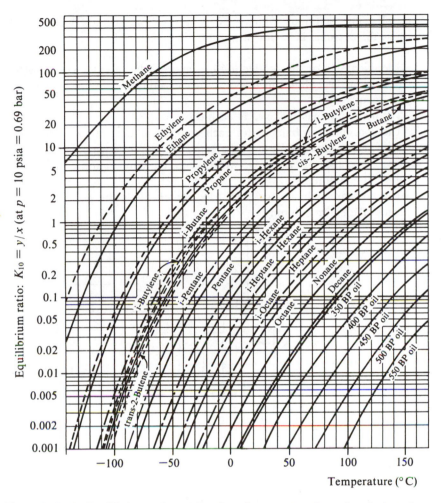

Figure 3.12.18. Equilibrium ratio as a function of temperature for various hydrocarbons.

are related. A description of the dynamic situation is, of course, more complicated. Balance equations of the types given by equations (3.12.5) and (3.12.7) can be written for each tray, the reboiler, the condenser, and the accumulator. Many studies and mathematical models of distillation columns have been reported since the first paper appeared in 1955 (Teager, 1955). Some of the important contributions are: Williams and Harnett (1957); Campbell (1958); Rademaker and Rijnsdorp (1959); Buckley (1964); Luyben (1969); Gould (1969); Rosenbrock (1962); Toijala and Fagervik (1972); Rademaker, Rijnsdorp, and Maarleveld (1975); Shinskey (1977/84); Gilles and Retzbach (1980); Dahlqvist (1981); McAvoy (1983); Buckley, Luyben, and Shunta (1985); and Deshpande (1985).

The following conditions, which are described with respect to Figure 3.12.19, determine the dynamic behavior of a distillation column:

1. The dynamics of the production of vapor in the reboiler system at the bottom of the column in response to the addition of primary energy to the reboiler (u_1).
2. The hydraulic mass balance at each tray.
3. The concentration balance for each component at each tray. This gives the concentration profile over the entire column for each component.
4. Energy balance at each tray, which gives the temperature profile across the entire column.
5. The pressure balance for each tray, which gives the pressure profile across the entire column.

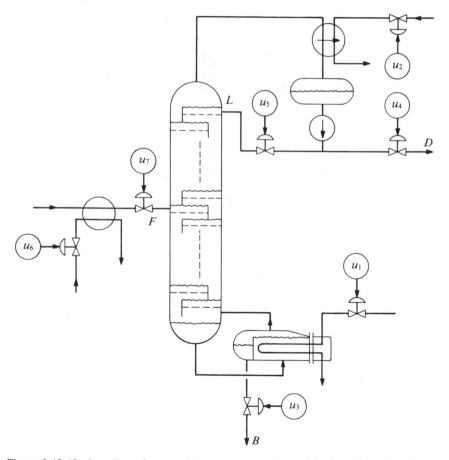

Figure 3.12.19. Location of the most important control variables in a binary distillation column.

6. The dynamic description of the condenser, which gives the amount of distillate produced in response to energy removed (u_2).
7. Dynamics of the accumulator.

These phenomena have differing bandwidths as determined by velocities, time constants, and so on. The vapor pressures and temperatures in a column are relatively rapid phenomena, while concentration changes are the slowest.

The pressure profile across the column will be monotonic, with the highest pressure at the bottom. The temperature profile, however, will often have a definite "knee," especially in binary distillation, between the boiling point for the bottoms product (highest temperature) and the boiling point for the distillate. A mathematical model that describes the motion in the vertical direction of such a temperature knee may give an effective description of the distillation column (Gilles and Retzbach, 1980). The concentration profile, or profiles in the case of multicomponent distillation, can, because of the relationship with the temperature, also often show a knee when the pressure drop over the column is not too high. A mathematical model that describes the movement in the vertical direction of the concentration knee may also give an excellent description of the distillation column.

Figure 3.12.19 indicates the seven most important control variables:

u_1: heat power added to the reboiler.
u_2: heat power removed from the cooler at the top.
u_3: flow of bottom product from the column.
u_4: flow of distillate from the column.
u_5: reflux flow to the column.
u_6: power to the heat exchanger for adjustment of the enthalpy of the feed.
u_7: feed flow (often not an independent variable).

It will be useful to consider which phenomena these control variables affect:

- Variable u_1 directly affects the production of vapor, which in turn directly affects the level of liquid in the bottom of the column. It influences the pressure in the column and the temperature and concentration profiles. The power added to the reboiler is, under normal conditions, proportional to the production rate of the column. Therefore, it is reasonable to use feedforward from the feed stream to the primary power supply for the reboiler.
- Variable u_2, the power removed from the condenser, will first influence the pressure at the top of the column. One would like to control this pressure, and this is often done by means of condenser cooling. The energy balance in the column requires that the energy added balance the energy removed. If the energy added to the reboiler is increased, the pressure in the column will increase, so the energy removed by the condenser must be increased.

- Variable u_3, the flow of bottom product removed from the column, must maintain a mass balance with the liquid flow into and the vapor flow out of the bottom section. In the type of reboiler shown in Figure 3.12.19, the mass balance can be measured by means of the level in the left-hand chamber because it is the overflow from the center chamber. This bottom section design leads very naturally to level control in the left chamber of the reboiler by means of the control variable u_3. If the feed stream to the column increases, the flow to the bottom section will increase, although it will be delayed by the hydraulic dynamics of the M trays below the feed tray. This causes an increase in the liquid level, which will call for an increase in u_3. When this occurs, evaporation in the reboiler must be increased, which means that the upward vapor flow will increase so that eventually the reflux will increase and cause an increase in the liquid flow down to the feed tray. From this we see that there is a significant dynamic difference between a change in the feed rate and the liquid flow downward measured below the feed tray with respect to that above the feed tray.

- Variable u_4, the flow of distillate out of the top of the column, affects the volume of distillate in the accumulator. However, this volume is also affected by the variable u_5, which controls the reflux flow. Since the ratio of reflux to distillate can be very large, especially for columns having many trays for distillation of components with low relative volatilities, it is usually most logical to control the level in the accumulator by means of the reflux flow u_5. If the distillate is sent to a subsequent process such as another distillation column that is sensitive to rapid flow changes, it may make sense to introduce flow control by means of u_4. Because the flow of distillate out of the system determines the mass balance over the entire column ($F = B + D$), a flow control loop on the distillate must have a higher level set point control with feedforward from the feed stream flow and/or from the reflux flow.

- Variable u_5, the reflux flow, plays a decisive role in the operation of a distillation column. This is so because the external reflux ratio, $R = L/D$, determines the slope of the operating line in the column's upper part [see equations (3.12.9) and (3.12.13)]. An increase in the reflux flow will result in a decrease in the liquid level in the accumulator and an increase in the liquid level at the top tray. This gives an increase in the liquid flow moving down the column as well as a decrease in the temperature at the top. This, of course, also decreases the pressure. Since operation of the column is based on the reflux ratio, it is common to use automatic ratio control on L and D. For a column with a given number of trays operating under steady-state conditions, the concentration of distillate, x_D, and bottom product, x_B, will be determined by the concentration of the feed, the thermal state (the q-line), and the reflux ratio.

The best separation possible occurs at total reflux; that is, $R = \infty$ or $D = 0$, but, as we have seen, under this situation the column produces no distillate. When the reflux ratio is reduced so that D is no longer 0, the difference between x_D and x_B will also be reduced. If we evaporate all of the bottom product, so $B = 0$, then the distillate will have exactly the same composition as the feed! Similarly, the bottom product B will have the same composition as the feed if we make $D = 0$.

In cases where the feed cannot be manipulated, perhaps owing to process considerations in a unit upstream of the column, then the column must take the feed as it comes. In this case, the concentration of the top product and the reflux flow must be used to maintain the material balance in the accumulator. The total material balance in the column is managed by the fact that the bottom product B is determined by the level in the bottom section, but the concentration of the bottom product will vary. A similar situation occurs if the concentration of the bottom product is specified. The boil up rate (vapor reflux) can be controlled so that the level of liquid in the bottom section is constant, while the flow of bottom product can be controlled by its concentration (given approximately by the temperature). If the concentrations of both top and bottom products are specified, then operation becomes a matter of balancing the reflux flow against the boil up rate.

- Variable u_6 adjusts the enthalpy of the feed. When the feed comes from an upstream process, the mass flow and concentration of the various components are usually known. The thermal state of the feed, however, often needs modification. As shown in Figure 3.12.19, this can be done with a heat exchanger that makes it possible to heat or cool the feed to adjust the enthalpy as desired. This is especially important if the upstream process delivers a feed stream in which the ratio of liquid to vapor changes considerably. If the feed is cold enough so that the temperature is below the boiling point, there will be less vapor and less liquid on each tray above the feed tray than below it. This means that one gets a "flood" in the stripping section of the column. Overheated feed, consisting of vapor only, will cause the liquid reflux at the top to increase in order to get liquid in the stripping section. This can lead to "flooding" of the rectification section. Usually, the feed stream is partially vaporized ($0.7 < q < 1$). The control variable u_6 makes this possible.

- Variable u_7 is a feed flow variable. Normally the feedstream cannot be manipulated independently of the upstream process, but if it is possible to store the feed stream between the column and the previous process, large flow fluctuations in feed can be evened out. For example, level control of the storage volume can be set to produce the smallest possible variations in the feed flow. Control variable u_7 makes this possible.

One can think of other control variables for special cases. This is especially true in multicomponent distillation columns where product is withdrawn from trays intermediate between the top and bottom. The flows removed from these trays are possible control variables. In some columns the feed stream is distributed among several trays at varying ratios, corresponding to a shift up or down of the feed location. This is another independent control variable.

Disturbances

A distillation column is subject to many disturbances that make automatic control absolutely necessary in order to obtain satisfactory operation. The most important disturbances are these:

- v_1: If one assumes that the other characteristics of the feed are constant, variations in the feed flowrate are the most obvious disturbances that will lead to changes in both the material and energy balances in the column, and to the need for different control actions in other places in the column. If fluctuations in the feed flowrate are due to the behavior of processes upstream of the column, there is usually little that can be done other than to damp these changes by means of level control of a buffer surge tank, or capacitance, between the process and the column. The transfer function between the liquid flow into and out of such a tank will be first order, with a time constant determined by the ratio between the area of the tank and the proportional gain of the controller (if one assumes a proportional controller is to be used). A large tank area and a small controller gain will give large damping. On the other hand, the changes in level will be inversely proportional to the gain, and the size of the tank is a matter of economics. A buffer tank is not especially practical, except for small processes operating at atmospheric pressure.
- v_2: Variation in the concentration of the different components in the feed (single-variable in binary distillation, but multivariable in multicomponent systems) is an especially common type of disturbance that affects nearly all the variables in the column. An increase in concentration of the most volatile component in binary distillation will lead, if the other variables are assumed to be constant, to an eventual increase in the flow of distillate, among other things. However, a properly designed control system can make use of the holdup capacity of the column to even out some of these variations. It is possible to feed some of the top and bottom product, as well as intermediate products for the case of multicomponent distillation, back into the column to help smooth variations in feed concentration. The effect of this positive feedback will be an apparent large increase in the holdup capacity of the column. However, this is not often done in industrial columns. In order to realize component concentration control of the feed, measurement of the individual components is required, which can be achieved only by very complicated and expensive analytical equipment (a chromatograph, for example). The dynamics of the concentration dis-

turbances, in terms of bandwidth, power spectrum, and so on, will for the most part be determined by the time constants in the upstream processes. Some of these processes are very fast, such as in cracking or pyrolysis in the production of ethylene and other light hydrocarbons from ethane and propane.

- v_3: Enthalpy changes in the feed are also common disturbances to a distillation column, but, as mentioned earlier, they can be smoothed out and perhaps eliminated by the use of a heat exchanger, as shown in Figure 3.12.19. Whether or not a heat exchanger and its associated equipment can be justified depends on the loss in quality or production that would result from uncontrolled enthalpy disturbances.

- v_4: External atmospheric conditions also present process disturbances in some distillation columns. In the case of uninsulated, uncovered columns (especially those located outside), these disturbances might be caused by solar heating, as well as variable cooling due to rain, snow, wind, and so on. This is also true of condensers that use atmospheric air. If the entire column is exposed to these influences, virtually all the variables will be affected, but if only the condenser is affected, for example by a rain shower, then just the variables in the upper part of the column will be affected. Avoiding disturbances caused by atmospheric conditions is generally a matter of construction and insulation, rather than automatic control; so it is a matter of determining the economic balance between the advantages and disadvantages of eliminating these disturbances or living with them.

- v_5: Variations in flow and enthalpy of the heating media of the reboiler at the bottom and the condenser at the top are often important disturbances. However, these disturbances can, and should, be eliminated by means of local control systems that control the addition or removal of energy. Many problems in the operation of a complex system such as a distillation column can be avoided if the disturbances themselves, or their causes, are removed.

- v_6: Pressure from the downstream processes that receive the distillate and bottom product can affect these flows, but disturbances from these sources are easy to counter by means of flow control. Therefore, they are not considered to be serious.

- v_7: Many disturbances can be caused at various points in the process or its auxiliaries by pump failures, valve failures, and so forth. Such failures can lead to disturbances of virtually all the variables in the column.

State Variables

At least four state variables are required to describe the dynamics of each tray in a distillation column, at least two to describe the reboiler, and two more for the condenser/accumulator stage. This gives a total of $4(N + 1)$ state variables for a column having N trays. The state variables are:

- Concentration of the most volatile component in the liquid phase at each tray.
- Concentration of the most volatile component in the vapor phase at each tray.
- Mass of liquid at each tray.
- Mass of vapor at each tray.

A reasonable simplification of the model is to neglect the dynamics of the vapor phase, as they are much faster than the dynamics that describe the concentration variations and the liquid volume. The differential equations that describe the changes in the mass of vapor thus can be replaced by algebraic relationships. If we assume constant relative volatility, the relation between concentration in the liquid phase to that in the vapor phase of the most volatile component can be described with two state variables per tray [see equations (3.12.5) through (3.12.9)].

On this basis a dynamic state space model will have the form

$$\dot{\mathbf{x}} = \mathbf{f}(\mathbf{x}, \mathbf{u}, \mathbf{v})$$

where:

$\mathbf{x}$ = vector of the state variables on each of the plates, at the bottom, and at the top

dim $\mathbf{x}$ = $2(N + 2)$

Many such models have been developed, and they have served as simulation bases for development of other models and for the testing of different types of control structures. Frequently, very detailed models are proposed for use with real-time state estimators for computation of the most important variables in the column, but so far these models have required so much computation that a total solution has been too complex. One can expect, however, that as the cost of computation goes down and the speed goes up, application of a total model will become more attractive for control.

Transfer Functions

One can derive many forms of reduced, or simplified, models on the basis of a detailed description of a distillation column. The choice of state variables often depends on the variables one thinks should be controlled and the ease with which the variables can be measured. These variables are:

- Concentration of the most volatile component in the distillate.
- Concentration of the most volatile component in the bottom product.

- Pressure at the top of the column.
- Pressure difference between the top and bottom of the column.
- Temperature at the top (related to concentration and pressure).
- Temperature at the bottom (related to concentration and pressure).
- Temperature profile over the column (related to the pressure and the con-centration profile).
- Volume of liquid at the bottom.
- Volume of liquid at the top (accumulator).

For further discussion of this subject, the reader is referred to Toijala and Fagervik (1972); Edwards and Jassim (1977); Taiwo (1980); Buckley, Luyben, and Shunta (1985); and Deshpande (1985).

Many authors have discussed the following form for the transfer function matrix for a distillation column:

$$\mathbf{y}(s) = H(s)\mathbf{u}(s) = \begin{bmatrix} h_{11}(s) & h_{12}(s) \\ h_{21}(s) & h_{22}(s) \end{bmatrix} \mathbf{u}(s) \qquad (3.12.23)$$

where:

$$\mathbf{y}(s) = \begin{bmatrix} \theta_N(s) \\ \theta_1(s) \end{bmatrix}$$

and:

$\theta_N(s)$ = temperature at the top (indicative of the top product concentra-tion x_D)

$\theta_1(s)$ = temperature at the bottom, corrected for pressure difference (indica-tive of the bottom product concentration x_B)

and:

$$\mathbf{u}(s) = \begin{bmatrix} q_R(s) \\ Q_1(s) \end{bmatrix}$$

where:

$q_R(s)$ = reflux flow
$Q_1(s)$ = heat flow to the reboiler

A number of authors have supplied terms for the transfer function matrix:

$$h_{11}(s) = \frac{-k_{11}e^{-\tau_{11}s}}{1 + T_{11}s}$$

$$h_{12}(s) = \frac{k_{12}}{(1 + T'_{12}s)(1 + T''_{12}s)} \qquad (3.12.24)$$

$$h_{21}(s) = \frac{-k_{21}e^{-\tau_{21}s}}{1 + T_{21}s}$$

$$h_{22}(s) = \frac{k_{22}}{1 + T_{22}s}$$

Typical values for the time parameters in these transfer functions are:

$$T_{11} \approx T_{22} \approx T'_{12}$$

$$\tau_{11} \approx 0.03T_{11} \rightarrow 0.1T_{11}$$

$$T''_{12} \approx 0.05T_{11} \rightarrow 0.1T_{11}$$

$$T_{21} \approx 2T_{11}$$

$$\tau_{21} \approx 0.1T_{11} \rightarrow 0.2T_{11}$$

The numerical values of the time constants (T_{11}, for example) depend completely on the size of the column and the materials being distilled, but one can expect time constants in the range of several minutes up to several hours.

No numerical value for the gain factor in equation (3.12.24) is given here because it depends on which units are used; but as all the factors k_{ij} are positive, we can tell which direction the response will take.

Several comments can be made with regard to the transfer functions given by equation (3.12.24). A variation in the energy supply to the reboiler gives a response in the bottom section that is characterized by first-order dynamics with a fairly large time constant. The top temperature also responds to an energy change in the reboiler, but it will show a second-order response with one time constant about the same as that of the bottom temperature, and the other (T''_{12}) about one order of magnitude less. This is explained by the fact that the transient in the vapor phase moves relatively quickly through the column.

The two responses that follow changes in the reflux flow are both characterized by a combination of a time constant and a transport delay. The transport delay is largest for the temperature in the lower end of the column, but it is still small compared with the dominant time constant. This means that even if these transfer functions have more phase shift than simple first-order processes, the time constants, not the transport delays, dominate the bandwidth that can be obtained in the control systems.

Another useful and more elaborate model form developed by Toijala and Fagervik (1972) is presented in equations (5.6.2) through (5.6.16). Still another structure, proposed by Gilles and Retzbach (1980), is given by equation (5.6.22). Finally, one should recall the approach indicated by Figures 2.13.4 through 2.13.6, which represent a structure suitable for a discrete model consisting of transfer functions separated by time delays.

Various authors have proposed methods for estimating the parameters of the transfer functions as functions of the column parameters. Examples can be found in Wahl and Harriott (1970) and Rademaker, Rijnsdorp, and Maarleveld (1975).

3.13. REFRIGERATION PROCESSES

Refrigeration is another important thermodynamic unit process. It has broad application from the smallest refrigerator and freezer in the home or laboratory to large industrial plants for the production of liquid gases, cooling for low-temperature distillation, and so on.

The refrigeration process is based on the fact that a liquid that evaporates will recover its heat of evaporation from the surrounding medium. If the liquid pressure is such that the boiling temperature is below the temperature one wishes to obtain by cooling, then heat will flow from the surroundings into the refrigerant medium.

Figure 3.13.1 is an enthalpy–pressure diagram for a typical refrigerant. The curved line on the left is the condensation curve, and that on the right is the evaporation curve. Liquid and vapor are in equilibrium between these curves, and the temperature will be fixed for a given pressure. Only liquid exists to the left of the condensation curve, and only vapor exists to the right of the evaporation curve. The refrigeration cycle drawn on the diagram of Figure 3.13.1 is realized by the equipment shown in Figure 3.13.2. At the bottom of this figure, we see an evaporator partly filled with refrigerant and containing heat exchanger tubes through which passes the liquid to be cooled. Vapor from the evaporator goes to the low-pressure side of a compressor that increases the pressure from

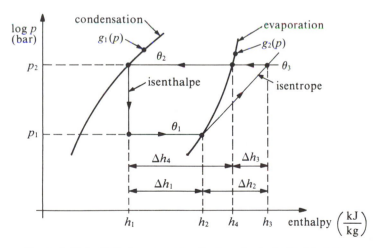

Figure 3.13.1. Enthalpy–pressure diagram for typical refrigerant.

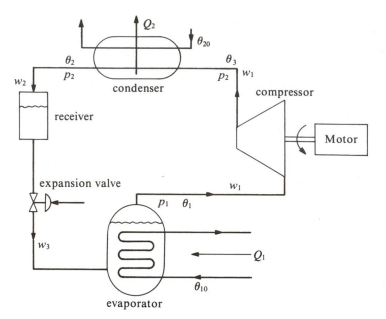

Figure 3.13.2. Main components of a typical refrigeration cycle.

the evaporator pressure p_1 to the condenser pressure p_2. If we assume the compression is adiabatic, the vapor absorbs only the enthalpy due to the mechanical power of compression (constant entropy). This is represented in Figure 3.13.1 by the isentropic line from the evaporation curve upward toward the right until it reaches the condenser pressure. The isentrope will be nearly straight because of the logarithmic pressure scale. Heat is removed at the condenser, so the temperature decreases until it reaches the evaporation temperature. Furthermore, heat is given off at p_2 until the vapor is completely condensed. The condensate is collected in a tank following the condenser, and if the tank is assumed to be well insulated, the temperature (θ_2) of the liquid in the tank will be determined by the pressure p_2. The heat flow Q_2 leaving the condenser can be used for process heat if desired. The refrigerant flows from the condensate tank through an expansion valve whose opening must be controlled. Upon expansion, the pressure drops from the condenser pressure p_2 to the evaporator pressure p_1. It is reasonable to assume that this expansion is adiabatic, which is to say that expansion occurs at constant enthalpy. In Figure 3.13.1, this is represented by the vertical line from the point at the condensation curve given by the condensation pressure p_2. This completes the refrigeration cycle. If the four corner points of the cycle in Figure 3.13.1 are identified by their enthalpies (h_1, h_2, h_3, and h_4) we can define the enthalpy differences as follows:

$\Delta h_1 = h_2 - h_1$: enthalpy change due to evaporation.
$\Delta h_2 = h_3 - h_2$: enthalpy change due to compression.

$\Delta h_3 = h_3 - h_4$: enthalpy change due to cooling of the superheated vapor.
$\Delta h_4 = h_4 - h_1$: enthalpy change due to condensation.

Since enthalpy has dimensions (kJ/kg), the total heat power is found by multiplying the enthalpy by the mass flow of the refrigerant. The mass flow of refrigerant leaving the evaporator is given by w_1 (kg/s), and w_2 is the mass flow leaving the condenser as condensate. The mass flow of condensate through the expansion valve is w_3.

If we assume that there is no heat loss to the environment, the total heat delivered by the condenser at temperature θ_2 will be the sum of the heat gained by the evaporator from the coolant and the mechanical energy supplied to the compressor. If the heat produced by the condenser can be used profitably, then the system will be energy-efficient because it will produce more energy than is used by the compressor.

Most gases can be used in refrigeration cycles because they have enthalpy–pressure diagrams similar to that of Figure 3.13.1. Nitrogen is particularly important as the refrigerant for the production of liquid oxygen. The enthalpy–pressure diagrams for nitrogen and oxygen are given in Figures 3.13.3a and 3.13.3b.

A number of refrigerants are produced that have characteristics well suited for use in refrigeration. These refrigerants can provide any desired vaporization temperature at different pressures. One of the most common is R-12 (dichlorodifluoromethane), which has the enthalpy–pressure diagram shown in Figure 3.13.4. For illustration, a refrigeration cycle is drawn on this figure where θ_1 and θ_2 are, respectively, $-30°C$ and $+50°C$. The corresponding pressures p_1 and p_2 will be about 1 bar and 12 bar. Notice that at the ideal conditions assumed here, the enthalpy difference for evaporation will be $338 - 250 = 88$ kJ/kg. In the condenser the enthalpy difference will be $384 - 250 = 134$ kJ/kg. Now, assume for example that the cooling capacity is specified to be 10 kW, or 10 kJ/s. We can calculate the required mass flow of refrigerant through the evaporator to be $w_1 = 0.114$ kg/s $= 6.82$ kg/min. At this flow rate of refrigerant the power delivered from the condenser will be $Q_2 = 0.114 \times 134 = 15.28$ kW. Therefore, the theoretical power added at the compressor must be 5.28 kW. In these calculations we have not allowed for any heat losses or other phenomena that might reduce the efficiency of the system. Therefore, a real system would not lead to quite the same numbers. A *heat pump* uses the fact that, as we have just seen, if we apply about 5 kW of power through the compressor, the power out of the condenser is increased about threefold.

Figure 3.13.5 shows the temperature–pressure relationship for several different refrigerants at liquid–vapor equilibrium. We see that the curves are nearly linear, which makes them simple to use in mathematical modeling.

A mathematical model of a refrigeration system of the type shown in Figure 3.13.2 can best be illustrated by means of a block diagram (Figure 3.13.6).

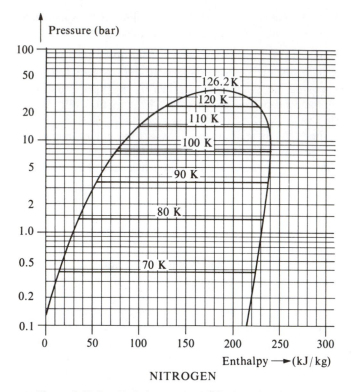

Figure 3.13.3a. Enthalpy–pressure diagram for nitrogen.

At the bottom center of the diagram is the block for the heat balance for the evaporator. The heat input Q_1 is proportional to the temperature difference between the evaporator (θ_1) and the primary medium (θ_{10}) times a heat transfer coefficient G_1. Heat coming in with the refrigerant is given by $w_3h_1 = w_3g_1(p_2)$. Heat leaving with the vaporized gas is $w_1h_2 = w_1g_2(p_1)$.

The block representing the condenser is located at the top of the figure. The heat Q_2 leaving the condenser is proportional to the difference between the condenser temperature θ_2 and the temperature of the secondary medium, θ_{20}, times a heat transfer coefficient G_2. Heat enters the condenser, w_1h_3, with the vapor from the compressor. An amount of heat w_2h_1 leaves with the condensate.

The pressure in the evaporator, p_1, and that in the condenser, p_2, are given by the temperature–pressure function for saturated vapor of the refrigerant. That is, $p = f_1(\theta)$.

The block diagram of the compressor appears at the right of Figure 3.13.6. This block includes equipment for speed control of the compressor. If we assume 100% efficiency, the mechanical energy P that is added will be given by $P = n \cdot d$ where n is the speed of rotation, and d is the torque. If the mechan-

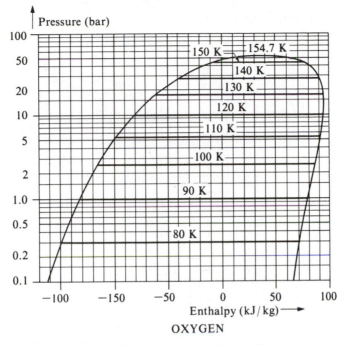

Figure 3.13.3b. Enthalpy–pressure diagram for oxygen.

ical energy is set equal to the enthalpy change of the refrigerant under isentropic conditions, we will have $P = w_1 \Delta h_2$.

The mass flow w_1 of vapor into the compressor is assumed to be a function of the rotational speed n and the pressure p_1. Thus, $w_1 = f_2(n, p_1)$.

In the center of Figure 3.13.6 we see how this simplified model describes the relationship between the mass flow into the condenser, w_1, and the mass flow out, w_2, as a first-order differential equation with a time constant T. At the lower left corner of the diagram, we have the mass balance in the evaporator tank, where the mass of refrigerant, W_1, is found by integrating the difference between the input flow w_3 and the output w_1. We assume here that the total mass controller operates the expansion valve such that the flow rate w_3 changes. The total mass is derived from the level, taking into account the swelling effect.

To the left in the middle of the diagram is the mass balance on the condensate tank, where the mass W_2 is obtained by integration of the difference between the condensate entering the tank, w_2, and that leaving, w_3. If the condensate tank is large enough, the condensate level in the tank will be of no interest although it will provide hydraulic decoupling between the condenser and the evaporator. However, there is a thermal coupling between these two elements of the system.

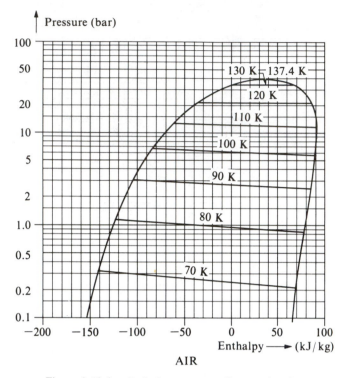

Figure 3.13.3c. Enthalpy–pressure diagram for air.

The condensation curve for Figure 3.13.6 is given by the function

$$h = g_1(p)$$

and the evaporation curve is given by

$$h = g_2(p)$$

The enthalpy change under adiabatic compression is

$$\Delta h_2 = g_3(p_1, p_2) \approx \frac{1}{k_2} \cdot \log\left(\frac{p_2}{p_1}\right)$$

where k_2 is the slope of the approximately straight isentrope of Figure 3.13.1.

The block diagram of Figure 3.13.6 contains five significant integrators. If W_1 is controlled with good sensitivity, we can set $w_3 \approx w_1$. If we also neglect the holdup time in the condenser, we can let $w_2 \approx w_1$, which leaves us with a third-order system.

In addition to the opening of the expansion valve, which is part of the mass control system in Figure 3.13.6, there are two important control variables in the process. These are the compressor speed set point n_0 and the heat flow out of

Pressure
bar

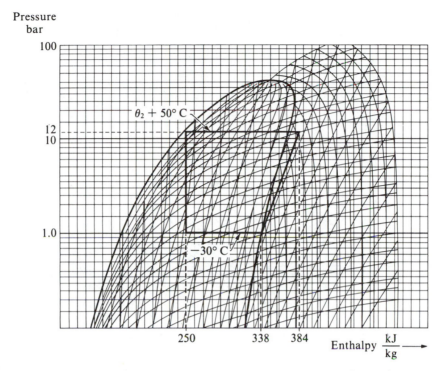

Figure 3.13.4. Enthalpy–pressure diagram for R-12, with cooling cycle.

the condenser, Q_2 (or possibly the secondary medium temperature, θ_{20}). The evaporator primary medium temperature, θ_{10}, is a disturbance. If the condenser condenses all of the vapor, the temperature θ_2 and the pressure p_2 will settle at levels determined by the energy flow from the compressor. In this case there will be just one control variable, n_0, and it will be natural to set up a controller relating the pressure in the evaporator to that variable. One can also let the compressor speed be controlled on the basis of the pressure in the condenser, p_2, which the temperature in the condenser will also follow.

Possible control structures for refrigeration processes will be developed in section 4.4.

The refrigeration cycle in Figure 3.13.2 is the most elementary of all refrigeration processes. With this as a foundation one can develop more complicated systems that might be more energy-efficient or provide a better fit of temperatures and pressures in different sections of the plant. An example of a more complex refrigeration system is shown in Figure 3.13.7, where there are two compressors, perhaps driven by the same motor, two evaporators, and four heat exchangers. The functions of the various elements of this system are indicated by the respective numbers appearing on the enthalpy–pressure diagram of Figure 3.13.7. A model for this process would include more state vari-

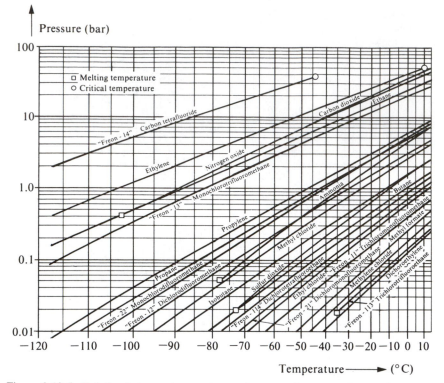

Figure 3.13.5. Relation between temperature and pressure in saturated vapor for various refrigerants.

ables, but it would have the same type of structure as that which is described by Figure 3.13.6.

3.14. CHEMICAL REACTIONS

All the processes discussed in the previous sections were characterized by pure physical interactions (mass transport, heat transfer, etc.) in which the materials experienced no changes other than phase changes. There was no chemical reaction involved. In this section we will discuss some basic considerations involving processes where chemical behavior leads to the creation of one or more new materials due to chemical reaction. This field is very extensive; here we can simply point out some of the features of these systems with respect to their relationships to automatic control (Gould 1969).

Many types of equipment and configurations are used for chemical reaction processes in industry. One can often choose between batch reactors and continuous reactors. In the first case, the reactants are added to the reactor as a batch and are held in the reactor until the reaction is complete. In the second case, the

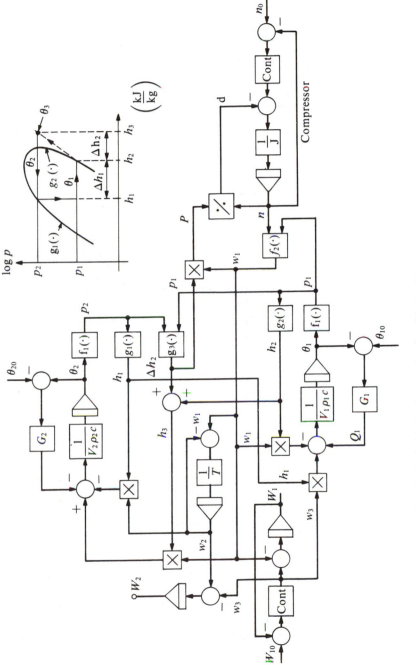

Figure 3.13.6. Block diagram for simplified refrigeration system.

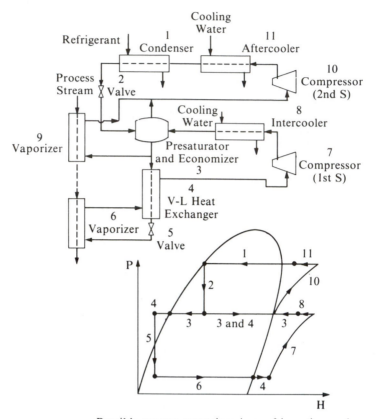

Possible process operations in a refrigeration cycle.

Figure 3.13.7. A two-stage refrigeration system (Barnes and King, 1974). (Reprinted with permission from *Industrial & Engineering Chemistry Process Des. Dev., 15*, 4, 495–504. Copyright 1974, American Chemical Society.)

reactants enter the reactor in a continuous stream, and the reactor delivers a continuous stream of product that is a result of the reaction occurring during a certain holdup, or residence time.

Chemical reactions often involve heat transfer, either because the reaction requires the addition of heat to make it "go" or because the reaction produces heat. Sometimes, both can occur. If the reaction requires the addition of heat, it is called endothermic. If it produces heat, it is called exothermic.

If the reaction takes place in a single physical phase (solid, liquid, or gas), it is known as a homogeneous reaction. Reactions that involve more than one physical phase are called heterogeneous. Heterogeneous reactions are often complicated to describe mathematically because the mass transfer between phases must be described in addition to the chemical reactions.

Some chemical reactions use catalysts, which do not enter into the reaction itself but influence the reaction, either promoting or retarding it, by affecting the reaction rate. Many reactions simply will not occur unless an appropriate catalyst is present.

An irreversible reaction is one in which the reactants convert completely to one product without any reverse reaction. This is often represented as $A \rightarrow B$, where A and B represent the reactant and product, respectively. On the other hand, a reversible reaction is one in which the reaction can proceed in either direction; that is, a product is created, but also the reactant can be created from the product. Usually, such a reacting system will stabilize at an equilibrium condition in which both reactant and product will exist. A reversible reaction of reactants A and B would be written $A \rightleftarrows B$.

When only one reactant and one chemical reaction are involved, only one stoichiometric equation is required to describe the reaction, but in the case of multiple reactions, more than one stoichiometric equation is required. Some types of multiple reactions are:

- *Consecutive* reactions, characterized by $A \rightarrow B \rightarrow C$, where, for example, B can be the desired product and C an undesired product.
- *Competing* reactions, which are characterized by $A \rightarrow B$ and $A \rightarrow C$, where B might be a more desirable product than C.
- *Mixed* reactions such as $A + B \rightarrow C$ and $C + B \rightarrow D$.

Very complicated systems can result from combinations of the above types of reactions. The net result can be a process that consists of a series of intermediate processes containing reversible, consecutive, competing, and mixed reactions.

Mathematical models of chemical reactions must include three major phenomena:

1. The stoichiometric relationships, which take the form of algebraic expressions. These are chemical formulas that describe the mass relationships between the different materials involved in the reaction.
2. The reaction rates, which determine the speeds at which the reactions take place and how quickly the various components change as functions of concentration, temperature, and so on.
3. Energy balance equations, which express the dynamic relationships between the addition, generation, loss, and storage of heat.

The rate at which a reaction takes place can be described by the derivative of the concentration of each individual component, expressed as the mole fraction. For the simple case of the reaction $A \rightarrow B$, we have:

$$\dot{c}_A = f_A(c_A, c_B) \tag{3.14.1}$$

$$\dot{c}_B = f_B(c_A, c_B) \tag{3.14.2}$$

An important factor implicit in the right side of equations (3.14.1) and (3.14.2) is the reaction rate r, which in this reaction can be assumed to be given by:

$$r = kc_A \tag{3.14.3}$$

where k is the reaction rate constant. For most materials the value of k obeys the well-known Arrhenius law, which states:

$$k = k_0 e^{-E/R\theta} \tag{3.14.4}$$

where:

k_0 = a constant factor
E = activation energy
R = the universal gas constant
θ = the absolute temperature

For a continuous reactor with constant volume and perfect mixing, equations (3.14.1) and (3.14.2) can be written:

$$\dot{c}_A = \frac{c_{A0} - c_A}{T} - k \cdot c_A \tag{3.14.5}$$

$$\dot{c}_B = \frac{c_{B0} - c_B}{T} + k \cdot c_A \tag{3.14.6}$$

where:

c_{A0} = concentration of A in the feed
c_{B0} = concentration of B in the feed
T = V/q_0, the holdup time in the reactor (time constant)
q_0 = volumetric flow into the reactor
V = volume of the reactor

The left side of the block diagram of Figure 3.14.1 represents the solutions of equations (3.14.5) and (3.14.6). According to equation (3.14.4), the rate constant k is dependent on temperature and will be constant for the case of an isothermal process.

The right side of the block diagram of Figure 3.14.1 describes the energy (heat) balance of the reactor. The heat due to the reaction is proportional to the reaction rate and the specific heat of reaction (ΔH) and must be added to any heat supplied (Q_0) or heat brought in by the materials to determine the overall rate of change of temperature. If the reactor has no temperature control (such as that shown by the dashed line in the figure), there will be a nonlinear feedback

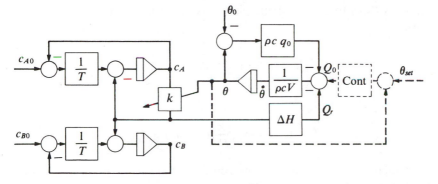

Figure 3.14.1. Elementary block diagram for first-order irreversible reaction.

from the energy balance to the reaction constant k. In an exothermic reaction, where $\Delta H < 0$, this feedback will be positive.

Notice that in Figure 3.14.1 there is no cross-coupling between the concentration of c_A and that of c_B. This means, among other things, that in order to study the stability of the reactor under nonisothermal conditions (that is, when there is no temperature control), only two state variables need be included, namely c_A and θ.

We then have:

$$\dot{c}_A = \frac{1}{T}(-c_A + c_{A0}) - k(\theta)c_A \qquad (3.14.7)$$

$$\dot{\theta} = \frac{1}{T}(-\theta + \theta_0) + \frac{1}{\rho c V}[-\Delta H k(\theta)c_A + Q_0] \qquad (3.14.8)$$

where:

θ_0 = temperature of flow into the reactor
ρ = density of material in reactor
c = specific heat of material in reactor

The steady-state conditions, where $\dot{c}_A = 0$ and $\dot{\theta} = 0$, can be found from equations (3.14.7) and (3.14.8). These conditions are designated as c_{As} and θ_s, respectively. First, we solve equation (3.14.7) for c_{As} as follows:

$$c_{As} = \frac{c_{A0} - c_{As}}{T \cdot k(\theta_s)} \qquad (3.14.9)$$

Inserting this solution into (3.14.8) and setting the right side equal to 0, we get:

$$\frac{1}{T}(-\theta_s + \theta_0) + \frac{1}{\rho c V}\left(\frac{-\Delta H(c_{A0} - c_{As})}{T} + Q_0\right) = 0 \qquad (3.14.10)$$

Equations (3.14.9) and (3.14.10) can be solved in terms of $(c_{A0} - c_{As})$, which gives:

$$(c_{A0} - c_{As}) = \frac{c_{A0} \cdot Tk(\theta_s)}{Tk(\theta_s) + 1} \qquad (3.14.11)$$

and:

$$(c_{A0} - c_{As}) = \frac{\rho c V}{\Delta H}(\theta_0 - \theta_s) + \frac{Q_0 T}{\Delta H} \qquad (3.14.12)$$

Equation (3.14.11), which is the solution of (3.14.9), gives the S-shaped curve shown in Figure 3.14.2. This curve approaches the limiting value

$$\frac{c_{A0} Tk_0}{Tk_0 + 1}$$

as the value of θ_s increases.

Equation (3.14.12) produces a straight line that, for an exothermic process with $\Delta H < 0$ will slope toward the upper right corner as shown in Figure 3.14.2. The figure shows a situation in which the two curves intersect at three different points, a, b, and c. These are singular points for the differential equations (3.14.7) and (3.14.8). Further investigation of these points will immediately show that points a and c are stable; that is, the eigenvalues of the system lin-

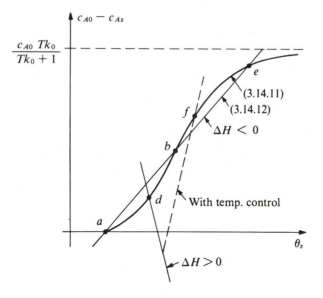

Figure 3.14.2. Determination of the steady-state solution for the reactor of Figure 3.14.1.

earized around these points will lie in the left half-plane. The singular point b will be unstable, which simply tells us that the reactor will have no stable equilibrium condition at that point.

If $\Delta H > 0$ (the case for an endothermic reactor), the straight line will angle down to the right as shown in Figure 3.14.2, and there will be only one point of intersection. It is easy to see that this point is stable.

If automatic temperature control is applied, as shown by the dashed line of Figure 3.14.1, the situation for an exothermic reactor will change significantly. There will then be only one intersection of the curves, as shown by point f in Figure 3.14.2. By this means, stable operation can be obtained at temperatures that are acceptable from the point of view of operation of the reactor. A closer analysis of the conditions of stability in such a system can be made by means of Liapunov's method, but this will not be considered here (Perlmutter, 1972).

For the case of a *reversible* reaction, $A \rightleftarrows B$, there will be two reaction rates, one for the forward reaction $A \rightarrow B$:

$$r_f = k_f c_A \tag{3.14.13}$$

and one for the backward reaction $B \rightarrow A$:

$$r_b = k_b c_B \tag{3.14.14}$$

The differential equations for this process are:

$$\dot{c}_A = \frac{1}{T}(-c_A + c_{A0}) - k_f c_A + k_b c_B \tag{3.14.15}$$

$$\dot{c}_B = \frac{1}{T}(-c_B + c_{B0}) + k_f c_A - k_b c_B \tag{3.14.16}$$

As before, the two reaction rates are functions of temperature according to the Arrhenius law:

$$k_f = k_{f0}e^{-\frac{E_1}{R\theta}} \tag{3.14.17}$$

$$k_b = k_{b0}e^{-\frac{E_2}{R\theta}} \tag{3.14.18}$$

A reversible reaction is shown in the block diagram of Figure 3.14.3. In this diagram, the reaction rates are indicated as being functions of temperature only.

Even if there are two reactions, there exists a linear dependence, defined as a reaction invariant, $z = c_A + c_B$, which can be found from a new differential equation obtained by adding equations (3.14.15) and (3.14.16):

$$\dot{z} = \frac{1}{T}(-z + z_0) \tag{3.14.19}$$

This differential equation is independent of the reaction rate and describes only the mass balance in the reactor without any effects due to reaction.

This principle has been generalized by Asbjornsen and Fjeld (1970) and has an important connection with the controllability of models of chemical reactors in general.

If we assume isothermal conditions in the reactor of Figure 3.14.3 (that is, k_f and k_b are constant), we can write equations (3.14.15) and (3.14.16) in state space form:

$$\dot{\mathbf{c}} = A\mathbf{c} + \frac{1}{T}\mathbf{c}_0 \tag{3.14.20}$$

where:

$$\mathbf{c} = \begin{bmatrix} c_A \\ c_B \end{bmatrix}$$

and:

$$A = \begin{bmatrix} -\left(\dfrac{1}{T} + k_f\right) & k_b \\ k_f & -\left(\dfrac{1}{T} + k_b\right) \end{bmatrix}$$

The Laplace transformation is used to find the transfer function matrix used in

$$\mathbf{c}(s) = H(s)\mathbf{c}_0(s)$$

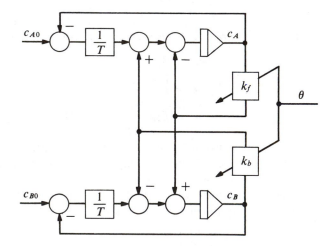

Figure 3.14.3. Elementary block diagram of first-order reversible process.

where:

$$H(s) = \begin{bmatrix} \dfrac{(1 + k_b T)}{1 + (k_f + k_b)T} \cdot \dfrac{1 + T_2 s}{(1 + Ts)(1 + T_1 s)}, & \dfrac{k_b T}{1 + (k_f + k_b)T} \cdot \dfrac{1}{(1 + Ts)(1 + T_1 s)} \\[3mm] \dfrac{k_f T}{1 + (k_f + k_b)T} \cdot \dfrac{1}{(1 + Ts)(1 + T_1 s)}, & \dfrac{(1 + k_f T)}{1 + (k_f + k_b)T} \cdot \dfrac{1 + T_3 s}{(1 + Ts)(1 + T_1 s)} \end{bmatrix}$$

$$(3.14.21)$$

and:

$$T_1 = \frac{T}{1 + (k_f + k_b)T}$$

$$T_2 = \frac{T}{1 + k_b T}$$

$$T_3 = \frac{T}{1 + k_f T}$$

Of special interest is the transfer function between the variations in concentration of the feed stream $c_{A0}(s)$ and variations in the product concentration $c_B(s)$. This appears in the lower left corner of equation (3.14.21) and is characterized by two time constants. One (T) is the hydraulic holdup time in the reactor, and the other (T_1), which is smaller, is determined by the sum of the reaction rate constants $(k_f + k_b)$.

If the temperature cannot be assumed to be constant, an energy balance must be developed, just as for the case of the irreversible reaction of Figure 3.13.1. Now, however, heat is produced (or used) by two different reactions. For an exothermic reactor, the heat generated by the reactions will be the sum of two contributions, as shown in Figure 3.13.1. An analysis similar to that used for Figure 3.13.2 can be used in this case as well, but it will not be done here.

If two *consecutive* reactions occur in the reactor, we write:

$$A \xrightarrow{k_1} B \xrightarrow{k_2} C$$

where k_1 and k_2 are the reaction rate constants for the two reactions, which have reaction rates $r_1 = k_1 c_A$ and $r_2 = k_2 c_B$. The reaction rate constants, as before, will be temperature-dependent.

$$k_1 = k_{10} \cdot e^{-\frac{E_1}{R\theta}}$$

and:

$$k_2 = k_{20} \cdot e^{-\frac{E_2}{R\theta}}$$

In these expressions, E_1 may or may not be equal to E_2, depending upon the particular reactions. The following differential equations can be written for this system:

$$\dot{c}_A = \frac{1}{T}(-c_A + c_{A0}) - k_1 c_A \tag{3.14.22}$$

$$\dot{c}_B = \frac{1}{T}(-c_B + c_{B0}) + k_1 c_A - k_2 c_B \tag{3.14.23}$$

$$\dot{c}_C = \frac{1}{T}(-c_C + c_{C0}) + k_2 c_B \tag{3.14.24}$$

In the same manner as before, we can now find a transfer function matrix for the system

$$\mathbf{c}(s) = H(s)\mathbf{c}_0(s)$$

where:

$$H(s) = \begin{bmatrix} \dfrac{1}{(1 + k_1 T)} \cdot \dfrac{1}{(1 + T_1 s)} & , & 0 & , & 0 \\[3ex] \dfrac{k_1 T}{(1 + k_1 T)(1 + k_2 T)} \cdot \dfrac{1}{(1 + T_1 s)(1 + T_2 s)} & , & \dfrac{1}{(1 + k_2 T)} \cdot \dfrac{1}{(1 + T_2 s)} & , & 0 \\[3ex] \dfrac{k_1 k_2 \cdot T^2}{(1 + Ts)(1 + T_1 s)(1 + T_2 s)} & , & \dfrac{k_2 T}{(1 + k_2 T)} \cdot \dfrac{1}{(1 + Ts)(1 + T_2 s)} & , & \dfrac{1}{1 + Ts} \end{bmatrix} \tag{3.14.25}$$

with:

$$T_1 = \frac{T}{1 + k_1 T} \qquad T_2 = \frac{T}{1 + k_2 T}$$

If B is the desired product, and C is an unwanted loss, it may be desirable to maximize the steady-state yield of B by determining the temperature that gives the highest concentration c_B, since temperature affects k_1 and k_2 as discussed above. Because the system is linear, this maximization is the same as finding the temperature that gives the maximum gain for the transfer function $h_{21}(s) = [h_{BA}(s)]$, which is the first term of the second row in $H(s)$.

If $E_1 = E_2 = E$, the optimal temperature is given by:

$$\theta_{\text{opt}} = \frac{E}{R \cdot \ln(T\sqrt{k_{10}k_{20}})} \tag{3.14.26}$$

This gives:

$$h_{21}(s)_{\text{opt}} = \frac{k_{10}}{(\sqrt{k_{10}} + \sqrt{k_{20}})^2} \cdot \frac{1}{(1 + T_1 s)(1 + T_2 s)} \tag{3.14.27}$$

where:

$$T_1 = \frac{T}{\sqrt{k_{10}} + \sqrt{k_{20}}} \cdot \sqrt{k_{20}}$$

$$T_2 = \frac{T}{\sqrt{k_{10}} + \sqrt{k_{20}}} \cdot \sqrt{k_{10}}$$

The ratio between the two time constants will be inversely proportional to the square root of the ratio between the respective reaction rate constants. Usually, that ratio will be $k_{10}/k_{20} \gg 1$, which means that $T_1/T_2 < 1$. If $k_{10}/k_{20} = 1$, the transfer function will be:

$$h_{21}(s)_{\text{opt}} = \frac{0.25}{\left(1 + \frac{T}{2}s\right)^2} \qquad (3.14.28)$$

For the reaction $A + B \rightarrow C$, the molar reaction rate per unit volume (at constant volume) will be:

$$r = kc_A c_B$$

where:

$$k = k_0 e^{-E/R\theta}$$

Clearly, we now have a nonlinear (multiplicative) coupling between the differential equations for c_A and c_B.

For the more complicated reaction $A + B \rightarrow C + D$, there will be two rate constants k_{fC} and k_{fD} for the forward reactions that produce C and D, respectively, and two rate constants, k_{bC} and k_{bD}, for the backward reactions from C and D. For example, the forward reaction rate for the production of C is:

$$r_{fC} = k_{fC} c_A c_B$$

and the reaction rate for the backward reaction that converts component C is:

$$r_{bC} = -k_{bC} c_C c_D$$

Again, we have a nonlinear multiplicative coupling between the various differential equations, but it is even more complicated than before.

The point of the previous discussion has been to illustrate some simple cases of chemical reactions taking place in well-mixed stirred tank reactors. Once the desired operating temperature has been determined, one usually wants to operate the reactor according to those considerations that will give the best possible yield in some respect. The control problem consists of control of the concentrations by manipulation of the flowrates of the various components in the reactor feed and temperature control of the reactor. Most difficult is control of temperature when the reaction is exothermic and one wants to operate at a temperature for which the uncontrolled reactor is unstable.

Tubular Reactors

A second type of common industrial chemical reactor is the tubular reactor. In this type of reactor the reaction takes place as the reactants pass through one or more tubes (or similar vessels). When a heat exchanger is placed around the tubular reactor, it becomes possible to control the temperature profile along the direction of reactant flow, thereby achieving control of the reaction in a manner unobtainable in a continuous stirred tank reactor. The so-called fixed bed reactor is a form of tubular reactor. In this reactor gas or liquid reactants are put in contact with a bed of solid material that can either participate in the reaction (as in the reduction of iron ore and pyrites) or act as a catalyst to promote various reactions.

As the reactants move through the reactor, they react according to the principles described in the previous section, but because the extent of the reaction is related to the position of the material in the reactor, the system is now a distributed system, which must be described by partial, rather than ordinary, differential equations. This also applies to the generation (or consumption) of energy; that is, the energy balances must also be formulated as partial differential equations.

As a simple example of the description of a tubular reactor, consider the case of the reaction $A \rightarrow B$. For simplicity, also assume that the flow conditions in the reactor are such that the cross-sectional velocity profile in the reactor is constant; that is, plug flow exists. We can then formulate the following partial differential equations for concentration and temperature:

$$\frac{\partial c_A(\tau, z)}{\partial \tau} = f_1(c_A(\tau, z), \theta(\tau, z)) - \frac{\partial c_A(\tau, z)}{\partial z} \tag{3.14.29}$$

$$\frac{\partial \theta(\tau, z)}{\partial \tau} = f_2(c_A(\tau, z), \theta(\tau, z), h, \theta_0) - \frac{\partial \theta(\tau, z)}{\partial z} \tag{3.14.30}$$

where:

$$f_1(\cdot) = -k_0 \cdot e^{-\frac{E}{R\theta(\cdot)}} c_A(\cdot) T \tag{3.14.31}$$

$$f_2(\cdot) = k_0 \cdot e^{-\frac{E}{R\theta(\cdot)}} c_A(\cdot) T \frac{\Delta H}{\rho c_p} + \frac{TSh}{A\rho c_p}(\theta_0 - \theta(\cdot)) \tag{3.14.32}$$

and:

$\tau = t/T$ = dimensionless time
$T = L/v$ = reactor holdup time
v = transport velocity in reactor

L = length of reactor
$z = l/L$ = dimensionless axial coordinate
l = axial coordinate
$c_A(\tau, z)$ = concentration of component A as a function of time and position
$\theta(\tau, z)$ = absolute temperature in reactor as function of time and position
S = reactor surface area per unit length
h = average heat transfer coefficient
A = cross-sectional area of reactor
θ_0 = absolute temperature of environment (possibly a heat exchanger or refrigeration medium)

If the heat transfer coefficient can be used as a control variable, the value of h in the last term of equation (3.14.32) can be expressed as:

$$h = h_s\left(1 + \frac{h - h_s}{h_s}\right)h_s(1 - u) \tag{3.14.33}$$

where:

h_s = average heat transfer coefficient for a given steady-state flow of heating (or cooling) medium

$u = \dfrac{h - h_s}{h_s}$ = control variable, the relative change of heat transfer coefficient with respect to the steady-state value

The temperature and concentration of the feed to the reactor are boundary conditions for the differential equations (3.14.29) and (3.14.30).

One approach to the solution of the partial differential equations is to approximate them by discretizing the spatial variable z. (This procedure is detailed in the example developed in Appendix L.) For example, if we divide the length of the reactor into five discrete segments, each segment having two state variables (concentration and temperature), the total reactor model will contain ten state variables related by a system of ten coupled, first-order, nonlinear ordinary differential equations. It is sensible to segment the heat exchanger that is to control the temperature profile of the reactor such that it has the same segment boundaries as the segments of the reactor model. The value of u, which determines the value of h according to equation (3.14.33), is applied to the system as shown in Figure 3.14.4. The resulting control vector will have as many independent elements as there are segments in the heat exchanger.

This method of handling distributed systems has proved to be satisfactory in most cases, and many successful applications have been reported (Balchen, Fjeld, and Olsen, 1971; Ray, 1981).

Several of the control structures that have been applied and/or proposed for chemical reactors of various types will be discussed in section 5.8. The most important variable for the control of chemical reactors is the transfer of heat, either to or from the reactor. This is so because the behavior of the reaction de-

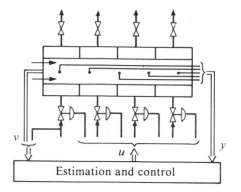

Figure 3.14.4. Tubular reactor with four individual control elements for control of the temperature profile.

pends upon the temperature; thus, the temperature profile in a tubular reactor, or the temperature–time history in a tank reactor, is important to the reaction behavior. Since it is difficult to measure the concentrations of the reaction products in a reactor, it is necessary to derive control actions from accessible quantities such as temperature. Furthermore, state estimation may possibly be applied, with slow updating based on analytical instrumentation.

3.15. OTHER PROCESSES

In this chapter we have considered many of the most common types of unit operations that are used as elements of larger, more complex industrial processes, but the discussion cannot be considered complete, as other important industrial unit operations have been omitted. However, most of them will be similar to those described here, so they can be approached with the same principles of model development as were used in this chapter. Specific examples of many other processes will be presented in Chapters 5 and 6.

Chapter 4
PROCESS CONTROL OF BASIC FUNCTIONS

In the following chapters we will use our knowledge of various process control principles and the behavior of elementary processes to examine how these processes and combinations of them can be equipped with automatic control. In this chapter we will examine the control of elementary functions such as level, pressure, flow, and energy. In Chapter 5 we will look at process control oriented toward unit operations and systems of relatively limited size. In Chapter 6 we will consider large industrial complexes.

4.1. LEVEL CONTROL

One of the most common control problems is control of the level of liquid in a tank. Although there are many reasons for level control, in terms of process dynamics and operation these stand out:

1. A fixed volume of a liquid may be maintained as a buffer, or reserve, to prevent shutdown of a continuous process in the event of failure of an upstream or downstream element in the process. There is no need for especially precise level control in this case. However it should be noted that maintaining a low level provides reduced reserve capacity for continuing production in the case of an upstream failure, and holding a high level provides reduced storage capacity for use in case of a downstream process failure.

2. Many unit processes function best when the volume of material in the process is constant. Typical examples are the bottom section and the accumulator at the top of a distillation column, the volume of solids in a ball mill, the level in a mixing tank, a batch chemical reactor, and so forth. These processes usually require that the levels be held within a few percent of the set point.

3. Level control may be used to smooth fluctuations in flows in cascade systems where the output of one unit is the input to the next. An example of this situation is the feed to a distillation column where, in order to have good operation of the column, the feed should not vary. However, the feed is often one of the products of a previous distillation column or another process that, if equipped with very sensitive level control (high gain, small proportional band) can provide wide flow variations. Here, a properly tuned level-controlled surge tank will damp the flow fluctuations and thereby improve operation of the downstream column.

Figure 4.1.1 shows several different structures for control of a tank.

The principle of differential pressure is used to measure the level in Figure 4.1.1a (LT), but other methods of measurement can also be used. The measurement signal is sent to a controller (LC) that operates a control valve in the discharge from the tank. The two feed streams, A and B, are uncontrolled. In Figure 4.1.1b the structure of the control is the same except that flow out of the tank is controlled by a flow controller (FC). This arrangement reduces sensitivity to variations in the forces opposing the flow that may result from changes in downstream process elements, changes in liquid viscosity, and so on. Figure 4.1.1c indicates how feedforward control from one of the feed streams, A, can be used to adjust the set point of the output flow controller so that the level controller need only make small adjustments. This solution is useful if one wishes to have very precise level control with high bandwidth, and when one of the feed streams is dominant with respect to the other. Figure 4.1.1d illustrates a layout in which one of the feed streams is controlled, and the output flow is allowed to run free. Feedforward could be added to this system as shown by the dashed line.

For most practical purposes, the solution shown in Figure 4.1.1b is equivalent to the solutions given by Figures 4.1.1a and 4.1.1d.

If the objective is to hold the level within narrow limits, a high gain is necessary in the level controller. However, the flow control system (or perhaps the

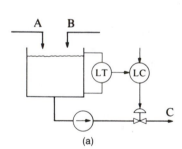
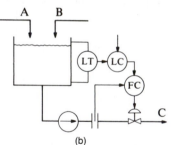

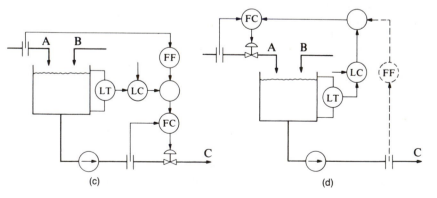

Figure 4.1.1. Different systems for level control in a tank.

control valve itself) has a limited bandwidth determined by a time constant (T_R), and since the process (in this case, the tank) is described by a pure integration of the net input flow to give the level, the level control loop will have a resonance [$|N(j\omega)|_{max} > 1$] if the controller gain is too high. This will be especially detrimental if the level control system includes much noise, such as that due to waves, bubbles, splashing, and so forth, in the tank. If the controller is a proportional controller, the gain in the loop that yields relative damping $\zeta = \sqrt{2}/2$ (no resonance) will produce a crossover frequency in the loop of:

$$\omega_c \approx 0.5 \frac{1}{T_R}$$

The transfer function between the flow variations into the tank and the level is:

$$\frac{h}{q_A}(s) = \frac{2T_R}{A} \cdot \frac{1 + T_R s}{1 + 2 \cdot \frac{\sqrt{2}}{2} \cdot \frac{s}{\omega_0} + \left(\frac{s}{\omega_0}\right)^2} \left(\frac{m}{m^3/s}\right) \qquad (4.1.1)$$

where:

$$\omega_0 = \frac{1}{\sqrt{2}} \frac{1}{T_R}$$

From equation (4.1.1) we see that the gain between the flow fluctuations and the level fluctuations is proportional to the time constant T_R and inversely proportional to the cross-sectional area A of the tank.

If precise control of the level is required, it may be necessary to use a proportional + integral (PI) controller. The loop will then have two integrations, which can cause difficulty if changes are so severe that either the valve or the measuring device saturates. In order to assure stability and an acceptable resonance characteristic, the ratio between the integral time of the controller and the time constant of the flow loop should be chosen to be approximately:

$$T_i/T_R = 10$$

The crossover frequency will then be:

$$\omega_c \approx 0.3/T_R$$

The transfer function between variations in flow into the tank and tank level is:

$$\frac{h}{q_A}(s) = \frac{\sqrt{10}\,T_R}{A} \frac{10T_R s}{1 + 10T_R s} \frac{1}{1 + 2\zeta\frac{s}{\omega_0} + \left(\frac{s}{\omega_0}\right)^2} \qquad (4.1.2)$$

We see that the level fluctuations will be about 50% greater at the middle frequencies, while there will be little response at low frequencies. The level response to a step change in flow will die out with a time constant of $10T_R$.

Many control valves suffer from hysteresis, usually due to dry friction in the packing around the valve stem. Its major effect is that when a signal is applied to the motor, the valve does not move either immediately or truly proportionally to the signal. In addition, it is unlikely to behave the same way in both directions of motion. This nonlinear behavior can be analyzed by the "describing function" technique, discussed in Balchen (1967) and Atherton (1982). The system may exhibit a cyclic behavior known as limit cycles. The amplitude of the cycle can be larger than the hysteresis width, and the frequency will be somewhat lower than the crossover frequency. If one needs integral control action in a flow control system, it is usually necessary to install a valve positioner on the valve to reduce hysteresis to a minimum. Certainly, there are exceptions to this rule, but these exceptions imply careful maintenance. The rule applies to most industrial applications.

Figure 4.1.2 illustrates the characteristic behavior of a control valve displaying both saturation and hysteresis due to friction. The describing function B_1 for this element is shown in Figure 4.1.3. Details of the construction of describing functions may be found in Balchen (1967).

When the frequency response $h_0(j\omega)$ of the other linear elements of a control loop is plotted with $|h_0|$ (dB) versus $\angle h_0$ together with the inverted describing function, $-1/B_1$, as shown in Figure 4.1.4, the occurrence of limit cycles (stable oscillations) or instability can be determined immediately. Two cases of proportional control (P control) and one case of PI control of the level in a tank are shown in Figure 4.1.4. It is assumed that the system contains both valve and measurement dynamics. The locus of the inverted describing function,

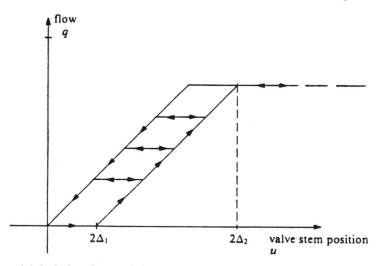

Figure 4.1.2. Valve characteristic showing saturation and hysteresis due to friction.

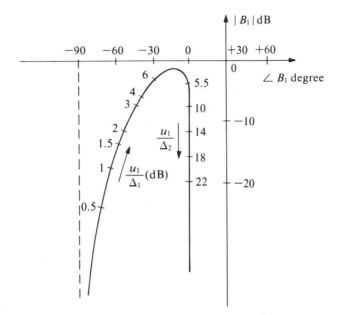

Figure 4.1.3. Describing function for control valve with saturation and hysteresis.

$-1/B_1$, plays the same role as the critical point (0 dB, 180°) in determining the stability of the system.

If the locus of $h_0(j\omega)$ crosses the locus of $-1/B_1$, the system will either be unstable (with increasing oscillations) or develop limit cycles (stable oscillations). In the case shown for the P controller with high gain, there will be a limit cycle having a frequency determined by the ω scale on the h_0 locus, and an amplitude u_1 determined by the u_1/Δ_1 scale on the $-1/B_1$ locus. The frequency of oscillation will be somewhat lower than the crossover frequency ω_c of the linear loop. The amplitude will be somewhat higher than the hysteresis $[(u_1/\Delta_1) \approx 4$ dB$]$ depending on the phase margin.

If the proportional gain is reduced as shown in Figure 4.1.4, the crossing between the two loci will move upward and to the right, indicating a reduction in frequency toward zero and a reduction in amplitude toward $(u_1/\Delta_1) = 0$ dB. This means that the valve will drift slowly within the hysteresis band.

If PI control is installed to remove steady-state deviation in the level, the limit cycle will become very pronounced. As shown in Figure 4.1.4, the amplitude and frequency will be relatively independent of changes in the gain. However, a new phenomenon related to saturation appears. The loci of $h_0(j\omega)$ and $-1/B_1$ approach each other near the vertical axis, indicating the possibility of large low frequency oscillations. This situation will occur if the system is given an initial upset sufficient to drive the valve to saturation. If the locus $h_0(j\omega)$ has the form shown by the dashed curve, the system will become totally unstable

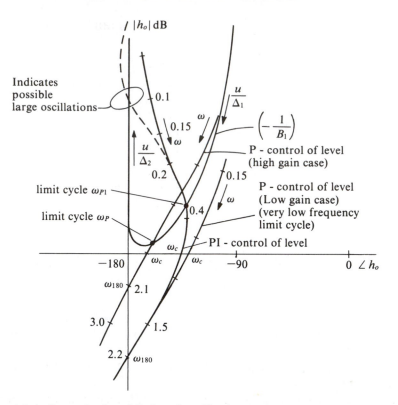

Figure 4.1.4. Determination of limit cycles of level control systems by means of the describing function technique.

when a large upset is introduced. However, this type of locus is not very likely to appear.

This method of analysis is very useful in many other cases of process control of systems exhibiting nonlinear behavior.

If the purpose of the system shown in Figure 4.1.1b is to smooth out fluctuations in the feed stream q_A in order to reduce variations in the product stream q_C, the level control system must be different from the one just described. Assuming that the level controller is a proportional controller and that the flow loop has, as before, a time constant T_R, the transfer function between variations in the feed flow and the outlet flow will be:

$$\frac{q_C}{q_A}(s) = \frac{h_0(s)}{1 + h_0(s)} = M(s) \tag{4.1.3}$$

where:

$$h_0(s) = \frac{K_p}{As(1 + T_R s)} = \text{loop transfer function} \tag{4.1.4}$$

If the proportional constant is small enough, the transfer functions for the system can be written:

$$\frac{q_C}{q_A}(s) \approx \frac{1}{\left(1 + \dfrac{A}{K_p}s\right)(1 + T_R s)} \tag{4.1.5}$$

and:

$$\frac{h}{q_A}(s) \approx \frac{1}{K_p} \cdot \frac{1}{1 + \dfrac{A}{K_p} \cdot s} \tag{4.1.6}$$

If the flow loop is fast, which implies a small T_R, then in order to obtain a certain damping factor M_A for a specific perturbation frequency ω_A, we must use a proportional constant given by:

$$K_p = A\omega_A M_A \tag{4.1.7}$$

which, substituted into equation (4.1.6), gives:

$$\frac{h}{q_A}(s) \approx \frac{1}{As} \tag{4.1.8}$$

Equation (4.1.8) shows that efficient damping at low-frequency flow perturbations leads to large level variations, as one should expect.

4.2. PRESSURE CONTROL

The pressure of a gas or vapor is one of the most common process variables subject to automatic control, either as an end in itself or as a means of controlling a more complicated system. The pressure of a liquid that is not in contact with a vapor or gas phase is also frequently controlled, commonly in hydraulic pump systems. Pressure in solids is difficult to identify in terms of the control problem, but changes in either tension or compression of solid construction materials such as steel are often suitable for measurement and for use as a step in the control of variables that are related to the strain in the material.

We will concentrate our discussion on pressure in a medium that has a significant steam or gas phase. Since a gas is simply a superheated vapor in which there is no liquid left, thermodynamic differences will distinguish between the phases.

The thermodynamic state of a vapor is completely defined by pressure, volume, and enthalpy. As an example, Figure 4.2.1 shows the enthalpy–pressure diagram for water and water vapor.

When a vapor is sufficiently superheated, its behavior will approach that of an ideal gas, which obeys the gas law:

$$pV = nRT \tag{4.2.1}$$

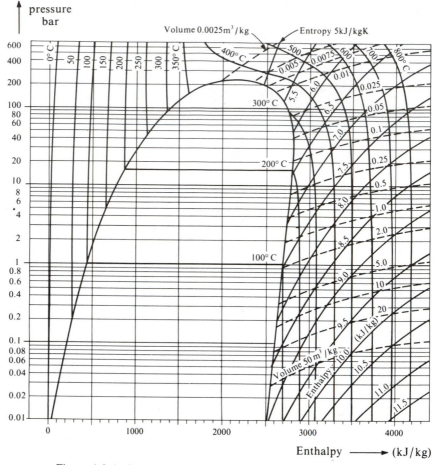

Figure 4.2.1. Enthalpy–pressure diagram for water/water vapor.

where:

p = pressure (bar)
V = volume (m^3)
n = number of moles of gas
R = universal gas constant ($8.3144 \cdot 10^{-5}$ m^3 bar/mole °K = 8.3144 J/mole °K)
T = absolute temperature (°K)

Under isothermal conditions (constant temperature), which occur when there is perfect heat exchange with the surrounding environment, the pressure will be inversely proportional to the volume. If the system is perfectly insulated, the conditions will be adiabatic; that is, no heat will be gained or lost by the system. In this case, the temperature, indicated on the right side of equation (4.2.1), will change. When the system is not perfectly insulated, there will be some heat exchange, and the system is said to be under polytropic conditions.

For simplicity, let us assume adiabatic conditions. In this case, a given amount of gas that undergoes a volumetric change will experience a related energy change, given by:

$$dW = -pdV \qquad (4.2.2)$$

For the case of an ideal gas, we can also express the energy change for a corresponding temperature change as:

$$dW = nc_V dT \qquad (4.2.3)$$

where c_V is the specific heat at constant volume. If we solve equation (4.2.1) with respect to p, put the result into equation (4.2.2), and use equation (4.2.3), we find that:

$$\frac{dV}{V} = -\frac{c_V}{R}\frac{dT}{T} \qquad (4.2.4)$$

Integrating both sides of equation (4.2.4), we find that:

$$\ln\frac{V_2}{V_1} = -\frac{c_V}{R}\ln\cdot\frac{T_2}{T_1} \qquad (4.2.5)$$

or, if we prefer:

$$\frac{T_2}{T_1} = \left(\frac{V_2}{V_1}\right)^{-R/c_V} \qquad (4.2.6)$$

The relation between the specific heat at constant volume, c_V, and the specific heat at constant pressure, c_p, is given by:

$$c_p = c_V + R \qquad (\text{J/mole} \cdot \text{°K}) \qquad (4.2.7)$$

and the ratio between them is:

$$k = c_p/c_V$$

Substituting into equation (4.2.6) gives:

$$\frac{T_2}{T_1} = \left(\frac{V_2}{V_1}\right)^{1-k} \qquad (4.2.8)$$

or:

$$TV^{k-1} = \text{constant} \qquad (4.2.9)$$

Using equation (4.2.1), we can also find, for the adiabatic case:

$$\frac{T_2}{T_1} = \left(\frac{p_2}{p_1}\right)^{(k-1)/k}$$

(4.2.10)

or:

$$Tp^{1-k} = \text{constant}$$

(4.2.11)

and:

$$\frac{p_2}{p_1} = \left(\frac{V_1}{V_2}\right)^k$$

(4.2.12)

or:

$$pV^k = \text{constant}$$

(4.2.13)

Equation (4.2.13) can be compared with equation (4.2.1) for the isothermal case, which expresses:

$$pV = \text{constant}$$

(4.2.14)

If we add a gas to a tank that has a constant volume, and if we hold the temperature constant, equation (4.2.1) can be applied to yield:

$$\frac{dp}{dt} = \frac{dn}{dt} \cdot \frac{RT}{V}$$

(4.2.15)

which says that the pressure changes at a rate that is proportional to the net flow of the gas, measured in moles, into the tank; proportional to the temperature; and inversely proportional to the volume of the tank.

Let us introduce:

$$dn/dt = w_1 - w_2$$

(4.2.16)

where:

w_1 = mass flow into the tank (mole/s)
w_2 = mass flow out of the tank (mole/s)
We now can write:

$$dp/dt = [RT/V](w_1 - w_2)$$

(4.2.17)

Therefore, the pressure is characterized by an integration from the mass flow to pressure.

The pressure in the tank can be controlled by manipulating the mass flowrate of either the feed stream or the output stream. The characteristics of the final control element (control valve, pump, compressor, etc.) determine whether the pressure control system will have difficulties in terms of dynamics.

Other important process-dynamics characteristics of gases and vapors are discussed with respect to automatic control in section 3.4 and in Appendix F.

In the previous discussion, we examined the pressure behavior of a gas from the point of view of lumped parameters, and ignored the force that is required to accelerate the mass of gas as it flows through a pipe. This approach is not acceptable for the study of flow in long pipes, where one must treat the process as a distributed system, which is similar in many respects to the propagation of transmission waves in electric transmission lines. Campbell (1958) has developed equations that describe changes of pressure and mass flows along a pipe under various conditions. Using the transfer functions found in Campbell, one can get an overview of what the dynamic behavior will be for pressure control in a long pipe.

A pipe section is "long" if it is longer than the wavelength of sound generated in the gas as the result of a pressure change. The wavelength is given by:

$$\lambda = 2\pi \frac{v_\ell}{\omega} \tag{4.2.18}$$

where:

v_ℓ = velocity of sound (m/s)
ω = frequency of the pressure change (rad/s)
The velocity of sound in an ideal gas is given by:

$$v_\ell = v_{\ell o} \sqrt{\frac{T}{293}} \tag{4.2.19}$$

where $v_{\ell o}$ is the velocity of sound (m/s) at 20°C.

Pressure waves can be reflected in a long pipe if it is not terminated by an acoustic impedance that is approximately equal to the characteristic impedance z_o of the pipe. If there are no reflections, it is easy to achieve control of pressure in the pipe by means of a restriction (control valve), as shown in Figure 4.2.2a or 4.2.2b. The choice between these two solutions depends upon the characteristics of the pressure source and the load.

If the pressure and/or temperature is such that the gas is saturated (the condition at which condensate just forms), the gas will not behave as an ideal gas. We can see from Figure 4.2.1 that there is a region (the saturation region) in which the temperature is constant when the pressure is constant. This region lies between the condition at which all the vapor is condensed and the condition where all the liquid has been vaporized. In this area the isotherms are horizontal lines. Along such a line, the enthalpy changes, but nothing else. If more heat is added to the system, the vapor becomes superheated, and its behavior will approach that of an ideal gas. In the saturation region it is possi-

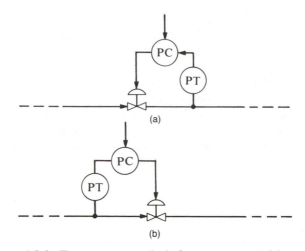

Figure 4.2.2. Two common methods for pressure control in a pipe.

ble to use pressure control to arrive at a temperature equal to the boiling point of the liquid. This phenomenon can be used to separate different condensed gases in a mixture by reducing the pressure stepwise in order to boil off the various components at their given boiling temperatures. A process frequently used in the petroleum industry is illustrated in Figure 4.2.3.

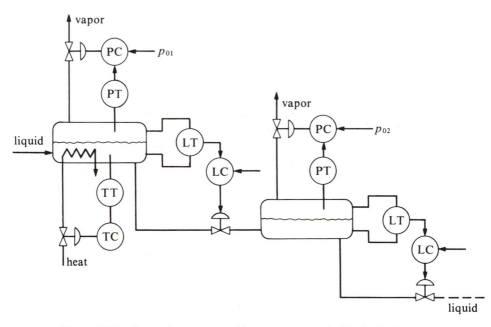

Figure 4.2.3. Separation process with pressure control of individual stages.

Here, the liquid entering at the left side of the system in Figure 4.2.3 is oil, which, at high pressure, contains certain volatile components that must be removed. If we assume that the temperatures in the different stages are the same, we will have a stepwise reduction in pressure from left to right. We see that this is managed by pressure control with the release of vapors of the various components, starting with the most volatile and ending with the least volatile component. Level control of the liquid in each separator tank is achieved by controlling the liquid flow out of each tank stage.

Because the boiling point temperature is dependent on the pressure (or vice versa), one must either choose the pressure reference as a function of the existing temperature, or else be sure that the temperature remains constant by means of addition or removal of energy, as shown for the case of the first tank in Figure 4.2.3.

The volatile components from each separator tank must be recovered and stored after condensation or, perhaps, compression followed by condensation.

Reduction of pressure is also used in stagewise steam heating processes in order to "flash" condensate at a lower pressure to remove further heat for process use. A common application of this principle is in the staged, sequential, series drying system on a paper-making machine.

Let us examine a pressure control system. Assume that the process has a pure integration as given by equation (4.2.17). Here, the element that is used to manipulate the gas flow will determine whether or not there will be a control problem. If the flow of the feed gas stream, w_1, comes from a pump or compressor that is driven by a motor or turbine having a controllable speed, and we wish to accomplish pressure control in the tank after the pump, then a control problem may result. In the scheme shown in Figure 4.2.4, the speed of the motor on the pump is controlled first (SC) and the flow of the feed gas, w_1, is controlled by the flow controller (FC) through the set point of the speed controller. The set point for the flow controller is made up of a combination of feed

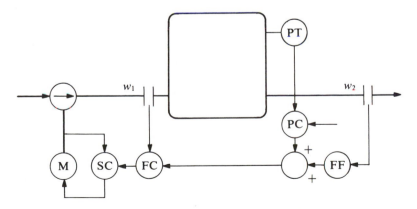

Figure 4.2.4. A system for pressure control in a tank.

forward from the output gas flow, w_2, and feedback from the pressure controller (PC). Whether all these control functions are necessary will depend upon the actual process, which in this case could be, for example, a chemical reactor that is to operate at constant pressure.

4.3. CONTROL OF VOLUME FLOW AND MASS FLOW

Many flow control schemes are used in modern industrial applications. They are influenced by the nature of the flowing material, which might be gas, liquid, suspended solids, dry solids, or virtually any combination of materials. The control scheme is also influenced by the type of final control element to be used. This element may be a control valve that restricts a pipe, or pump (either a centrifugal pump or a gear pump) with varying rotation speed, a restriction on the pressure side, or a variable bypass rate. Finally, the behavior of a flow control system depends upon the means to measure the flow.

In this section we will consider flow control systems in which the final control element is a control valve. We will consider control of compressors and pumps in section 5.7.

The most common difference in the control of liquids and gases is due to the difference in bandwidth of the process itself. This difference is related to the velocity of the medium in response to a change in valve opening. When the medium is a vapor or gas, the bandwidth is usually so high that it can be considered to be instantaneous compared with the other elements in the loop such as the measuring element, the controller, and the final control element. However, if the medium is a liquid, the mass effect (inertia) may cause a relatively slow reaction of the stream to changes in the valve opening. As a result, the bandwidth of the reaction may be comparable to the bandwidths of other elements in the control loop.

The situation becomes more complicated if the liquid is partially compressible, as it would be if it contained entrained gas. Then, in addition to the friction effect due to pipe resistance and the inertia due to the mass of the liquid, there will be a capacitive effect due to compressibility. As a result, the piping system will behave analogously to an electric circuit containing resistance, capacitance, and inductive elements, and will be likely to be described by transfer functions with complex poles. However, if the compressibility is small (the gas volume is very small compared with the liquid volume), the two poles will be real and distinct. In the limit, the pole determined by the mass and the friction loss will dominate.

Assume that the system shown in Figure 4.3.1 is filled with liquid, the length of the pipe is L, and the pressure drop over the length L is Δp. This will produce a nonlinear differential equation that describes the volumetric flow (q) as:

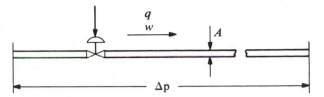

Figure 4.3.1. Pipe with control valve.

$$\dot{q} = -\left(\frac{f}{2Ar_h} + \frac{A}{Lg(u)^2C_V}\right)q^2 + \frac{A}{L\rho}\Delta p$$

$$= -h(u)q^2 + \frac{A}{L\rho}\Delta p \qquad (4.3.1)$$

where:

f = Fanning friction factor (Figure 4.3.2)
A = cross-sectional area of the pipe (m^2)
r_h = hydraulic radius of the pipe (m)
L = length of the pipe (m)
u = control signal to the control valve
$g(u)$ = valve characteristic
C_V = valve coefficient (m^2)
Δp = total pressure drop (N/m^2)
ρ = liquid density (kg/m^3)

Equation (4.3.1) can be used directly when Δp varies, as is the case when the system uses a variable-speed pump or compressor. However, if the control valve opening is changed and the pressure drop Δp in the system is assumed to remain constant, it becomes advantageous to linearize equation (4.3.1). For the linearized system, the time constant between the change in valve opening and the change in volumetric flow is:

$$T_w = \frac{1}{\left.\dfrac{\partial(h(u)q^2)}{\partial q}\right|_{q=\bar{q}}} = \frac{1}{2h(u)\bar{q}} = \frac{L\rho\bar{q}}{2A\,\overline{\Delta p}} = \frac{L\bar{w}}{2A\,\overline{\Delta p}} \qquad (4.3.2)$$

where:

$\bar{q}$ = average volumetric flow (m^3/s) obtained at $\dot{q} = 0$
$\bar{w}$ = average mass flow (kg/s)
$\overline{\Delta p}$ = average pressure drop across the system (N/m^2)
T_w = time constant (s)

We see that the time constant is proportional to the length and the average mass

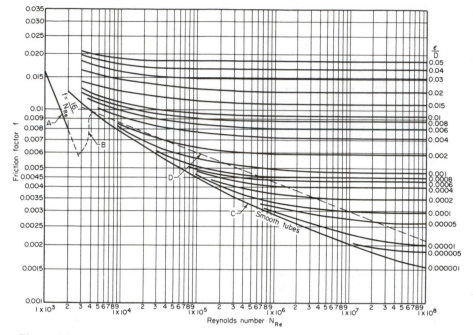

Figure 4.3.2. Fanning friction factor as function of pipe roughness. (Based on Moody, 1944. Trans. Am. Soc. Mech. Engr., *66*, 671.)

flow, and inversely proportional to the pressure drop and the cross-sectional area of the pipe.

The most common arrangement for liquid flow control is shown in Figure 4.3.3. In addition to the dynamics of the response of the process itself (the response of the liquid stream following a change in valve opening), there will be dynamics in the measuring element, the signal transmission system between the measuring device and the controller, the controller, the signal transmission between the controller and the valve, and the valve motor. In pneumatic control systems, which are widespread in industry because of either tradition, cost, or

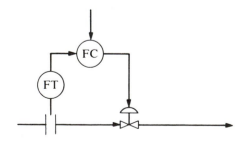

Figure 4.3.3. The most common arrangement for flow control.

the need for explosion-proof instrumentation, controllers were often placed in control rooms at some distance from the control valves and measuring elements. This required long pneumatic signal transmission lines whose dynamics could completely dominate the dynamics of other elements. A major advantage of electronic instrumentation is the elimination of these signal transmission lags even when the control room is quite remote from the measuring and final control elements.

Flow is often measured as a function of the pressure drop across an orifice, venturi, or other fixed restriction in the line. The differential pressure across a restriction is proportional to the square of the mass flowrate. If one wishes to design a flow control system that operates on the flow rather than the square of the flow, the signal from the differential pressure device must be converted by a square root extractor. Modern devices, including pneumatic types, can extract the square root very rapidly; thus, they contribute virtually no dynamic characteristic to the system. Therefore, in a modern flow control system, the major dynamic elements are the control valve, the controller, and the process itself.

A pneumatic diaphragm valve motor with an electro-pneumatic valve positioner (which assists the diaphragm motor by feeding back a positioning signal from the actual position of the valve stem) can, with reasonable accuracy, be treated as a first-order system with a single time constant T_v. The value of the time constant is determined by the volumetric air capacity of the diaphragm chamber and the rate at which the electro-pneumatic positioner can deliver air to the diaphragm chamber.

From the preceding discussion, we see that the portion of the system from the output of the controller to the measured flow will be second-order. There will be two time constants, one for the valve (T_v), and one due to the inertial effect of the mass of the liquid flow (T_w).

Precise flow control requires a PI controller. Derivative control action should be avoided because of the inevitable presence of noise generated by differential pressure measurement in a line with turbulent flow. The integral time in the control loop will be chosen to be approximately the same as the largest time constant in the system. If the inertial effect in the pipe is small, the time constant of the valve motor will dominate. Since this time constant is usually relatively constant, the integral time can be set to a constant. The controller is then tuned by adjusting the proportional constant rather than the integral time. This can proceed until the crossover frequency of the system becomes:

$$\omega_c \approx 1/T_w$$

If the inertial effect dominates the system, then one must choose the integral time for the controller such that $T_i = T_w > T_v$. The bandwidth for the system will then be:

$$\omega_c \approx 1/T_v$$

Control valves are designed to generate various functional relationships between valve opening and flow. A semilogarithmic or equal percentage characteristic is often used in order to provide good control for both large and small flowrates. This characteristic, shown in Figure 4.3.4, is described by:

$$q = R^{u-1} \qquad (4.3.3)$$

where:

R = ratio between the maximum and minimum average flow
$0 < u < 1$ is the normalized stem position
$0 < q < 1$ is the normalized liquid flow

We see for this characteristic that when $R = 50$ and $u = 0$, $q = 0.02$.
   The gain of a semilogarithmic valve is:

$$dq/du = q \ln R \qquad (4.3.4)$$

and is therefore proportional to the flowrate. For $R = 50$, we have:

$$dq/du = 3.9q$$

The characteristic curve of Figure 4.3.4 applies when the pressure drop across the valve is constant. The pressure drop will not be constant in the system shown in Figure 4.3.3 if the pipe is long or if it discharges into a process that presents resistance to the flow. If the minimum pressure drop across the valve, Δp_{min}, is small relative to the maximum drop in the system, Δp_{max}, the semilogarithmic valve characteristic will approach linear behavior over much of the valve stem travel. This is shown in Figure 4.3.5, where we see a quite linear

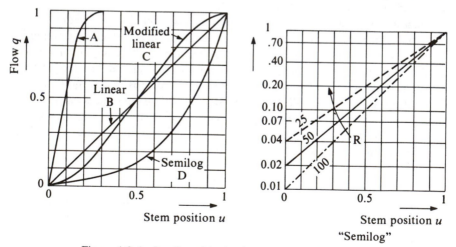

Figure 4.3.4. Semilogarithmic characteristic valve curve.

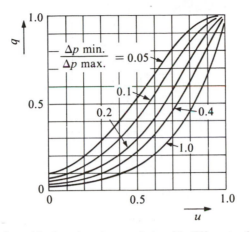

Figure 4.3.5. Semilogarithmic valve characteristic with different valve pressure drops.

curve for the case when the pressure drop across the valve is only 5% of the total pressure drop in the system. When all of the drop occurs at the valve, we can see from the figure that the characteristic is semilogarithmic.

Another important application of the semilogarithmic valve characteristic is in control loops where the process gain is *inversely proportional* to the flow in the process. This is very often the case. In such cases the property of the valve that the gain is *proportional* to the flow means that the total loop gain will be *independent* of the flow.

Flow control loops are frequently cascaded with pressure, temperature, concentration, or consistency control loops. The advantage of this is that local disturbances such as variations in the driving force (or pressure) in a pipe, variations in viscosity, and so on, can be eliminated as and where they occur, and prevented from appearing downstream in the process.

When ratio control is used to maintain a desired stoichiometric ratio between two streams in a mixture, it is important to maintain precise control of the stream flowrates. However, in such a case it is not the actual volumetric flowrate that is important; it is the "concentration" (species, or component in the stream) flow that matters. This is, of course, the volumetric flow multiplied by the component concentration in that flow. A control scheme for such a system is shown in Figure 4.3.6. Here we see that the concentration measurement is multiplied by the linearized flowrate measurement, and the product is combined with the set point in the controller to provide the signal for the control valve.

Frequently the device that makes the concentration measurement will have a significantly slower time response than the flow measuring device, but the concentration will in fact usually change much more slowly than the flowrate. When this is the case, the system shown in Figure 4.3.6 will be satisfactory.

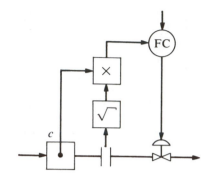

Figure 4.3.6. Control system for a single component in a stream.

It is often necessary to distinguish between *volume* flow and *mass* flow. Many flow measuring devices work on principles that provide a signal as a function of volumetric flow (m³/s). This applies, for example, to displacement meters, turbine meters, and magnetic meters. Other devices are based on principles that require knowledge of the density of the material in the stream. Typical of the latter are pressure-drop devices such as orifices and venturi meters. For the latter:

$$w = \rho_1 \cdot q_1 = K\sqrt{\Delta p \cdot \rho_1} \tag{4.3.5}$$

where:

w = mass flow rate (kg/s)
q_1 = volume flow upstream of the measurement (m³/s)
ρ_1 = density of the medium upstream of the measurement (kg/m³)
Δp = pressure drop across the measuring device (N/m²)
K = a constant

Equation (4.3.5) shows that it is necessary to know the density of the medium in order to measure either mass flow or volumetric flow. If the density can vary, it is necessary to measure it and either multiply it by the differential pressure (for measurement of mass flow) or divide the differential pressure by the density (for volumetric flow) before taking the square root. This is shown in Figure 4.3.7 for a mass flow control system.

For control of a thermal process such as a heat exchanger, flow control is often cascaded with temperature control. If the control valve is located on the primary side of the heat exchanger, it is of less interest to control the flow of the primary medium than it is to control the heat flow, which is the product of the specific heat, mass flow, and temperature. With this product, which can be generated as described above, good control of the energy flow in the heat exchanger can be obtained. This control scheme is illustrated in Figure 4.3.8.

In addition to cascade operation, flow control systems are often integrated with other control systems, such as concentration control, for example. There

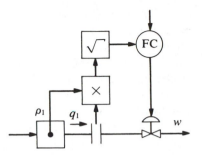

Figure 4.3.7. System for controlling mass flow.

will then be significant interaction that will affect the dynamics of both systems. Examples of this were discussed in section 2.8.

Control of the flow of pulverized dry solids has many applications in industrial processes. A common objective of such systems is to meter a solid stream at a given ratio to the mass flow of some other stream, perhaps a liquid, in order to obtain a mixture of uniform quality. One flow control system that can be used for pulverized solids and liquid was discussed in detail in section 2.7 (see Figure 2.7.3). It calculates the mass flow with the help of an estimator, which is updated by the weight of the contents of the silo containing the material to be metered. This method can give very precise mass flow control.

Use of an estimator includes a model of the flow process that must be updated by appropriate measurements. This is probably the best technique for mass flow control in cases where there is no simple applicable measuring instrument such as one has for the measurement of liquids.

4.4. CONTROL OF ENERGY: TEMPERATURE AND ENTHALPY CONTROL

Many of the unit processes discussed in Chapter 3 involve the transfer of various forms of heat energy, to heat solids, liquids, or gases or to evaporate

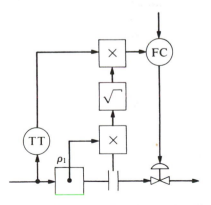

Figure 4.3.8. System for energy flow control in a liquid stream.

liquids or condense vapors. In that chapter, mathematical models were developed to show, either in the state space or as transfer function matrices, how the system reacts to various control variables.

In Chapter 5, we will consider in detail the control of various thermal processes such as heat exchange, evaporation, drying, boiling, and so forth. Therefore, in this section we will present only some elementary introductory observations regarding heat control.

In heat energy processes, the energy state is controlled by manipulating the addition or removal of energy from the system. The energy state of the process can be characterized by temperature for either solids or liquids. For superheated steam and for cooled condensate (liquid water), the energy state is generally given by the temperature with little dependence on pressure. However, for a saturated vapor, or a mixture of vapor and liquid, the temperature and the relative amounts of liquid and vapor must be known to account for the heat of evaporation.

The enthalpy of a material is always stated relative to an arbitrary reference value, so the *change* in enthalpy is of interest when a control system is to be designed to maintain a constant energy state.

Addition and removal of energy can take place either as heat conduction or radiation or by the addition or removal of mass with its enthalpy. Saturated steam or the saturated vapor of liquids other than water is often used as an energy source. In a process such as a distillation column, the medium for heat transfer may be a process liquid that contains energy, but in most processes water and steam are used. In Figure 4.2.1, we see clearly that at a given pressure, the saturated steam will have a certain temperature, and the energy that can be delivered per unit mass (the enthalpy change), during condensation in a heat exchanger, for example, is given by the specific heat of vaporization multiplied by the mass flow. The specific heat of vaporization depends on the pressure and is largest at low pressure and zero at the critical pressure. In order to control the energy flow, it is best first to control the steam pressure. This is preferred over temperature control because it is faster and cheaper. Obviously, when the pressure of saturated steam is reduced, the steam becomes slightly superheated. However, as can be seen in Figure 4.2.1, the enthalpy increment of the superheated steam is small compared with the heat of vaporization. The mass flow is then controlled as shown in Figure 4.4.1. Mass flow multiplied by the enthalpy change gives the energy flow. The mass flow of steam or a gas can be measured by the pressure drop over a restriction (such as an orifice), and, for an ideal gas, it is:

$$w = K_1 \sqrt{\Delta p \cdot \frac{p_1}{T_1}} \qquad (4.4.1)$$

where:

p_1 = pressure upstream of restriction (N/m^2)
T_1 = temperature upstream of restriction (°K)

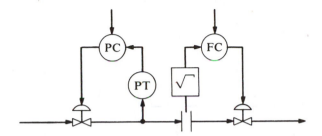

Figure 4.4.1. System for control of energy flow in saturated steam.

Δp = pressure drop across restriction (N/m^2)
w = mass flow (kg/s)
K_1 = constant

We have already applied pressure control to the saturated steam, so equation (4.4.1) reduces to:

$$w = K_2 \cdot \sqrt{\Delta p} \qquad (4.4.2)$$

where K_2 is a constant. In order to properly calibrate the system, numerical values for K_2 must first be determined on the basis of the type of steam (or saturated vapor of a liquid other than water) and the parameters of the measuring system.

Once the above relationships are known, the amount of heat delivered per unit time when the steam is completely condensed will be proportional to the set point of the flow controller. This solution is quite satisfactory, and is widely used in industrial processes.

When one can control the input of heat to a process or unit operation, it is also useful to measure the heat leaving the system if possible. One can use this information to automatically adjust the set point (in a feedforward sense) of the heat supply to the system, with the result that control of the temperature of the process, which is a measure of its energy state, will require only small adjustments. A measure of the heat leaving a heat exchanger is the product of the mass flow and the temperature difference between the input and output of the secondary medium. Figure 4.4.2 shows a heat exchanger where the primary medium is saturated steam, and the secondary medium is a liquid that can have large variations in both flowrate and input temperature. The system shown in Figure 4.4.1 is used for control of the steam flow, and the set point for the flow rate is determined by the amount of heat leaving the system, computed as stated above. The feedforward action is updated by a temperature control loop that processes the output temperature by means of an ordinary temperature controller. The system shown in Figure 4.4.2 is a realization of the general scheme discussed in section 3.8. It is relatively complicated, but it can be simplified in various ways if one is willing to accept poorer performance under some operating conditions. The complexity one would use depends on the relative values of various levels of performance.

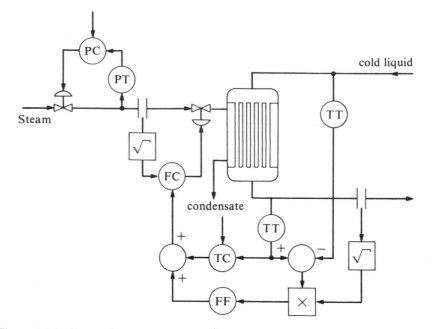

Figure 4.4.2. System for control of heat exchanger based on balancing of input and output energy flows.

Temperature control is a common means of maintaining the desired operating conditions in a reactor. A typical situation is one in which a chemical reactor must be held at a certain temperature in order to control the reaction. In a stirred tank reactor with good mixing, there will be one temperature that must be controlled, as in the systems shown in Figures 3.14.1 and 3.14.2. In a tubular reactor, a temperature profile along the length of the reactor must be established and maintained by means of individual temperature controllers on a number of zones, as was seen in Figure 3.14.4. These individual temperature control systems have strong mutual coupling that must be considered in the design of the system.

In an exothermic reactor it is possible that the heat due to the reaction can increase faster than the natural removal of heat, as shown in Figure 3.14.2. This requires that automatic temperature control have such a strong negative feedback that one gets a characteristic and stable equilibrium point as shown by the dashed line in Figure 3.14.2. In terms of dynamics, a pole (eigenvalue) of the linearized system that lies in the right half-plane must be moved to the left half-plane in order to achieve stability.

To illustrate this problem, let us consider the simple reactor illustrated in Figure 3.14.1 and described by the differential equations (3.14.7) and (3.14.8). Linearization around a singular point (c_{As}, θ_s) gives the following differential equations:

$$\frac{d}{dt}\begin{bmatrix} \delta c_A \\ \delta \theta \end{bmatrix} = A \begin{bmatrix} \delta c_A \\ \delta \theta \end{bmatrix} + \begin{bmatrix} 0 \\ \dfrac{1}{\rho c V} \end{bmatrix} Q_0 + \begin{bmatrix} \dfrac{1}{T} \\ 0 \end{bmatrix} c_{A0} + \begin{bmatrix} 0 \\ \dfrac{1}{T} \end{bmatrix} \theta_0 \qquad (4.4.3)$$

where matrix A is the Jacobian matrix derived by taking the partial derivatives of the right sides of equations (3.14.7) and (3.14.8) with respect to c_A and θ. We find:

$$A = \begin{bmatrix} -\dfrac{1}{T} - k(\theta), & -c_{As} \dfrac{\partial k(\theta)}{\partial \theta} \\[2ex] -\dfrac{\Delta H}{\rho c V} k(\theta) \ , & -\dfrac{1}{T} - \dfrac{\Delta H c_{As}}{\rho c V} \cdot \dfrac{\partial k(\theta)}{\partial \theta} \end{bmatrix} \qquad (4.4.4)$$

The eigenvalues of matrix A can be found from the characteristic equation:

$$|\lambda I - A| = \lambda^2 + \lambda \left(\frac{2}{T} + k(\theta) + \frac{\Delta H c_{As}}{\rho c V} \cdot \frac{\partial k(\theta)}{\partial \theta} \right) + \frac{1}{T} \left(\frac{1}{T} + k(\theta) + \frac{\Delta H c_{As}}{\rho c V} \cdot \frac{\partial k(\theta)}{\partial \theta} \right)$$

$$= (\lambda - \lambda_a)(\lambda - \lambda_b) = 0 \qquad (4.4.5)$$

where λ_a and λ_b are eigenvalues. If the process is strongly exothermic, $\lambda_a < 0$ and $\lambda_b > 0$; that is, there will be one eigenvalue in the left half-plane and one in the right half-plane. The transfer function between $Q_0(s)$, which is the heat removed by cooling, and $\delta\theta(s)$, which is the change in reactor temperature, is:

$$\frac{\delta \theta}{\delta Q_0}(s) = \frac{1}{\rho c V} \cdot \frac{s + \dfrac{1}{T} + k(\theta_s)}{(s - \lambda_a)(s - \lambda_b)} \qquad (4.4.6)$$

This can be rearranged to give:

$$\frac{\delta \theta}{\delta Q_0}(s) = \frac{K \cdot (1 + T_1 s)}{(1 + T_a s)(-1 + T_b s)} \qquad (4.4.7)$$

In addition to a controller, a control system such as that shown in Figure 3.14.1 must include a heat exchanger as a means of removing the heat generated by the reaction. At the very least this exchanger will have a first-order transfer function. We assume it will have a time constant T_v. Figure 4.4.3 is a Bode diagram for the temperature control loop for the case where the controller is a simple proportional controller. The dashed line shows the result when a PI controller is used. With only proportional control, the condition for stability, with respect to the controller gain, becomes:

$$K_p > (-\Delta H) c_{As} \cdot \frac{\partial k(\theta)}{\partial \theta} \cdot \frac{1}{1 + Tk(\theta_s)} - \frac{\rho c V}{T} \qquad (4.4.8)$$

(Note that ΔH is negative for an exothermic process.)

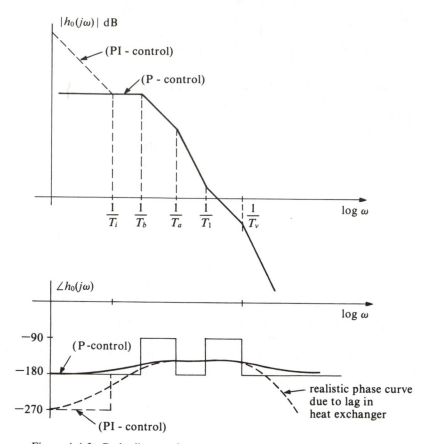

Figure 4.4.3. Bode diagram for temperature control of exothermic reactor.

If, for simplicity, we assume that $T_a \approx T_1$, we can see from Figure 4.4.3 that a requirement for stability is that $T_v < T_b$. This means that heat must be removed from the reactor faster than it is generated by the exothermic reaction.

Because it is necessary to use a PI controller in order to get good temperature control, the integral time T_i must be chosen large enough to provide a satisfactory phase margin at the crossover frequency, which must fall between $1/T_b$ and $1/T_v$.

It is often difficult to measure the temperature of a material that is to be temperature-controlled. This is the case, for example, when one would like to control the interior temperature of a steel block by changing the surface temperature. The problem also occurs when it is difficult to maintain the measuring element in a hostile environment such as the highly corrosive electrolyte in an aluminum electrolysis cell. It is then feasible to use a mathematical model for calculation of the temperature rather than direct measurement. Such a model must be updated by measurements of some sort, and these measurements must have one or more known relationships to temperature that provide observability sufficient to make the estimate converge. If there are not sufficient measurements in the process, the estimator will be ballistic; that is, one must rely on the model without updating. Even this can lead to remarkable results. An example of this occurs in the "blowing" of oxygen in an LD converter for the production of steel, where the temperature in the converter is difficult to measure continuously. This estimator must be based on calculation of the energy balance.

4.5. CONTROL OF CONCENTRATION ON THE BASIS OF pH MEASUREMENT

In many industrial processes it is necessary to meter one or more materials to obtain a mixture that has a particular concentration with respect to one or more components. The concentration of hydrogen ions, which defines the acidity of a solution, is a frequently controlled characteristic. The measure of acidity, pH, is defined as:

$$pH = -\log(H^+)$$

where H^+ is the hydrogen ion concentration (moles/l). Water that has a hydrogen ion concentration of 10^{-7} moles/l and a hydroxyl ion concentration of 10^{-7} mole/l is said to have a pH of 7. An acid has an excess of hydrogen ions and a pH < 7. A base has an excess of hydroxyl ions and a pH > 7.

pH is measured by means of a special type of electrode that generates a voltage as a function of the hydrogen ion concentration in the liquid. Because pH is defined as a logarithmic function, the relationship between the concentrations of various components and the pH measurement is highly nonlinear. This can be illustrated by the example of a mixture of hydrochloric acid (HCl) and sodium hydroxide (NaOH) in water at concentrations of x_A and x_B respectively. Hydrochloric acid is a strong acid in that it readily gives up its hydrogen atoms to water. Similarly, sodium hydroxide is a strong base because it readily "grabs" hydrogen atoms. Acids that are not completely ionized influence pH less than strong acids. The relationship of the difference between x_A and x_B, the concentrations of a strong acid and a strong base, and pH is shown in Table 4.5.1. The table is given in graphic form in Figure 4.5.1, where we can easily see the tremendous change that occurs at about pH = 7 for this mixture.

This change generates a very large process gain in a loop that measures pH and controls either the acid or the base. The extent of the control problem depends on the strength of the acid and the base. Neutralization, or modification to any desired pH value, is easier to accomplish if one controls the flow of the weaker component, either acid or base. This is so because the flow of the weaker component will be greater than that of the stronger, and therefore easier to control with higher precision and less severe stability problems than might result from the very high gain characteristic of the mixture itself.

Table 4.5.1 Relation between pH and concentration difference $x_A - x_B$ of HCl and NaOH.

pH	$x_A - x_B$
0	1
2	10^{-2}
4	10^{-4}
6	10^{-6}
7	0
8	-10^{-6}
10	-10^{-4}
12	-10^{-2}
14	-1

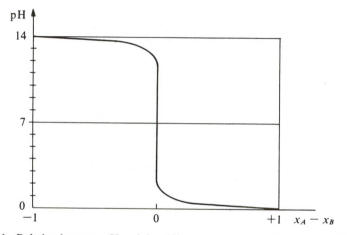

Figure 4.5.1. Relation between pH and the difference in concentration between HCl and NaOH control systems.

Figure 4.5.2 illustrates how the pH of different acids changes when mixed with a flow of 1.28 N caustic (NaOH). We see that the slope when the weak citric acid is being neutralized is considerably less than is the case for the strong sulfuric acid.

In systems where a strong acid is to be neutralized by a strong base, it is advisable to arrange the process in at least three stages in order to gradually raise the pH to neutrality (pH = 7). An example of such a system is shown in Figures 4.5.3a and 4.5.3b (McMillan, 1984). Figure 4.5.3a shows a system that cannot work because it has only one stage to achieve pH control. The same system, modified, is shown in Figure 4.5.3b. First, a relatively fast pH loop (pHIC-1) controls the first mixing stage to a low pH value. This loop may tend to oscillate at a rather high frequency, but the effect of this oscillation is filtered by the sump. The pH is measured, and the second pH controller adjusts the set point of pHIC-1. Finally, the pH is increased to 7 in the last stage, which includes a relatively large mixing tank. The pH controller in this loop (pHIC-3)

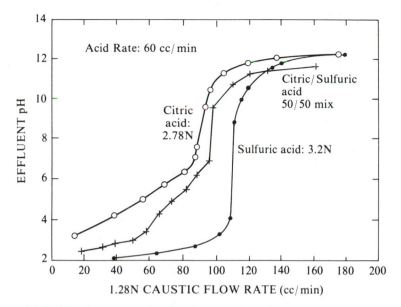

Figure 4.5.2. Titration curves showing how process gains (the slopes of the curves) change with pH set point and with effluent composition. (From Piovoso and Williams, 1985.)

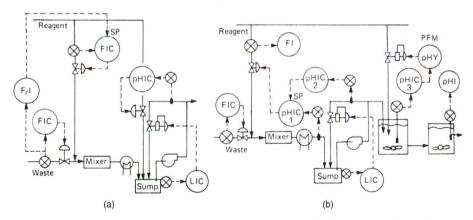

Figure 4.5.3. Unsuccessful (a) and successful (b) schemes for the control of pH. (From McMillan, 1984.)

contains a nonlinear characteristic (pHY) so as to reduce the gain around pH = 7. The valve is equipped with a positioner to achieve precise control.

Many applications of pH control arise because chemical reactions often require certain pH conditions. An example of such a process is shown in Figure 4.5.4. This is a two-stage process for the oxidation of sodium cyanide (NaCN) in a waste treatment system where sodium hydroxide and chlorine are

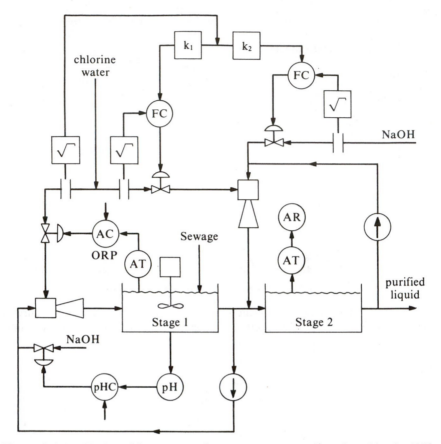

Figure 4.5.4. pH control in a process for waste treatment. (Based on Liptak, 1973. From *Instrument Engineers' Handbook: Process Control* by Bela G. Liptak. Copyright 1985 by the author. Reprinted with permission of the publisher, Chilton Book Company, Radnor, Pa.)

mixed in the tank at the left to produce the reaction:

$$Cl_2 + 2NaOH + NaCN \rightarrow NaCNO + 2NaCl + H_2O \qquad (4.5.1)$$

This reaction requires that the pH be at least 8.5. We see that the addition of sodium hydroxide is controlled by a pH controller, while the addition of chlorine is based on measurement by a special ORP (oxidation-reduction-potential) electrode. The ORP unit measures the amount of oxidation that has taken place. In addition to the indicated control functions, the level in the left-hand tank must also be controlled. Notice especially that the sodium hydroxide and the chlorine are mixed together in a stream of recirculated process liquid coming from the first tank, and that the first tank is equipped with a stirrer to provide good mixing.

The process liquid from the first tank is mixed with additional chlorine and sodium hydroxide as it enters the second tank. The chemical flows are precisely controlled in this case by means of flow control loops that are ratio-controlled based on preset ratios. The set points for the flow control loops are controlled proportionately to the measured flow of chlorine in the first stage. There is no feedback in the second stage from any measurement in the second tank. The ORP analyzer is used in this case only as a monitor, which the operator can use to determine whether the process is operating satisfactorily. For this structure to be satisfactory, it is necessary that the concentration of the sodium hydroxide be constant so that the ratio control system of Figure 4.5.4 can provide the desired pH in the second stage. If the concentration varies, it is necessary to place a pH meter in the second stage in order to adjust the ratio control system.

As mentioned earlier, pH control systems often suffer from extreme variations in loop gain if strong acids and bases are used, and the set point is in the region of pH = 6 to 8. Often, the result of this is that the pH control loop will exhibit limit cycles of appreciable amplitude. One way to ease this problem is to use a controller that has a variable gain. This type of controller can be programmed to have a low gain when the deviation is small, and a high gain when the deviation is large. If a PI controller is used, the integral action should be disconnected when the deviation is greater than some limit.

Another way to solve the problem is to use an adaptive control algorithm (recall section 2.14) that continuously estimates the process parameters and adjusts the controller parameters accordingly. This has been investigated by many authors (Buchholt and Kummel, 1979; Proudfoot, Gawthrop, and Jacobs, 1983; Piovoso and Williams, 1985), and positive industrial experience has been reported.

Furthermore, in recent years various new proposals for pH control based on the principles of state estimation have appeared in the literature (Waller and Gustafsson, 1983). Figure 4.5.5 is an illustration of how such a system is intended to work. A model of the pH meter and a numerical state space model describing the concentrations of the various chemical components of the process are in parallel with the process itself. The measurement model will be highly nonlinear, but the model of the reaction components will be approximately linear. As shown in the figure, the state vector $\mathbf{x}$, which describes the component concentrations, is estimated on the basis of feedback from the pH measurement. This feedback will be highly nonlinear because of the pH measurement, giving roughly the same situation as in the case of ordinary pH control. However, control of the process is based on the estimated component concentrations. These control loops will be approximately linear and can be designed with good robustness.

The performance of the estimation loops can be improved by preprogrammed variable gain as indicated above, and by adapting the parameters of the nonlinear pH measurement such that the innovation process ε (the difference between the true measurement and the estimated measurement) is minimized. When

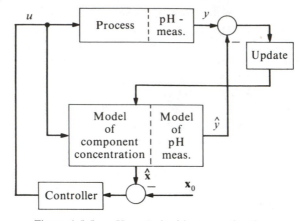

Figure 4.5.5a. pH control with state estimation.

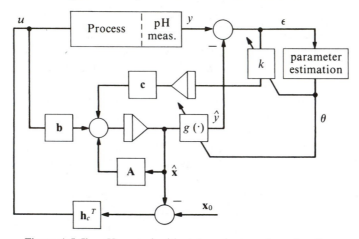

Figure 4.5.5b. pH control with state and parameter estimation.

these parameters are found, the gains in the estimation loops can be adjusted. This is shown as integral updating in Figure 4.5.5b.

This approach to solution of the pH control problem has not yet been widely applied, but there is reason to believe that it soon will be. An important lesson to be learned from the system shown in Figure 4.5.5 is that *the measurements are used only to update the estimator; control is based on the estimated states*.

A good general discussion of many practical considerations of pH control can be found in Shinskey (1973) and Waller and Gustafsson (1983).

4.6. CONTROL OF CONCENTRATION BY MEANS OF VISCOSITY MEASUREMENT

There are many industrial processes in which the concentration of one or more components in a suspension can be determined relative to either the viscosity or density of the suspension. Figure 4.6.1 illustrates a system for mass flow control with a linear flow signal (differential pressure and the square root element can be replaced with a linear magnetic flowmeter) that is multiplied by a linear density measurement obtained from the analyzer, A. In addition, a multiplication factor k is used to adjust the calibration of the density measurement on the basis of a more precise measurement made downstream in the process, perhaps in the laboratory. The product of the linear volumetric flow and concentration is controlled by the flow controller.

Figure 4.6.2 is a process segment that has a combination of precise ratio control of the flows of the components, A and B, in a mixing tank and control of the viscosity of the mixture. Viscosity is adjusted by heating, based on measurement of the viscosity of the flow from the tank. If the relationship between the viscosity and the temperature is well known, temperature control of the mixture would be a sufficient control. If this is not the case, viscosity control based on the actual measurement of viscosity should be implemented in cascade with the temperature control, as shown in the figure.

4.7. CONTROL OF THE COMBUSTION PROCESS

The important fundamentals of the combustion process were described in section 3.8. In this section we will examine some of the automatic control structures that are commonly applied to this process.

It is absolutely essential that the flows of fuel and air be maintained at the proper ratio in the combustion chamber. This ratio is determined by the stoichiometric equations given in section 3.8. Because no combustion process is ideal, it is necessary to operate with some amount of excess air in order to prevent the presence of significant amounts of unburned hydrocarbons and carbon monoxide in the flue gas or in the later stages of the combustion chamber.

A common solution for the control problem is shown in Figure 4.7.1. The input streams of both air and fuel are equipped with flow controllers, so that possible influences due to pressure changes in the lines or viscosity changes of

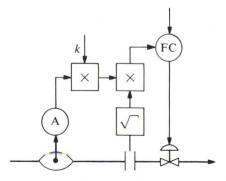

Figure 4.6.1. System for mass flow control of a component in a suspension.

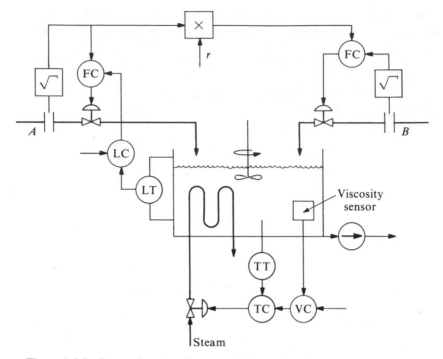

Figure 4.6.2. System for viscosity control in cascade with temperature control.

the fuel are eliminated. Ratio control between the two flows is designed so that the measurement of the the air flow is used to give the set point for the fuel flow. The reason for this is safety; if the air supply system fails for some reason, the fuel flow will be shut off automatically. Another possible solution is to let the measured flow of fuel control the set point for the air flow, but then some independent means of stopping the fuel flow when the air flow fails must be provided. The set point of the air flow controller in the system of Figure 4.7.1 is changed when the power output of the process is to be changed.

The figure indicates that measurement of the flue gas characteristics can be used to adjust the ratio between the fuel and the air flow. It is common practice to measure the oxygen in the flue gas. Often, an oxygen content of 2% in the flue gas is used as the set point for the controller. According to Figure 3.8.1, most fuels operate best at an air flow about 10% in excess of the theoretical value.

Depending somewhat on its design, a gas analyzer will generally have a transfer function of the form

$$h(s) = K\frac{e^{-\tau s}}{1 + Ts}$$

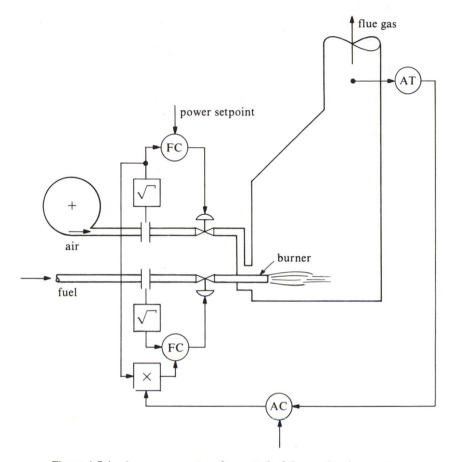

Figure 4.7.1. A common system for control of the combustion process.

where the transport delay τ will be of the order 10 to 40 seconds, and the time constant T will be about 5 to 20 seconds. This transfer function, combined with the dynamics of the combustion process itself (given by equation 3.8.4), leads to a control system with limited bandwidth. However, a high bandwidth is not required because variation in fuel quality is usually slow. Variation of fuel quality is the most common reason for control based on gas analysis.

An alternative approach to the analysis of flue gas is to measure the amount of unburned carbon monoxide (CO). This is usually done by means of infrared absorption. This is somewhat more expensive than measuring oxygen, an approach that is based on the paramagnetic characteristics of oxygen. However, control on the basis of CO measurements can give somewhat better fuel economy in large furnaces, and therefore is an economically justifiable solution.

There is a long list of details that determine the complete instrumentation scheme for the practical design of an industrial combustion system, but it will not be given here. These details concern the control of the fuel when different types of fuel are used at the same time (coal, oil, gas, etc.); different systems for air flow control (fan or pump speed, dampers in the inlet, dampers in the outlet, etc.); and various safety considerations (Liptak, 1970).

Chapter 5
PROCESS CONTROL OF THE COMMON UNIT
PROCESSES

It is difficult to distinguish between the basic functions described in Chapter 4 and the unit processes that will be considered in this chapter. For example, the combustion process described in section 4.7 is on the borderline between basic functions and unit processes.

It is important to organize the material as we have done because production systems in widely differing types of industries have many similar unit operations, where the requirements of the control systems will be nearly the same from industry to industry. Of course, many problems are special, or peculiar to various industries. These problems will be discussed in Chapter 6.

5.1. CONTROL OF MECHANICAL SEPARATION PROCESSES

In sections 3.4, 3.5, 3.6, and 3.7 we discussed several unit processes used for the separation of different phases (gas, liquid, and solid) from each other based on various differences in physical characteristics. These processes often must be controlled automatically to maintain the process variables within acceptable limits. The various types of separation processes have many different possible control variables, as discussed in Chapter 3. Because the separation of two or more components (phases) often leads to a need to balance separation quality and separation quantity (production rate), one can easily end up with an optimization problem. Many times the separation *quality* will be better if there is a large holdup time in the process, but in that case, the separation *quantity* will be less. A large holdup time in the system is easily obtained by recirculation of some of the product stream after classification (sorting or grading). As described in sections 3.5 and 3.7, this is a positive feedback that makes the system slower.

It can generally be said that the characteristics of the process dynamics of most separation processes lead to few dynamic control problems. The problems are centered more on obtaining measurements that give an accurate picture of the state of the process and on choosing control variables that give economically acceptable and safe operating conditions.

Figure 5.1.1 is an example of a relatively complete control system for a separation process. This process is a flotation unit of the type described by Figure 3.7.13.

Both feedback and feedforward can be used in the control of flotation processes. For practical and economic reasons it is not possible to apply automatic

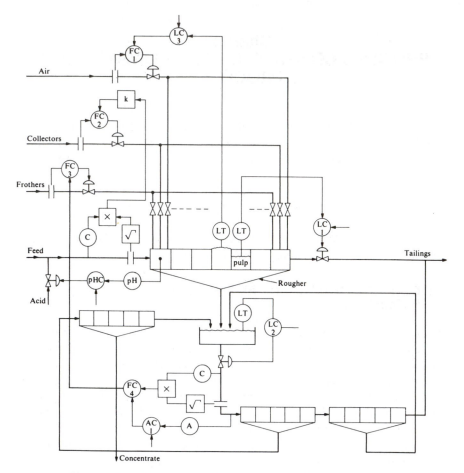

Figure 5.1.1. Flowchart of flotation process with control system.

control to each individual cell in a flotation system. Therefore, the automatic control system will consider the process as a whole, while the individual cells are "trimmed" by individual manual control valves on each cell.

The most important control variables for the flotation process, described in section 3.7, are listed here for convenience:

- Air flow into the cell
- Flow out of the cell (tailings)
- Flow of collectors
- Flow of frothers
- Water/solid ratio in the pulp

The most important measurements that can be made in the flotation chamber are:

- Flow of the various components (air, pulp, collector, etc.)
- Pulp level
- Depth of froth
- pH of the pulp
- Analysis of the mineral makeup and feed concentrate by means of an X-ray fluorescence analyzer
- Analysis of particle size distribution in the pulp

The flotation process is strongly multivariable, which means there are strong couplings between the various control variables and the important states of the process. The literature presents many different proposals for automatic control structures, but it cannot be said that one structure is necessarily better than another. The choice depends upon the particular design and application of the flotation system and the materials to be separated. The system shown in Figure 5.1.1 is based on principles presented in LeGuea (1975) and Gault et al. (1979). The system contains four flotation units similar to the one in Figure 3.7.13. To simplify the diagram, only the "rougher" unit in Figure 5.1.1 is equipped with automatic control, but in practice a similar control scheme is applied to the other units as well. The feed, from a ball mill having its own control system, enters the process at the left side of the diagram. In order to maintain constant pH in the pulp, which is important for good operation of the flotation process, pH is measured in the first cell. On the basis of this measurement it is possible to control the addition of weak acid (or a weak base if appropriate) to the feed stream. Addition of collector is controlled by feedforward from calculation of the mass flow of the feed stream. There is only one flow controller in the common collector feed stream, so distribution of this stream to the individual cells is adjusted by manual valves. The same principle is used for the addition of air and frothing agents.

The total flow of air added to the cells is controlled by feedback from a measurement of the froth level (defined as the height of froth above the pulp level). The pulp level is controlled by a valve in the outlet, often a butterfly valve in the last cell. The concentrate from the rougher stage is sent to a collection tank where it is mixed with return flows from the other stages. The level in this tank is controlled by the outlet flow. The mass flowrate of this stream is computed and, through flow controller FC-4, adjusts the set point of the flow controller in the frothing agent feed line. An X-ray fluorescence analyzer (A) determines the contents of the various components in the concentrate, and this information, possibly with additional calculations, controls the set point for the mass flow controller FC-4. This solution assumes a known relationship between the flow of frothing agent and the concentration of valuable minerals in the concentrate.

Because of the highly multivariable nature of the flotation process, there has been a great deal of effort to develop mathematical models and to experimentally determine the values of the parameters of these models (Olsen and Henriksen, 1976; Henriksen, Kaggerud, and Olsen, 1982). Using these models as a starting point, it has been possible to develop multivariable control strategies of the linear-quadratic type (Kaggerud, 1983). Further study of control structures should lead to still better improvements in control of these processes.

5.2. CONTROL OF HEAT EXCHANGERS

Heat exchange occurs in many forms in industrial processes and is one of the most important of all unit operations subject to automatic control. The dynamics of several heat exchange processes are discussed in section 3.8 and Appendix H, and in section 4.4 we discussed the fundamentals of energy and temperature control. In this section we will present a general overview of why automatic control of heat exchange (and transfer) is important and how the application of the various principles of automatic control takes place. We will discuss the special cases of evaporation in section 5.3, crystallization in section 5.4, and drying in section 5.5. All these processes involve heat exchange or transfer in some form.

Control of Heat Exchangers

Heat exchangers as separate process units can take on a number of forms, depending on which media are primary and secondary. This was discussed in section 3.8. The primary medium delivers heat to the secondary medium. In a condenser the primary side contains vapor and condensate, while the secondary will contain cold liquid or gas (air, perhaps). In an evaporator, the primary side will often contain steam or warm gas, but a warm liquid is also common. The secondary medium in an evaporator will be boiling liquid and vapor.

The purpose of a heat exchanger used as a stage in a larger process is usually to supply a product (process liquid or steam) with certain qualities such as temperature, pressure, mass flow, and enthalpy. It is useful first to consider how the objective of the heat exchanger should be specified, identify which control variables are possible, and determine which are the most important disturbances. These considerations give direct indications of which control structure should be chosen.

The basic heat exchanger process is simple enough that it is not necessary to generate a general, detailed theoretical development in order to arrive at a control structure that is completely satisfactory for most objectives. Certainly, dynamic optimal control based on estimation of nonmeasurable states obtained by means of a state space model of the process, combined with modeling and estimation of the disturbance states, can lead to "perfect" automatic control of the heat exchanger. However, at this point we will use some simpler observations to develop a control structure. Later, in connection with boilers and evaporators, we will use a more formal approach.

Let us now consider a series of system solutions. The control principles used in these solutions are:

- Feedback
- Feedforward
- Cascade control
- Computation of process variables (models, material and energy balances, etc.)

Figure 5.2.1 is a diagram of the simplest form of control for a liquid–liquid heat exchanger where the temperature of the secondary medium outflow is measured. If the driving pressure on the primary side is essentially constant, then this solution will usually give satisfactory results. Because of the dynamics in the heat exchanger (see section 3.8), it is necessary to use a PI (perhaps a PID) controller to obtain satisfactory precision and speed of response. However, this solution has several weaknesses that are related to the way the disturbances affect the process.

Figure 5.2.2 shows a cascade control scheme on the flow of the primary stream to eliminate flow variations in the primary loop. These changes result from pressure changes in the line when the valve opening remains the same. This solution is further extended in Figure 5.2.3 by the addition of temperature measurement of the primary stream input. The product of flow and temperature makes it possible to calculate the heat content of the primary flow, which can be

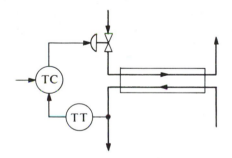

Figure 5.2.1. The simplest form of control of a heat exchanger.

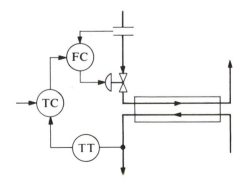

Figure 5.2.2. Heat exchanger with flow control of primary side.

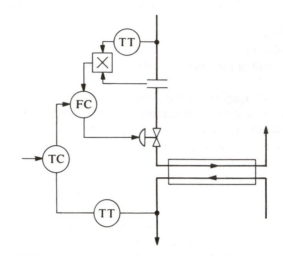

Figure 5.2.3. Heat exchanger with energy flow control of the primary medium.

controlled by means of a flow controller. When the set point of this flow is controlled by the temperature controller on the outlet of the secondary medium, the disturbances in both the primary temperature and pressure will be eliminated.

Figure 5.2.4 is an application of feedforward from the heat requirement of the secondary side (expressed as the product of the temperature difference between the input and output multiplied by the flowrate) combined with temperature feedback to the primary flow. The advantage of this arrangement is that disturbances in both temperature and flow of the secondary medium will be

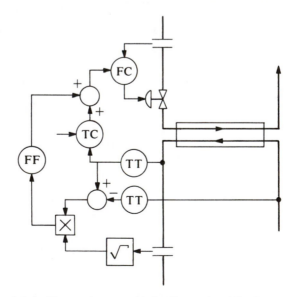

Figure 5.2.4. Heat exchanger with feedforward and feedback control.

quickly compensated by a corresponding change in the amount of heat added to the primary side. One could also use the more complicated solution of Figure 5.2.3 for control of the heat flow. The final temperature of the secondary medium is obtained via an ordinary temperature controller that integrates the effect of the temperature measurement in the outlet of the secondary stream. The feedforward, shown in Figure 5.2.4, can have a dynamic transfer function as developed in section 2.3 and based on the dynamics of the heat exchanger.

Figure 5.2.5 illustrates the most common arrangement when the primary medium is steam that condenses in the heat exchanger. Because saturated steam has a well-defined relationship to temperature and pressure, it is best to equip the primary side with a pressure controller that operates a valve on the supply side. This solution has such rapid response that it usually is not necessary to use cascade control with a flow controller as in Figure 5.2.2. Temperature control is accomplished by measuring the temperature of the outlet of the secondary medium and controlling the set point in the steam pressure controller by means of a temperature controller (TC) of either a PI or a PID type. The system in Figure 5.2.5 can be extended to include feedforward and heat flow control as shown in Figures 5.2.3 and 5.2.4.

Figure 5.2.6 shows a heat exchanger where the primary medium is condensed steam, and control is achieved by varying the surface of the tubes exposed to the steam. The means of control is to move the condensate level up or down as needed by means of a control valve on the outlet. This method gives a quite slow temperature response because the condensate level is proportional to the integral of the condensed steam flow. Nevertheless, for technical reasons this approach is used in certain cases.

Figure 5.2.7 is an example of parallel control, which was discussed in section 2.6. This is especially applicable when a liquid is to be heated in the heat exchanger, and the temperature and flowrate of the secondary medium can vary

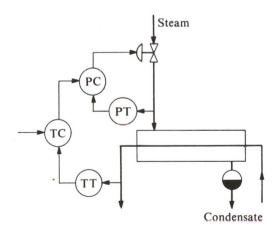

Figure 5.2.5. Heat exchanger with pressure control of saturated steam on the primary side.

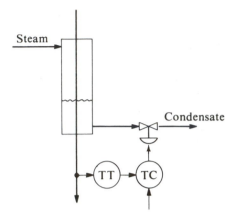

Figure 5.2.6. Heat exchanger with control from level of condensate on primary side.

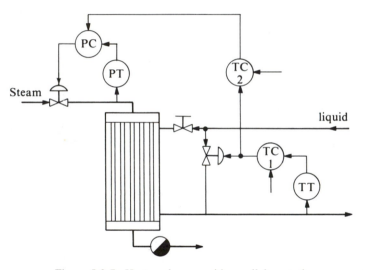

Figure 5.2.7. Heat exchanger with parallel control.

strongly. A bypass provides a stream in parallel with that flowing through the exchanger. Therefore, the liquid leaving the exchanger consists of a mixture of cold liquid from the bypass and hot liquid from the secondary side of the exchanger. The control system has two controllers, one of which measures the temperature in the outlet stream and controls the valve in the bypass. This unit must be either a P or a PD controller (see section 2.6). Because the opening of the control valve is completely dependent on the temperature of the stream in the exchanger, and because it is desirable to operate the valve around its mid-range, it is necessary to add a controller (TC-2) that takes its signal from the output of the first temperature controller (TC-1). This controller should be a PI type acting on the set point of the steam pressure controller on the primary side or some equivalent controller for control of the power addition to the system. The ratio between the flow through the heat exchanger and the parallel loop is adjusted by means of the valve in the parallel path. The advantage of this scheme is that it makes it possible to compensate very quickly for disturbances in flow and/or temperature, but perhaps even more important is its very quick response to changes in the set point of temperature controller TC-1.

The solution given in Figure 5.2.4, which shows the case for a liquid–liquid exchanger, can also be used when steam is the primary medium. Calculation of the enthalpy of the primary stream on the basis of measurement of temperature or pressure and mass flow can be made, corresponding to the layout given by the figure.

Many other arrangements are also possible depending on the control variables, measurements, and possibilities for feedback and feedforward. For control it is most important to keep in mind that, in heat exchangers, all the energy one wants to transfer to the secondary medium must come from the primary medium. Therefore, calculation of the heat requirement (or cooling demand) makes it possible to request the necessary addition of heat (or cooling). As always, one must balance the cost of more complicated control systems against the quality of the control required.

Boiler Control

One very special and widely used heat transfer process is the boiler, which actually involves many heat transfer processes. Boilers are used in industry to produce process steam for a wide variety of applications, including heating and driving turbines, which in turn drive pumps, compressors, engines, and generators. In thermal power generating stations, boilers can be very large units, using gas, oil, or coal as the primary energy source. Steam turbines are also used to propel ships. These turbines are driven by smaller boilers that must be suitable for shipboard installation and must allow an easy, rapid response that will provide good ship maneuverability.

The elements of a typical stationary boiler are shown in Figure 5.2.8. A combustible material, usually gas, oil, or coal, is burned with air in the firebox. The air is usually preheated by heat exchange with the flue gas in the upper area of the combustion chamber. A discussion of problems related to combustion control was given in section 4.7, and will not be repeated here. It is important that the ratio of air to fuel be carefully controlled in order to obtain good, safe, and efficient combustion.

The water that is to be evaporated is added to a drum, or horizontal cylinder, following preheating in the upper portion of the combustion chamber. The feedwater is often condensate that has been returned to the drum from the various steam consumers in the system. The level of water in the drum is an important variable in the control system. From the drum, the water goes down through the downcomers, which are located outside of the firebox. The water then goes

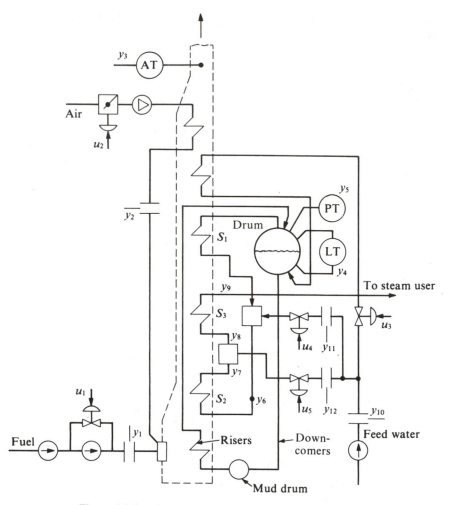

Figure 5.2.8. The elements of a typical stationary boiler.

into the risers, which are located in the hottest part of the furnace. Here, the water evaporates, and the steam rises and flows back up to the drum. The steam then goes from the drum through three superheaters, S_1, S_2, and S_3, where it is heated to a temperature higher than that obtained in its saturated state. In order to control the temperature and pressure in the superheaters, atomized water can be added between the first and second and second and third superheaters. This water immediately evaporates, rapidly affecting the temperature.

In addition to the fuel, u_1, and air, u_2, there are three other control variables for this boiler—the feedwater, u_3, and two water sprays, u_4 and u_5, between the superheaters. Frequently, only one water spray is used between the super-heaters, and if so, it will usually be located between the first and second super-heaters (u_4).

The most important measurements in the system of Figure 5.2.8 are:

y_1: fuel flowrate
y_2: air flowrate
y_3: flue gas analysis
y_4: water level in the drum
y_5: steam pressure in the drum
y_6: steam temperature before the second superheater
y_7: steam temperature after the second superheater
y_8: steam temperature before the third superheater
y_9: steam temperature after the third superheater
y_{10}: total feedwater flow
y_{11}: water added between first and second superheaters
y_{12}: water added between second and third superheaters
y_{13}: steam flow out of the boiler

Boiler control systems can be built up as a combination of conventional single-variable control loops, with or without feedforward, and computation of certain variables that cannot be measured directly (enthalpy, for example). One could also approach the problem from the standpoint of a state space model of the system and use a state estimator for calculation of the states that cannot be measured. From this, one can generate control variables where, in principle, all the states will affect all the control variables. In the following sections we will examine both approaches as they appear in the literature.

The Requirements of Boiler Control Systems   The function of a boiler is to deliver steam of a given quality (temperature and pressure) either to a single user, such as a steam turbine, or to a network of many users. The specific requirements of the system depend on the type of system. In the case of a steam turbine, the load on the boiler can change from zero to maximum in a very short time, a behavior typical of ship boiler systems. During this rapid change, the steam quality must be held within certain limits. In addition, the water level in the drum must remain within certain limits to prevent water from entering the turbine on the one hand, and to prevent overheating of the risers in the boiler on the other. Both of these situations represent serious safety hazards. The requirements for control of a stationary boiler that supplies steam to a network of users are not quite so stringent because the steam demand, or load on the boiler, is unlikely to undergo such wide, rapid changes.

Conventional Boiler Control.   Figure 5.2.9 is a somewhat simplified conventional boiler control system for the boiler described in Figure 5.2.8. The feedwater flow is controlled by the feedwater control valve u_3, and the total feedwater, including the desuperheater spray water, is measured by a single flowmeter y_{10}. An important function of the feedwater controller is to maintain

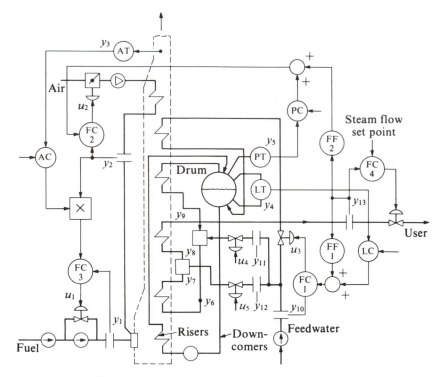

Figure 5.2.9. A conventional boiler control system.

the proper water–steam mass balance. A well-known phenomenon that affects this problem is the so-called shrink and swell effect in the drum. It causes the apparent water level (the measured value) in the drum to depend on the volume of bubbles in the drum and the risers. The volume of bubbles depends on the pressure; at low pressure, the bubbles are larger than at high pressure. An increase in the volume of the bubbles causes an increase in the total volume of the water–bubble mixture, which causes the level in the drum to rise even though the actual amount of water has not changed. Therefore, a mass balance based on measurement of the level and feedback to the feedwater control valve can give misleading results if one wishes to have a control system with high bandwidth.

A common solution is to use feedforward (FF-1) in the form of ratio control between the steam flow, y_{13}, and the feedwater flow controller, FC-1, and to update this slowly with a level controller, LC, operating on the basis of the measured water level in the drum. The feedforward from the steam flow should have some dynamics (a little low-pass filtering) in order to compensate for the effects of zeros in the right half-plane due to the decrease of the water level in the drum when the bubble volume temporarily drops upon the addition of feedwater. Problems of adjustment of the feedforward and the level controller have been the subject of much discussion and research. The first systematic solutions were obtained by use of the theories of decoupling and multivariable control, to be discussed further in the next section.

The power control of a boiler (fuel addition coordinated with air addition) is conventionally achieved by providing feedforward (FF-2) from the steam flow to the set point of the fuel flow controller, shown here as the set point for the air flow controller. This feedforward accounts for the fact that an increase in steam flow requires an increase in the energy supplied to the boiler. The feedforward can also be given dynamic action to compensate for the dynamics of the boiler itself. Pressure controller PC can update the feedforward on the basis of the pressure measured in the drum, a measurement that indicates the energy content of the steam. The pressure controller should be a PI type with the integral time matched to the dominant time constant of the boiler. Because a boiler can usually be approximated by a first-order transfer function between the fuel addition and the steam flow, a proportional controller can also give satisfactory results, especially when it is combined with feedforward as described above. If the steam flow from the boiler is not controlled directly but is manipulated only by a steam valve (as shown at the right of Figure 5.2.9), the steam flow will be a function of both the pressure and the valve opening. The feedforward from the steam flow to the set point of the fuel controller will then actually be a positive feedback. Therefore, we assume that there is a direct steam flow controller, FC-4, that controls the steam valve. Without this controller, the feedforward from the steam flow to the set point of the fuel flow control must be corrected with pressure such that the feedforward becomes transient.

Use of Decoupling for Boiler Control. One of the earliest control schemes, more advanced than that just described, was developed by Chien et al. (1958). This scheme used a multivariable controller to decouple the strong interactions occurring in the process transfer function matrix. Such a transfer function matrix can be derived from the state space model for a common boiler, presented in Appendix I. Figure 5.2.10 is a simplified diagram giving the transfer function matrices for the boiler. This diagram indicates the transfer function matrices

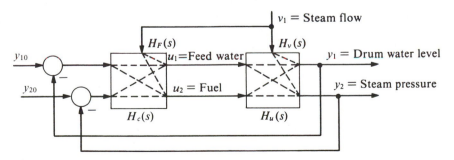

Figure 5.2.10. Boiler control with decoupling.

for the control variables $H_u(s)$, disturbances $H_v(s)$, and two controller transfer function matrices consisting of a feedback control matrix $H_c(s)$ and a feedforward matrix $H_F(s)$. We see here that the feedbacks are taken from measurements of the water level y_1 (notice the different notation from that used in Figure 5.2.8) and the steam pressure y_2. The two control variables are feedwater flow (u_1) and fuel flow (u_2). The dominant disturbance is the steam flow (demand) v_1, which is assumed to be independent of the control variables. Chien, et al. (1958) presented the following transfer functions as representative of those for a boiler from the standpoint of decoupling and feedforward control:

$$h_{u11}(s) = k_1 \cdot \frac{1}{s}$$

$$h_{u12}(s) = -k_2 \cdot \frac{(1 - 122s)}{(1 + 3s)(1 + 40s)}$$

$$h_{u21}(s) = -k_3 \cdot \frac{1}{(1 + 40s)}$$

$$h_{u22}(s) = k_4 \cdot \frac{1}{(1 + 40s)}$$

$$h_{v11}(s) = -k_5 \cdot \frac{(1 - 16s)}{s(1 + 3s)}$$

$$h_{v21}(s) = -k_6 \cdot \frac{1}{(1 + 40s)}$$

These transfer functions are simpler than those one would get by precise analysis, but they reflect the most important phenomena. It is unlikely that a more complicated model would add much to the analysis of the control system under discussion here. In these transfer functions, all the factors k_i are positive and are scale factors whose numerical values have no significance to our discussion. The time constants have the dimensions of seconds. We see that three transfer functions, those that describe the response of the steam pressure, are characterized by a single time constant. The transfer function between the fuel flow and the water level, $h_{u12}(s)$, has a dominant zero in the right half-plane because of the shrink and swell effect. The same effect appears in the transfer function between the steam flow and the water level. The zero in $h_{u12}(s)$ is closer to the origin than that of $h_{v11}(s)$, but the phase of the frequency responses of the two transfer functions will show about the same form in the vicinity of $-180°$.

Chien et al. (1958) proposed a control system consisting of feedforward $[H_F(s)]$, as shown by the dashed line in Figure 5.2.10, and feedback $[H_c(s)]$ which diagonalizes the loop transfer function matrix (decoupling). The theory for this was discussed earlier in sections 2.3, and 2.9, and 2.10. The feedforward required to eliminate the influence of steam flow as a disturbance to the water level and steam pressure is:

$$H_F(s) = -H_u^{-1}(s)H_v(s) \tag{5.2.1}$$

Substituting the various transfer functions, we get:

$$h_{F11}(s) = \frac{h_{u12}(s)h_{v21}(s) - h_{u22}(s)h_{v11}(s)}{h_{u11}(s)h_{u22}(s) - h_{u21}(s)h_{u12}(s)} \tag{5.2.2}$$

$$h_{F21}(s) = \frac{h_{u21}(s)h_{v11}(s) - h_{u11}(s)h_{v21}(s)}{h_{u11}(s)h_{u22}(s) - h_{u21}(s)h_{u12}(s)} \tag{5.2.3}$$

The next requirement is that the loop transfer function matrix

$$H_0(s) = H_u(s)H_c(s) \tag{5.2.4}$$

be a diagonal matrix in order to obtain decoupling so that changes in the level control loop influence the pressure control loop as little as possible, and vice versa. Setting the transfer functions into equation (5.2.4) gives:

$$h_{u11}(s)h_{c12}(s) + h_{u12}(s)h_{c22}(s) = 0 \tag{5.2.5}$$

$$h_{u21}(s)h_{c11}(s) + h_{u22}(s)h_{c21}(s) = 0 \tag{5.2.6}$$

$$h_{u11}(s)h_{c11}(s) + h_{u12}(s)h_{c21}(s) = h_{011}(s) \tag{5.2.7}$$

$$h_{u21}(s)h_{c12}(s) + h_{u22}(s)h_{c22}(s) = h_{022}(s) \tag{5.2.8}$$

Equations (5.2.5) and (5.2.6) contain the elements in equation (5.2.4) that do not lie on the diagonal, while equations (5.2.7) and (5.2.8) are the diagonal elements. The diagonal elements $h_{011}(s)$ and $h_{022}(s)$ must be specified in such a way that we obtain a unique solution (see section 2.9).

We find then:

$$h_{c11}(s) = \frac{h_{011}(s)}{h_{u11}(s)} \cdot \frac{1}{1 - Y(s)} \tag{5.2.9}$$

$$h_{c22}(s) = \frac{h_{022}(s)}{h_{u22}(s)} \cdot \frac{1}{1 - Y(s)} \tag{5.2.10}$$

$$h_{c12}(s) = -\frac{h_{u21}(s)}{h_{u22}(s)}h_{c11}(s) \tag{5.2.11}$$

$$h_{c12}(s) = -\frac{h_{u12}(s)}{h_{u11}(s)}h_{c22}(s) \tag{5.2.12}$$

where:

$$Y(s) = \frac{h_{u12}(s)h_{u21}(s)}{h_{u11}(s)h_{u22}(s)} \qquad (5.2.13)$$

When we insert the numerical values into the transfer functions, we get:

$$\frac{1}{1 - Y(s)} = \frac{(1 + 3s)(1 + 40s)}{(1 + 3s)(1 + 40s) - k(1 - 122s)s} = \frac{(1 + 3s)(1 + 40s)}{(1 + 6.4s)(1 + 35.9s)} \qquad (5.2.14)$$

where $k = (k_2 k_3)/(k_1 k_4) \approx 0.9$. Chien et al. (1958) chose:

$$h_{011}(s) = h_{022}(s) = \frac{k_7}{s(1 + 3s)} \qquad (5.2.15)$$

but it should be recognized that other suitable choices are possible.

Combining this with equations (5.2.9) through (5.2.12) gives:

$$h_{c11}(s) = k_8 \cdot \frac{1 + 40s}{(1 + 6.4s)(1 + 35.9s)} \approx k_8 \cdot \frac{1}{1 + 6.4s} \qquad (5.2.16)$$

$$h_{c22}(s) = k_9 \cdot \frac{(1 + 40s)^2}{s(1 + 6.4s)(1 + 35.9s)} \approx k_9 \cdot \frac{1 + 40s}{s(1 + 6.4s)} \qquad (5.2.17)$$

$$h_{c21}(s) = k_{10} \cdot \frac{1 + 40s}{(1 + 6.4s)(1 + 35.9s)} \approx k_{10} \cdot \frac{1}{1 + 6.4s} \qquad (5.2.18)$$

$$h_{c12}(s) = k_{11} \cdot \frac{(1 - 122s)(1 + 40s)}{(1 + 3s)(1 + 6.4s)(1 + 35.9s)} \approx k_{11} \cdot \frac{1 - 122s}{(1 + 3s)(1 + 6.4s)} \qquad (5.2.19)$$

Inspection of these results shows that the level controller, $h_{c11}(s)$, is a proportional controller plus a little damping (the time constant 6.4). The pressure controller, $h_{c22}(s)$, is a PI controller with the addition of a little damping and an integral time equal to the dominant time constant $T_i = 40$. The cross-coupling (interaction) terms, $h_{c21}(s)$ and $h_{c12}(s)$, have relatively simple transfer functions, and we notice especially that one of them has a zero in the right half-plane.

Similarly, insertion of numerical values into equations (5.2.2) and (5.2.3) gives feedforward from the steam flow to the respective control variables feedwater and fuel flow as:

$$h_{F11}(s) = k_{12} \cdot \frac{(1 + 0.1s)(1 + 40s)}{(1 + 6.4s)(1 + 35.9s)} \approx k_{12} \cdot \frac{1 + 0.1s}{1 + 6.4s} \qquad (5.2.20)$$

$$h_{F21}(s) = k_{13} \cdot \frac{(1 - 23.3s)(1 + 51.9s)}{(1 + 6.4s)(1 + 35.9s)} \approx k_{13} \cdot \frac{1 - 20s}{1 + 6.4s} \qquad (5.2.21)$$

The first of these is essentially proportional action with one pole (time constant of 6.4), while the other, even though it has the same pole, is dominated by a zero in the right half-plane.

Chien et al. (1958) also investigated the behavior of the boiler with ordinary PI controllers (from level to feedwater and from pressure to fuel flow) without feedforward or decoupling. The process was upset with a small step change in steam flow. The response of the drum pressure is shown in Figure 5.2.11a, and the response of the drum water level is shown in Figure 5.2.11b. For the easy comparison, the performance of the system with the feedforward and decoupled control scheme is shown by the dashed line. The improvement is obvious.

For the system described above to give the desired improved performance, the process parameters must be essentially constant over the operating range of the boiler. If this is not the case, the parameters in the various transfer functions must be adjusted as functions of, for example, the operating levels of the boiler. This represents a complication that might make the control scheme less attractive.

Boiler Control with State Estimation and State Feedback. There have been many recent attempts to develop control systems based on a combination of feedback from states generated by state estimators and feedforward from measured or computed disturbances. Tyssø et al. (1976) and Tyssø and Brembo (1978) developed such a system for a ship's boiler. A boiler that supplies steam to the turbines of a ship is sketched in Figure 5.2.12. We note that in addition to the fuel flow and the feedwater flow, the flow in the desuperheater system is also a control variable. Figure 5.2.13 is a block diagram of the associated multivariable control system. This includes a mathematical model as part of the state estimator that generates estimates of the process variables and provides feedback through matrix G. Multivariable integral action is obtained through a matrix of integrators G_2. There is feedforward between the various set points and the steam flow.

The mathematical model of the state estimator has eight state variables. These are:

x_1: enthalphy of the water in the drum and downcomers
x_2: mass of water in the drum and downcomers
x_3: average enthalpy of the water/steam in the risers
x_4: density of the saturated steam in the drum
x_5: temperature of the metal in the risers
x_6: temperature of the metal in the first superheater
x_7: temperature of the metal in the second superheater
x_8: water flow through the downcomers

The control variables are:

u_1: fuel oil flow
u_2: feedwater flow
u_3: relative flow of desuperheated steam

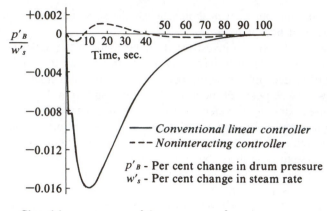

Closed-loop response of drum pressure for
a step-change in steam rate for the two controllers.

(a)

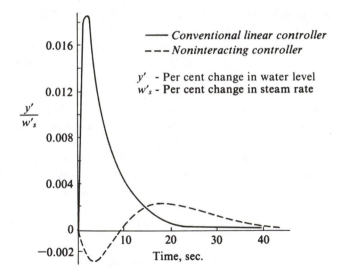

Closed-loop response of water level for a
step-change in steam rate for the two controllers.

(b)

Figure 5.2.11. Comparison of decoupling with conventional control of a boiler. (From *Control Engineering, 5;* 95–101, 1958.)

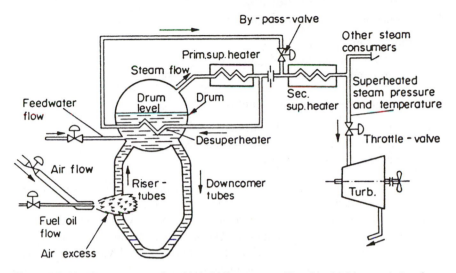

Figure 5.2.12. Components of a ship's boiler system. (Reprinted with permission from *Automatica, 14;* 213–221, Tyssø, A. and J.C. Brembo. Copyright 1978, Pergamon Press, Ltd.)

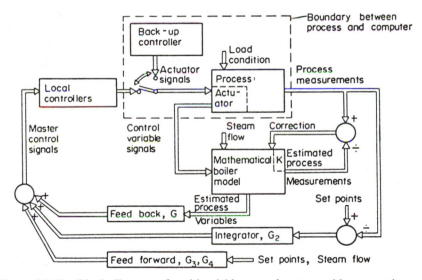

Figure 5.2.13. Block diagram of multivariable control system with state estimator. (Reprinted with permission from *Automatica, 14;* 213–221, Tyssø, A. and J.C. Brembo. Copyright 1978, Pergamon Press, Ltd.)

The measurements that can be used are:

y_1: pressure in the drum
y_2: water level in the drum
y_3: steam temperature after the second superheater

The most significant disturbances are:

v_1: steam flow from the boiler (demand or load)
v_2: excess air ratio in the combustion process
v_3: temperature of the feedwater

The multivariable control strategy is developed on the basis of optimal control theory (see sections 2.11 and 2.13). The deviations of the states from their nominal values are weighted by a quadratic form ($\Delta x^T Q \Delta x$) where the weighting matrix Q is chosen to be diagonal. Quadratic weighting is also applied to the variation of the control variables from their nominal values ($\Delta u^T P \Delta u$) where the weighting matrix P is also diagonal.

An important concern with a ship's boiler system that does not occur with stationary boilers is the wave motion of the ship. This motion can induce significant oscillations in the system, as shown for the case of conventional shipboard boiler control in Figure 5.2.14. Since the drum in a ship's boiler usually has a small diameter, the level control system must be extremely accurate and responsive. If the water level changes too much, the system will be tripped automatically. The result of this is power failure of the turbine and loss of maneuverability of the ship.

Figure 5.2.15 illustrates the behavior of a system with multivariable control when it is subjected to a realistic, large change in steam demand (from 28 kg/sec down to 4 kg/sec), as shown by the upper curve in the figure. The response of the same boiler with a conventional control system to the same demand change is shown in Figure 5.2.16. When the steam flow decreases suddenly, the pressure increases. As a result the level in the drum will decrease owing to a sudden compression of the bubble volume, and the feedwater controller will respond by increasing the feedwater flow. This leads to an excess of water in the drum; the safety limit (200 mm) is exceeded, and the turbine is decoupled from the system. The corresponding water level change in the system of Figure 5.2.15 is significantly less. A comparison of conventional and multivariable control is shown in Figure 5.2.17, where we also see a significant difference in the performances of the two systems with regard to feedwater flow control.

A disadvantage that has inhibited application of the more complicated multivariable controller is that up to now these systems have been much more expensive than conventional control systems because they require an on-line computer. Further, they can be difficult to tune and adjust because they have

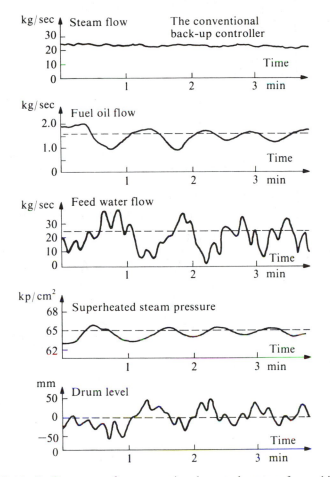

Figure 5.2.14. Performance of a conventional control system for a ship's boiler. (Reprinted with permission from *Automatica, 14;* 213–221, Tyssø, A. and J. C. Brembo. Copyright 1978, Pergamon Press, Ltd.)

many more parameters than a conventional control system has. However, there are many ideas, yet to be proved, that might make it possible to reduce the number of adjustable parameters in a multivariable controller to about the same number as for a conventional control system. As the cost of computer equipment decreases, these solutions will become more attractive.

The estimator of the system described above uses a model linearized around a nominal set of operating conditions. In some cases this may not be satisfactory. If it is not, one might use a nonlinear model that includes the dominant nonlinear effects but does not necessarily lead to a significant increase in the complexity of the system.

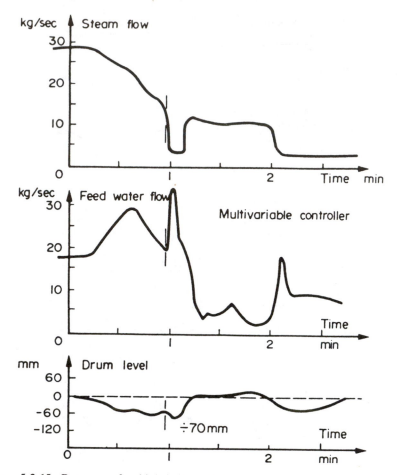

Figure 5.2.15. Response of a ship's boiler with multivariable control to a large demand change. (Reprinted with permission from *Automatica, 14;* 213–221, Tyssø, A. and J. C. Brembo. Copyright 1978, Pergamon Press, Ltd.)

5.3. CONTROL OF EVAPORATORS

The physical characteristics of an evaporator were described in section 3.9. Development of a mathematical model of an evaporator is given in Appendix J.

Multi-effect evaporators are very common in industry. We will discuss conventional control and then a multivariable structure based on optimal control theory for multi-effect evaporators. We will also compare the different solutions.

Conventional Control

Figure 5.3.1 is a diagram of a double effect evaporator equipped with a conventional control system. The first effect consists of a tank with a jacket for

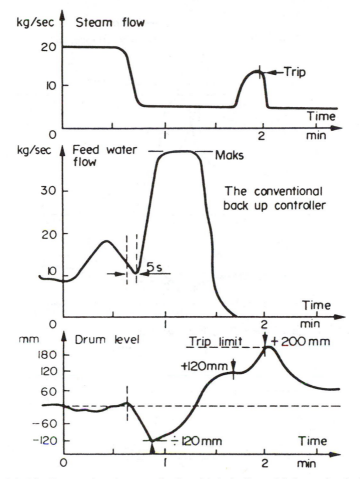

Figure 5.2.16. Conventional control of a ship's boiler with large load change. (Reprinted with permission from *Automatica, 14;* 213–221, Tyssø, A. and J. C. Brembo. Copyright 1978, Pergamon Press, Ltd.)

steam-heating the feed from which the most volatile component is to be evaporated (water, for example). The desired product is a concentrate of the feed. Vapor from the first effect goes to the second effect, where it causes additional boiling and evaporation in the vertical pipes through which the liquid is circulated before the new vapor is separated in an outer separator tank, as shown. The vapor from the last effect then goes to a cooler that operates at a slight vacuum to depress the boiling point somewhat.

The total mass balance is taken care of by two level control systems, which control the level in the first evaporator effect and that in the second effect separator. Both of these levels must be held constant. This is done by controlling the flows leaving each effect by means of cascade control of the set points of

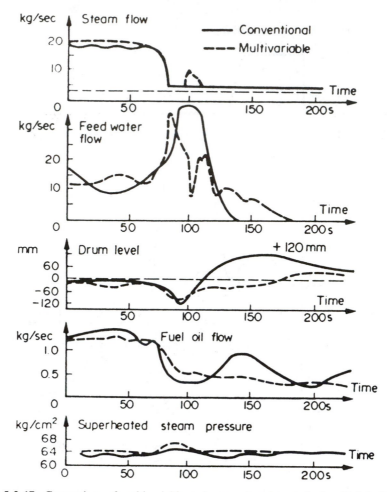

Figure 5.2.17. Comparison of multivariable and conventional control of a ship's boiler. (Reprinted with permission from *Automatica, 14;* 213–221, Tyssø, A. and J. C. Brembo. Copright 1978, Pergamon Press, Ltd.)

the flow control loops on each branch. It is assumed that the feed flow is determined by a process upstream of the evaporators, but certainly each evaporator has an upper limit on the amount of feed it can accept. This limit must be accounted for by the upstream process so that the evaporator is not overloaded. The advantage of flow control of the branches is that pressure changes in the vessels and viscosity changes in the concentrate will not affect the flow ratios.

The low pressure (or vacuum) in the second effect is controlled by a pressure controller. The flow of cold water to the condenser can also be controlled automatically, but here we will assume that the cooling is sufficient to give complete condensation.

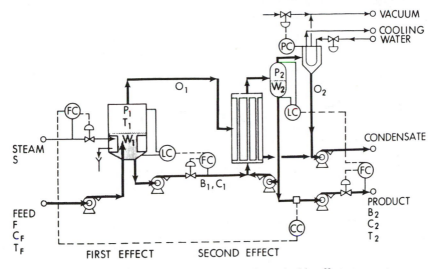

Figure 5.3.1. Conventional control system for a double effect evaporator.

The last, and very important, control system measures product quality (concentration, consistency, viscosity etc.) and provides signals from which the controller manipulates the set point of the steam flow controller on the inlet to the first effect. When the energy (steam input) is increased, production of vapor in the first effect will also increase, thereby increasing evaporation in the second effect as well.

The dominant disturbances in this system are variations in flow and concentration of the feed stream. Other disturbances, such as pressure variations in the steam and vacuum and temperature variations of the cooling water, can easily be eliminated by local, individual control systems. It is apparent that feedforward from the feed stream flow and concentration will be advantageous because the energy that must be added at the first effect by the steam is directly proportional to the amount of volatile liquid (water, for example) to be evaporated from the feed. Improvements of the system shown in Figure 5.3.1 should, therefore, include feedforward.

Because of the interconnections between the first and second effects, the system will exhibit a dynamic behavior with many interactions. Therefore, a multivariable control strategy might be advantageous.

Multivariable Control of a Double Effect Evaporator

Fisher and Seborg (1976) investigated a number of control strategies for the type of evaporator shown in Figure 5.3.1. Figure 5.3.2 gives a comparison of four different control solutions with respect to product concentration changes resulting from a 20% increase in feed rate. Curve A is the response for the control system shown in Figure 5.3.1. Curve B was obtained from a solution that

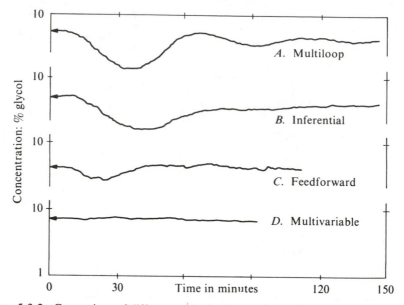

Figure 5.3.2. Comparison of different methods of control of a double effect evaporator.

used a simplified model (inferential control), and indicates some improvement. Curve C shows the effect of feedforward from the feed stream to the steam flow, and curve D shows the result obtained with a complete multivariable control strategy.

The multivariable control strategy uses a model with five state variables:

x_1: mass of the liquid in the first effect
x_2: concentration of the liquid in the first effect
x_3: enthalpy of the liquid in the first effect
x_4: mass of the liquid in the second effect
x_5: concentration of the liquid in the second effect

The control variables are the same as in Figure 5.3.1. The measurements are:

y_1: liquid level in the first effect
y_2: temperature in the first effect
y_3: liquid level in the second effect
y_4: concentration of the liquid in the second effect

The multivariable system studied by Fisher and Seborg (1976) did not include measurement of the flow and concentration of the feed stream, but the results, nevertheless, were good, as seen in Figure 5.3.2. This success was due in part to the fact that the feedback from the level in the first effect was relatively fast.

A process of the type described above has very strong coupling between the variables and represents a situation where multivariable control has apparent advantages. However, it is important that a state space model include the most important disturbances, at least as constants (modeled by integrators). Then the multivariable control strategy will include feedback from the process states and feedforward from the disturbances (or estimated disturbances) as discussed in section 2.13.

Standard methods of decoupling and multivariable feedforward have been used with good results by Palmenberg and Ward (1976). The methods they used are described in sections 2.9 and 2.10. The results will not be discussed further here.

5.4. CONTROL OF CRYSTALLIZATION

The practical design and application of crystallizers was discussed in section 3.10, and a mathematical model for a simple crystallizer is developed in Appendix K. In this section we will examine some structures for the control of practical crystallizers.

Figure 5.4.1 is an Oslo crystallizer similar to that of Figure 3.10.3. The main feed and the return line from the centrifuge at the product end of the process both enter the system at the tank at the left side of the diagram. Level con-

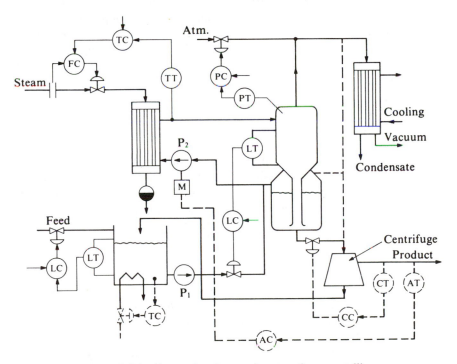

Figure 5.4.1. Conventional control system for a crystallizer.

trol on this tank causes the feed stream to adjust to the return stream. When the crystallizer is not operating properly, it produces too many fine particles, which are returned to the feed tank at a ratio to the feed stream much higher than when operation is normal.

From the tank, the liquid is pumped through a valve to the suction side of another pump where it is mixed with a stream that circulates from the lower to the upper section of the crystallizer. This flow goes through a heat exchanger where it is heated to a temperature high enough to cause evaporation in the upper chamber. On the primary side of the heat exchanger, steam is supplied under flow control, and the temperature of the secondary side is controlled by a temperature controller in cascade with the steam flow controller. The mass balance in the crystallizer is maintained by a level control system that measures the level in the upper stage and controls the feed rate out of the tank.

It is often advantageous to operate a crystallizer under a slight vacuum in order to obtain evaporation at a lower temperature. This requires pressure control, which is obtained by connecting the upper stage to a vacuum device and admitting air from the atmosphere through a valve opening controlled by the pressure controller.

The bottom section of the crystallizer and the centrifuge are valved to the same vacuum system.

Most of the crystallization process takes place in the lower section of the crystallizer. The unsaturated liquid is removed from the top, and the crystals are taken out at the bottom. The function of the centrifuge is to separate the large crystals from those that must be recirculated for further growth.

The basic control system of Figure 5.4.1 is used only to maintain suitable conditions for crystallization. Figure 5.4.1 shows no explicit measurement of product quality, but if the centrifuge is reliable, it provides an indirect indication of how well the process is operating. The measure of performance in this case is the ratio between the product stream and the reject stream from the centrifuge.

The degree of supersaturation is the factor that most affects the creation of nucleii and the growth of crystals in a crystallizer; so it would be useful to be able to measure this factor. However, there is little reference in the literature to successful systems for making this measurement (Gupta and Timm, 1971). Examination of the mathematical model of the crystallization process in Appendix K shows that the feed concentration and the feed flowrate are completely determinate for crystallization. In the system of Figure 5.4.1, the return flow from the centrifuge can be considered as an internal phenomenon that, if great, will result in a small net feed flow (q). In that case, there will be a long holdup time in the crystallizer. In Figure 5.4.1 the feed flowrate is determined by the return flowrate, which gives a sort of automatic control of the holdup time based on the separation in the centrifuge. Depending on the type of centrifuge used (see section 3.7), it may be possible to control the flow from the crystallizer to the centrifuge on the basis of a measurement of quality in either the product line or the return line. The dashed lines in Figure 5.4.1 show control of the flow to the centrifuge on the basis of some measured value (size distribution, density, viscosity, etc.) of the product. As a result, the return flow will change, forcing a change in the feed rate.

One control variable that can be used is the circulation rate through the heat exchanger, as determined by pump P_2. This flow will affect the degree of supersaturation because it removes the unsaturated mother liquor from the lower section of the crystallizer and replaces it with supersaturated mother liquor from the upper section. As we see in Figure 5.4.1, an increase in circulation through pump P_2 leads to an increase in the supply of steam (due to temperature control) so that production increases. Increasing the degree of supersaturation in the lower section, however, results in a strong tendency for the creation of nucleii. The speed of pump P_2 is, then, a possible means of control of the crystal size distribution in the crystallizer. This is shown by a dashed line, which represents a control scheme based on measurement of the size distribution of the product.

The feed will often have a temperature lower than that of the return stream from the centrifuge. If the return stream temperature varies much, then the temperature of the material added to the crystallizer through pump P_1 can vary. The dashed lines in the figure show a temperature control loop that will hold the outflow at the desired average temperature.

Control of the crystallization process directly by addition and removal of nucleii has been proposed and attempted. In this case there is an extra control variable that is not shown in Figure 5.4.1.

Feedforward Control

Han (1969) presented a simulation study designed to develop a method of feedforward from measurement of the feed concentration, to control the feed rate. From Figure K.2, Appendix K, it can be seen clearly that this is possible provided we have free manipulation of the feed rate. This cannot be done in the system of Figure 5.4.1, but it becomes possible if the valve in the feed line to the centrifuge is used to maintain the level in the crystallizer, and the valve after pump P_1 is used to manipulate the feed rate via a flow controller. Feedforward must be based on measurement of the concentration of the feed after pump P_1. Han (1969) obtained a favorable response to dynamic feedforward developed on the basis of linearization of the mathematical model presented in Appendix K.

Multivariable Control of a Crystallizer on the Basis of a State Space Model and State Estimation

As indicated in Appendix K, the crystallization process is characterized by five state variables: the degree of supersaturation, the number of particles, the average particle length, the average particle surface area, and the average volume of the particles. In addition, we could perhaps define at least one state variable related to the thermal condition (temperature or pressure) in the evaporation zone. For a system similar to that of Figure 5.4.1 where there is mixing of fresh feed and a return stream in the feed tank, we could define one or two additional state variables for this portion of the process. If we also consider the liquid levels in the crystallizer and the tank, we could easily have about ten state variables, seven control variables, and perhaps six measured variables. This encourages consideration of a multivariable control strategy based on a state estimator and multivariable feedback based on linearization around some average operating point (see section 2.13). This solution has been suggested by several investigators (Frew, 1973; Liss and Shinnar, 1976), but it has not become common in industrial practice. Further progress can be made in this area.

5.5 Control of Drying Processes

Some of the most common industrial drying processes and a theoretical basis for understanding the mechanisms involved were presented in section 3.11. A mathematical model of the dynamics of a drying process is presented in Appendix L, which provides a foundation for the design of various control systems.

As we saw in section 3.11, in any drying system the temperature difference between the environment and the material to be dried is the driving force that determines the rate of drying. In the design of a drying process, one must be sure that this difference is maintained at an appropriate level, and that the material to be dried is held in the drying zone long enough for the temperature difference to physically drive the moisture (and vapor) from the material. The drying process appears to be simple to control, but this is not always the case; the simplest control structures do not include enough satisfactory measurements of the state variables to achieve complete control of the process.

In the following paragraphs we will present some of the conventional and simplest control systems for drying processes. We will also show how they can be improved by various methods.

Figure 5.5.1 is a schematic diagram for a rotating drum dryer. The screw conveyor at the left carries the moist, solid feed material into the system at what we will assume to be a constant velocity. The moist material is transported from left to right by the rotation of the drum. Air, heated in a heat exchanger with steam as the primary medium, is forced into the drum at the right, and moves countercurrent to the flow of the solid material. The heat exchanger has a temperature controller that assures that the air enters the drum at constant temperature. The air flowrate is controlled by a motor-driven damper in the inlet. The dryer operates under slight pressure controlled by another motor-driven damper under the action of a pressure controller in the air outlet at the left. Therefore, it is necessary to have an air lock on the outlet for the dry material. This is shown at the bottom right of the figure.

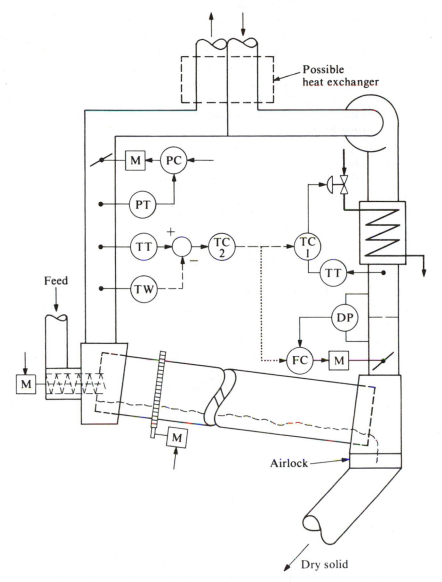

Figure 5.5.1. Conventional control system for rotating drum dryer.

The three controllers in this solution simply maintain conditions in which drying can occur. There is no feedback from any process variable related to the moisture content of the solid material being dried.

The process of Figure 5.5.1 is an example of a single-pass drying system, in that the hot air used to heat and remove the moisture passes through the dryer only once. This is very expensive, as the energy used to heat the air leaves with the air and is not reused. Because of the increasingly high cost of energy, it makes sense to reuse the energy (hot air) as much as possible. Therefore, the warm, moist air can be passed through a heat exchanger to preheat new air for use in the process. This decreases the steam demand, which is controlled by the temperature controller in the inlet to the dryer. Similarly, it might be reasonable to pass the hot, dry solid material through another heat exchanger to preheat the moist feed and thereby recover the heat remaining in that material. Whether these two heat-exchanging processes are economically justifiable will not be discussed further here.

The two most important distrubances to this process are variations in the flowrate and in the moisture content of the feed. Normally, the feed rate will never exceed the designed capacity of the dryer. However, the rate can be much less than the designed rate if, for example, there is a failure in the process upstream of the dryer. In the worst case this could lead to a final product much drier than desired if the control scheme of Figure 5.5.1 were used. Many times this is not a problem except that it represents a waste of energy, but some products cannot tolerate overdrying. A more sophisticated control system can eliminate this operating problem.

Without doubt, the most important disturbance is the moisture content of the feed. If the feed rate is held constant, then an increase in moisture content requires more energy for drying if the moisture content of the product is to remain the same. Since both temperature and flow of the inlet air are controlled, an increase in moisture content of the feed will soon cause a drop in temperature of the air leaving the dryer. The simplest way to compensate for this is to put a temperature controller on the outlet air with cascade control, as shown by the dashed line in Figure 5.5.1. The new temperature controller TC-2, if it has the integral mode (PI controller), will cause the set point of the temperature controller on the inlet, TC-1, to increase until the outlet air temperature once again attains its original value. This is shown in Figure 5.5.2. At the top, the temperature of the solid (θ_s) and the temperature of the gas (θ_g) are plotted versus the distance along the length of the dryer. The lower plot of Figure 5.5.2 shows the moisture content of the solid versus dryer length.

The dashed lines in Figure 5.5.2 give the steady-state temperature and moisture conditions following a slight increase of moisture in the feed for a system equipped only with temperature control on the gas (air) supply and with no feedback from the temperature of the gas leaving the dryer. We see that the zone in the dryer where the drying rate is constant is shifted somewhat to the right. The gas temperature is now always below the solid line, but it starts at the same value at the inlet. The moisture in the product at the outlet is slightly higher than before.

The dotted curves also show the steady-state conditions following an increase in feed moisture content, but in this case the system has feedback from the temperature of the gas leaving the dryer (temperature controller TC-2). We see that the temperature of the *outlet* gas is now equal to that of the solid curve, but the temperature is above the solid curve through the entire dryer because the inlet

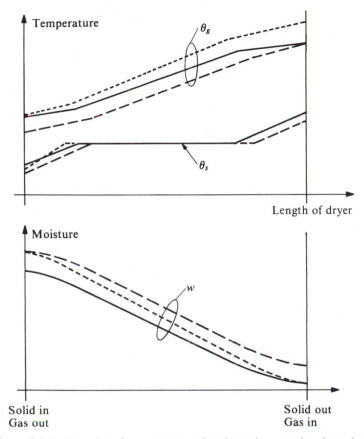

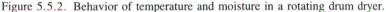

Figure 5.5.2. Behavior of temperature and moisture in a rotating drum dryer.

temperature is raised by the controller. The moisture, indicated by the dotted line in the lower figure, has a higher value at the left (the entrance to the drum), but drops more quickly because of the higher temperature difference. The final moisture is about the same as the value indicated by the solid line. The actual final value depends on the type of solid material, dryer design, and so on. Holding the temperature at the outlet constant, as the new control system does, is not necessarily sufficient to obtain the desired product moisture content.

Since the driving force for drying is the temperature difference between the gas and the average temperature of the solid material, it is highly desirable to be able to measure this temperature difference. The temperature of the solid can be approximately inferred from measurement of the wet-bulb temperature of the gas (the temperature measured by a bulb covered with a moist sleeve). This principle was described in section 3.11. In order to get a proper picture of the conditions in the dryer, the temperature should be measured at several points along the length of the dryer. This is not always practical, however; one must often work with measurements from the two ends of the dryer. Further, the use of wet-bulb temperatures presents many practical difficulties. The measurement often indicates a temperature higher than the actual value, and it is sensitive to dust in the gas flow. There are other ways to measure the moisture in the gas, one being to use a dew point measurement. Assuming that we can in fact obtain a wet-bulb temperature, the difference between the wet-bulb and dry-bulb temperatures can be used by the temperature controller (TC-2), as shown by the dashed lines in the center of Figure 5.5.1.

A possible control variable is the gas flow through the dryer. The set point for the flow controller on the gas leaving the dryer in Figure 5.5.1 is assumed to be constant. However, we could let the temperature controller on the outlet gas act on this set point instead of the set point of the temperature controller on the inlet flow. This is shown by the dotted line in Figure 5.5.1. Increasing the gas flow adds energy to the dryer, but this has the advantage that the temperature in the inlet will not need to be as high. This can be important in cases where the solid material is sensitive to high temperature.

Feedforward for control of a dryer requires a measurement of the moisture content of the feed, as that is the most important disturbance. Unfortunately, this measurement is usually very difficult and/or expensive to obtain. But, if such a measurement should be available, feedforward to the set point for temperature controller TC-1 or the air flow controller FC in the inlet could be very effective. In all countercurrent processes that include a significant transport time for the solid material flow, as is the case for a rotary drum dryer, there is the problem that the response of the moisture in the solid product to a change in moisture in the feed will be characterized by an essentially pure transport delay. Therefore, if the set point for the fresh gas temperature (or flow) is controlled by feedforward from the feed moisture through a corresponding transport delay, the result will not be especially effective. Determination of the "best" feedforward dynamics is a question of optimization that can be treated by the theory of optimal control, as considered later.

Materials that are especially sensitive to overheating should be dried in co-current dryers. In this case the hot, dry gas enters the process at the left side of Figure 5.5.1, and the moist gas leaves at the right. An advantage to this system is that the hottest gas contacts the moist solid. As a result, the initial warming zone for the solid is shorter, the zone of constant drying rate is longer, and the temperature of the material leaving the dryer is lower. A disadvantage is that a co-current dryer of a given length will usually have less capacity than it would if it were operated countercurrently. The control techniques are about the same except that the feedforward from the feed moisture of flowrate is easier to realize in this case because the manipulated variable (the air temperature) follows the solid as it passes through the dryer.

Multivariable Control with State Estimation of a Rotary Drum Dryer

The drying process is a good example of a process where a state space model updated by the available measurements can be used to estimate the most important dynamic states, which can then be used for multivariable control. Nevertheless, this method of attack has received little attention in the literature, probably because most dryers operate satisfactorily with the simple control schemes discussed previously. Thus, there may be insufficient motivation to develop significantly more complicated control schemes.

However, some drying processes are very critical, and the benefit of improved control of these processes can be substantial. In these cases, complex control schemes can be justified.

Before developing a state estimator, one must choose the model structure; that is, the variables that determine the process dynamics must be recognized. Since a rotating drum dryer includes mass transfer (the solid moving through the drum), physical phase changes (moisture evaporating to vapor and gas), and heat transfer (from gas to solid to water), a state space model that can describe the transport processes is preferred. In Appendix L we develop a highly simplified mathematical model where the dryer is divided into several segments along the axial direction. These elements are short enough to make reasonable the assumption that the state variables (moisture in the solid w, temperature in the solid θ_s, and temperature in the gas θ_g) do not vary over the length of the element. When we assume that the velocity of the solid moving through the drum is constant, we can discretize the time variable such that the time step is $\Delta t = \Delta z / v$ where Δz is the length of the element in the axial direction, and v is the transport velocity of the solid material.

As we see in Appendix L, this leads to a discrete time model with $3N$ state variables where N is the number of elements in the axial direction of the dryer. The more elements used to discretize the dryer, the greater the required computational effort. On the other hand, a larger number of elements gives a better approximation, up to a point. Because the basis of the model is an approximation, and because the system will be equipped with feedback, it usually makes little practical sense to use more than about five elements, and certainly there should be fewer than ten. The control variables are temperature and flow of the hot gas entering element $(N + 1)$. The dynamics of the heat exchanger used to heat the gas can be modeled separately as they are independent of the activity in the dryer. Further, we can assume that the pressure controller is perfect so that it does not affect the system.

The most important disturbance is the moisture in the feed entering element number 0. If one also wants to observe the variations of mass flow, one must define a state variable in addition to the three given above. This state variable will be the mass of dry material (m) in each element along the dryer.

On the basis of the discrete model, a discrete state estimator can be formulated to update the calculated states of the model using any available process measurements. Figure 5.5.3 is a block diagram for such a state estimator and the control system that goes with it, designed with the help of optimal control theory.

The two estimates of the state vector, the *a priori* estimate [$\overline{\mathbf{x}}(k)$] and the *a posteriori* estimate ($\hat{\mathbf{x}}(k)$] which contain the $3N$ elements (possibly $4N$),

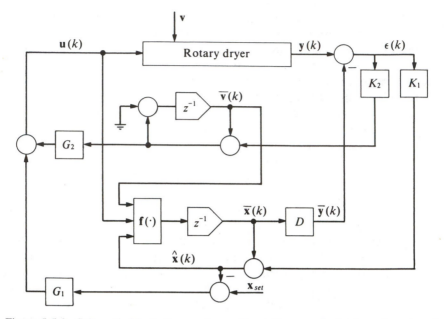

Figure 5.5.3. Schematic block diagram for multivariable control of a drum dryer based on state estimation.

are computed repeatedly by means of the model presented in Appendix L with updating through the matrix K_1 (assumed to be constant) from the difference between the actual measurements $[\mathbf{y}(k)]$ of temperature and moisture in the gas outflow and temperature of the product dry material and the corresponding calculated measurements $[\overline{\mathbf{y}}(k)]$.

The scheme in Figure 5.5.3 includes a disturbance vector $\mathbf{v}(k)$ that has the same dimensions as the measurement vector $y(k)$. Two of the elements of this vector will be the dominant disturbances, feed moisture and flow. If we have more than two independent measurements, we can add a corresponding number of fictitious disturbances. These could be, for example, process parameters. As shown in the figure, a disturbance modeled as a constant can be updated from the measurements through the matrix K_2.

The control vector (temperature and flow of hot gas) is generated by a linear combination of the estimated state vector's deviation from the reference (set point) vector and the disturbance vector through matrices G_1 and G_2, respectively. These matrices are assumed to be constant. The matrices K_1, K_2, G_1, and G_2 are calculated off-line ahead of time using the computation routines for optimal control and optimal state estimation.

The difference $\Delta\mathbf{x}(k)$ between the specified state profile $\mathbf{x}_{\text{ref}}$ and the estimated state profile $\hat{\mathbf{x}}(k)$ enters into an objective functional for computation of the optimal feedback through the quadratic form $\Delta\mathbf{x}^T(k)Q\,\Delta\mathbf{x}(k)$, where the vector matrix Q is chosen to be a diagonal matrix. The elements of this matrix will have different numerical values, depending on the weight one chooses to put on the deviations of the various states in relation to the specified values for those states along the dryer. It is reasonable that the weight assigned to the deviation of the moisture from the specified values should be greater toward the end of the dryer than at the inlet.

As we have said, the control scheme shown in Figure 5.5.3, which is considerably more complex than that in Figure 5.5.1, can be justified only if the requirements of the process are substantially more stringent than those that can be satisfied by the simplest solution. The cost of realizing the solution of Figure 5.5.3 is essentially the cost of the computing equipment. If we ignore the cost of development, the rapid decrease in the cost of computer equipment will make the multivariable solution attractive for the more demanding drying systems.

Yliniemi and Uronen (1983) have presented a simplified solution of the kind described above. In their solution the state vector contains only three variables, all of which are measurable. They are moisture of the dried solid (w_N), temperature of the dried solid (θ_{sN}), and temperature of the gas at the outlet (θ_{g1}). This system gave quite satisfactory control performance.

An alternative to the model structure chosen for the state estimator above, which was based on division of the axial direction of the dryer into elements with three or four state variables in each element, is to model the different state variable profiles by a piecewise straight-line approximation as shown in Figure 5.5.2. The break points and angles for these lines will be new "(state variables)" that can give an effective representation of the process dynamics and can be updated by measurements using the same principles as shown in Figure 5.5.3. This method gives a model with fewer state variables than the model described previously, but special care must be used to get a proper model of the transport delay phenomenon in the dryer. This method has been used by Gilles and Retzbach (1980) for control of a distillation column having a pronounced temperature break, with very good results. The method will be discussed further in section 5.6.

The previous discussion considered a rotary drum dryer, which is the most widely used drying system in industry today. Drying by means of a fluidized bed has many distinct advantages if it is suitable for the particular materials to be dried. It is easier to obtain good control of the fluidized bed than of the drum dryer because there is no transport delay in the fluidized bed, and because the fluidized bed provides nearly perfect contact between the hot gas and the solid.

A fluidized bed with a simple control scheme is shown in Figure 5.5.4. Gas, usually air, enters at the left driven by a fan, and then passes through a steam-heated heat exchanger to the bottom of the dryer, which is a vertical cylinder. At the point where the cylinder begins, the horizontal distributor has a large number of orifices that increase the velocity of the gas and cause the solid particles to float. The feed enters through the angled tube at the left, and the dry product leaves through the corresponding tube at the right. A special lock ar-

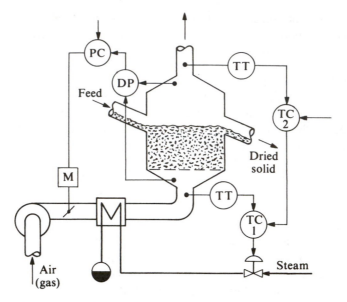

Figure 5.5.4. A typical system for control of a fluidized bed.

rangement on the outlet prevents the gas from leaving the system with the product; the gas leaves the dryer at the top. The gas flow is controlled by a damper after the fan, and the setting of the damper is controlled on the basis of the differential pressure across the bed. The dynamics of this loop are complicated, but a PI controller is a good solution and is easily adjusted. Temperature control is about the same as that described for the system of Figure 5.5.1, where the steam flow to the heat exchanger is controlled by a temperature controller operating on the measured temperature of the gas at the inlet to the dryer. The set point for this temperature controller is controlled by a temperature controller operating on the basis of the temperature of the gas at the outlet. It is also possible in this case to measure the wet-bulb temperature and let the difference between it and the dry-bulb temperature set controller TC-2. Furthermore, feedforward from a measured feed flow can be added to the set point of TC-1.

In this system it is also possible to exchange the heat of the warm gas leaving the bed with the cold gas ahead of the fan in order to get better heat and energy economy. The result of this will be reduced steam consumption, and no particular change in the structure of the control system is needed.

5.6. CONTROL OF DISTILLATION COLUMNS

One of the most important industrial unit processes, distillation can be used as an element in complex combinations in many kinds of industries. The basic techniques of distillation were described in section 3.12. Many different types of mathematical models have been developed for distillation processes, and these models offer various possibilities for control, depending on the type of control system one wishes to design. Some models use a state space description with many state variables, perhaps four per plate. These models provide a starting point for the development of reduced models of lower order. Models have also been presented in the form of transfer function matrices, often based on linearized detailed state space models. The transfer functions in these matrices usually have one or two time constants and a transport delay (see section 3.12). In some cases it is possible to use a model that represents the column in terms of stylized profiles of temperature and concentration, for example, where the "state variables" are characteristic points on these profiles (Gilles and Retzbach, 1980).

Some Typical Control Structures

Let us now look at some of the common ways of controlling a binary distillation column. We will look first at methods based on conventional control principles including isolated (individual) loops, cascade control, ratio control, and feedforward. Then we will discuss some methods based on state estimators and optimal control. Finally, we will examine some results obtained with adaptive control. We will provide commentary and explanation as needed.

The most important problem in a distillation column is control of the material balances. This has to do primarily with control of the various flowrates, but it is also strongly coupled to the energy balances, especially those related to the addition of energy to the reboiler and removal of energy from the condenser. The next most important problem is control of product quality, evaluated as product concentration.

There are three possible arrangements for control of the material balance in a simple column having only a bottom product and a distillate, as shown in Figures 5.6.1a, b, and c. In Figure 5.6.1a, the level in the accumulator is controlled by the distillate flow, and the level in the reboiler is controlled by the flow of bottom product. In Figure 5.6.1b, the accumulator level is controlled by the flow of distillate, but the level in the reboiler is now controlled by the addition of heat to the reboiler. The level of the accumulator in Figure 5.6.1c is controlled by the reflux flow, and the bottom product flow controls the level in the reboiler. In addition to the systems for material balance control, Figure 5.6.1 also indicates different structures for controlling product quality. The quality of the top product in Figure 5.6.1a is controlled by means of a controller cascaded to the reflux flow (we know from section 3.12 that increasing the reflux flow gives higher purity to the top product, but at a reduced production rate). Concentration of the bottom product is controlled by a controller in cascade with the energy flow to the reboiler, which changes in the vertical flow of vapor through the column. The top concentration in Figure 5.6.1b is controlled the same as in Figure 5.6.1a, but the bottom concentration is now controlled by changing the set point of the bottoms flow controller. The result of

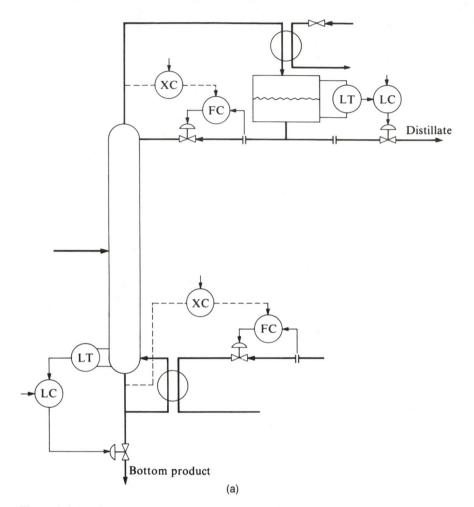

(a)

Figure 5.6.1. Three structures (a, b, and c) for control of a simple distillation column.

this will be about the same as for the system of Figure 5.6.1b because a large flow of bottom product out of the system affects the liquid level in the bottom section and this, in turn, acts on the energy flow to the reboiler. In Figure 5.6.1c, the top product concentration is controlled by controlling the set point in the flow controller for the distillate. This solution can give a considerably more stable distillate flow than the system shown in Figure 5.6.1a because the reflux flow is usually significantly greater than the distillate flow. This is important if the distillate is used as the feed to a subsequent unit process where it is desirable to have a feed that varies as little as possible.

Control of the level in the bottom section by using the flow of heat input as shown in Figure 5.6.1b is not used as often as the schemes shown in Fig-

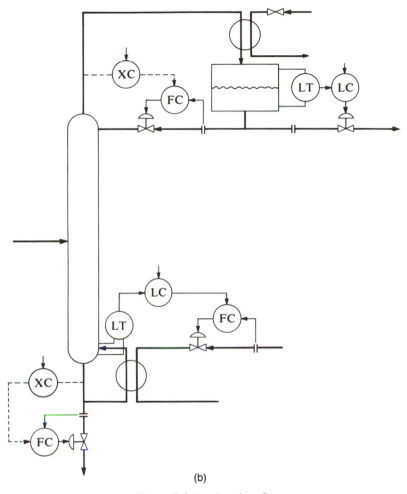

(b)

Figure 5.6.1.   (continued)

ures 5.6.1a and 5.6.1c because this method has difficult dynamics that can lead
to undesirable variations in the vapor flow in the lower part of the column. It is
most common to use the flow of the bottom product for level control.

Figure 5.6.2 shows how the pressure in the top of the column is controlled
by controlling the energy leaving the condenser. The objective is to maintain a
constant pressure or pressure drop across the column. The figure also shows a
means of control of the internal reflux ratio, based here on measurement of the
reflux flowrate and the top vapor flowrate. Alternatively, the reflux ratio control could be based on measurement of the distillate flowrate instead of the top
vapor flow [considering the relationship given by equation (3.12.13)]. The concentration of the top product is controlled by the internal reflux ratio.

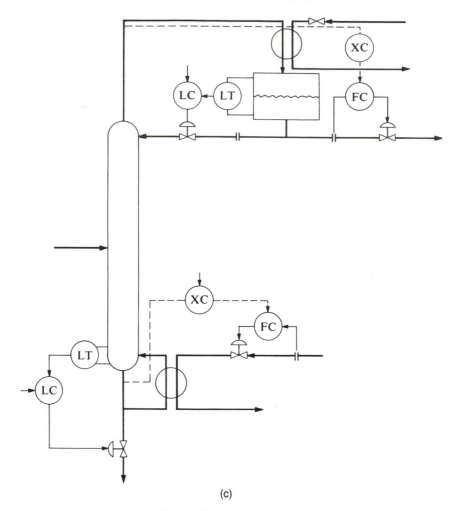

(c)

Figure 5.6.1. (continued)

Figure 5.6.3 illustrates a case where the top product is removed in vapor form, and pressure control is obtained by means of a valve in the vapor line. Flow of the cooling medium to the condenser is controlled by a ratio controller from the top product. The flow of condensate will be proportional to the flow of cooling medium, and the condensate goes to the accumulator, which has level control using the reflux flow; so there will be a known ratio between the top product flow out and the reflux flow. This ratio is adjusted on the basis of analysis of the concentration of the top product. As discussed in section 3.12, it is the reflux ratio that best controls the top product concentration at the cost of distillate (top product) flowrate.

When the top product is condensed, the reflux ratio, relative to the distillate flow, can be controlled as shown in Figure 5.6.4. The level in the accumulator

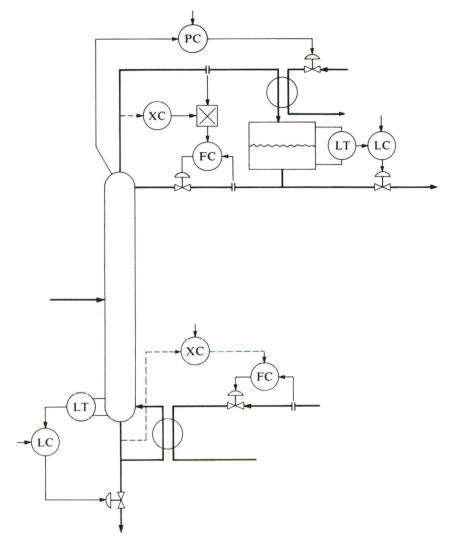

Figure 5.6.2. The most common method for pressure control in a distillation column; an example of control of the reflux ratio.

controls the set point of a flow controller on the distillate stream, and the measurement of the distillate flow is sent to the ratio controller on the reflux stream flow. The ratio setting is automatically controlled on the basis of the measured concentration (possibly temperature) in the top product vapor or distillate. Figure 5.6.4 also shows how the bottom section can be controlled by the boil-up ratio, which is similar to the reflux ratio. The vertical flow of vapor, which is proportional to the heat supplied to the reboiler, is made proportional to the flow of bottom product. The boil-up ratio is controlled on the basis of bottom product flow.

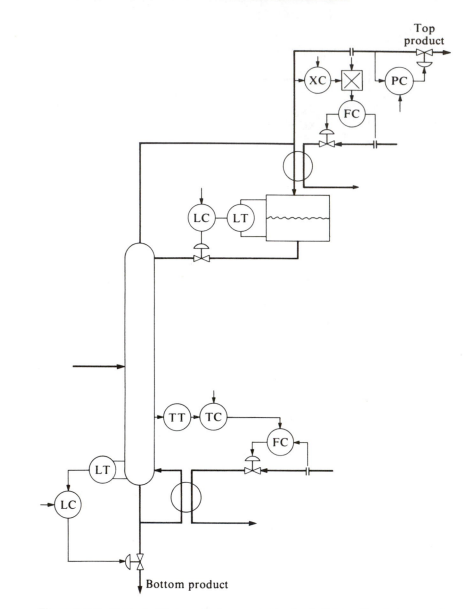

Figure 5.6.3. Control of top product concentration by indirect reflux ratio control.

Without doubt, the most frequent disturbance to a distillation column is flow variation of the feed to the column. If we assume that the concentration and temperature of the feed are constant, then an increase in feed flowrate will appear first as an increase of the liquid flow downward in the lower part of the column. If the heat supply to the reboiler is held constant during this change in

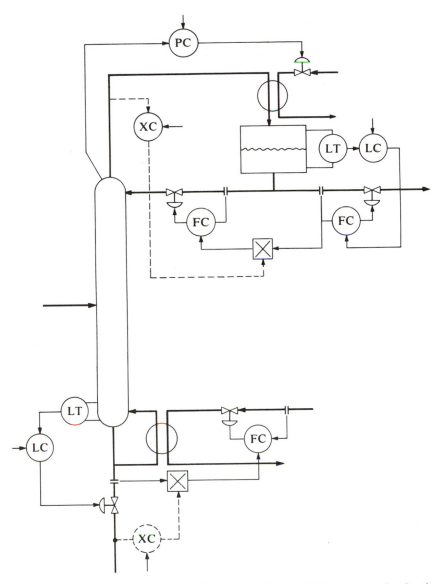

Figure 5.6.4. A typical arrangement for control of top product concentration by the reflux ratio and bottom product concentration by the boil-up ratio.

feed, the vertical flow of vapor will not change much. As a result, the flow of bottom product will increase because the action of the level controller in the bottom section will maintain the mass balance in that section of the column. The mass balance based on the level controller also tells us that if the bottom product concentration is to stay the same, an increase in feed flow must also lead to an increase in the flow of vapor up through the column, a condition that in turn requires an increase in heat to the reboiler. An increase in feed flowrate will also cause an increase in the vertical liquid flow above the feed entry point due to the reflux flow that follows the vapor flow.

These observations lead immediately to consideration of feedforward from the measured feed flow to the set point for the heat flow controller, to the reboiler, and to the reflux flow controller, as shown in Figure 5.6.5. These feedforwards must have correct dynamic characteristics based upon the transfer

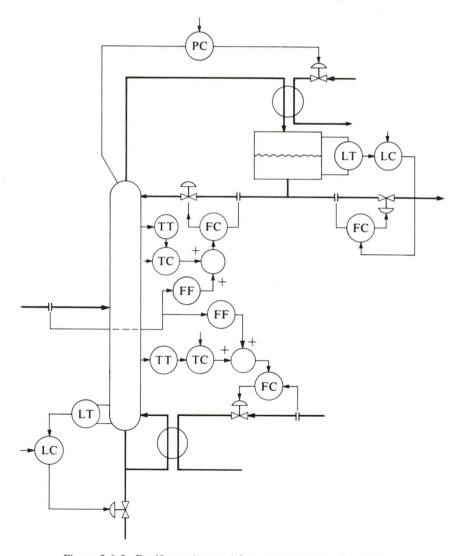

Figure 5.6.5. Feedforward control from measurement of feed flow.

functions of the column in both directions. The feedforwards need not be concerned with the steady state (zero response at zero frequency) because this is controlled by the concentration controllers (shown as the two temperature controllers). Alternatively, feedforward to the set point of reflux flow, shown in Figure 5.6.5, can be moved to the distillate flow if the reflux flow is used for level control of the accumulator. It is also possible to let the feedforward from the feed flow act on the set point for the bottoms flow controller, but this has little advantage over the solution shown in Figure 5.6.5.

Variations in the thermal state of the feed, if ignored, will change the conditions in the entire column. Except for the cost of installation, it is feasible to remove these changes by means of a heat exchanger in the feed line, as shown in Figure 5.6.6. The heat exchanger can provide either heating or cooling, depending on the situation, and control is attained on the basis of a measurement or calculation of enthalpy.

Condensate of the top product can easily be overcooled such that the reflux liquid will have a lower temperature than the temperature of the top tray in the column. This is an unfortunate situation, as it causes the internal reflux (the liquid flow from the top plate downward) to be different from the external reflux, which we control with a valve. Therefore, we want to design a control system that controls the internal reflux on the basis of a simple estimator that calculates the actual internal reflux flow by means of a few measurements. A schematic drawing of the top plate in the column is given in Figure 5.6.7 (Perry and Chilton, 1973). The different variables are designated with the same notation as used in section 3.12, where V and L are mass flows of vapor and liquid, respectively, and R is the mass flow of the reflux. If h_V and h_L give the enthalpies of

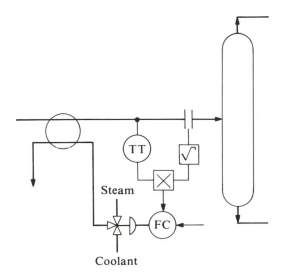

Figure 5.6.6. Enthalpy control of the feed.

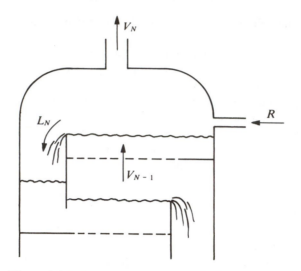

Figure 5.6.7. Detail of top plate in a distillation column.

the vapor and liquid, the following two equations can be written to describe the enthalpy and mass balances on the top plate:

$$V_N h_{VN} + L_N h_{LN} = V_{N-1} h_{V(N-1)} + R h_R \tag{5.6.1}$$

$$V_N + L_N = V_{N-1} + R \tag{5.6.2}$$

It is a reasonable approximation to assume that:

$$h_{VN} = h_{V(N-1)} = h_V \tag{5.6.3}$$

and:

$$h_V = h_{LN} + \lambda \tag{5.6.4}$$

where λ is the heat of evaporation, and:

$$h_{LN} = h_R + c_p(\theta_{LN} - \theta_R) \tag{5.6.5}$$

where:

θ_{LN} = temperature of the liquid at the top plate
θ_R = reflux temperature
c_p = specific heat of the liquid

Substituting equations (5.6.3) through (5.6.5) into equations (5.6.1) and (5.6.2), and solving for L_N, which is the internal reflux, we get:

$$L_N = R\left(\frac{c_p(\theta_{LN} - \theta_R)}{\lambda} + 1\right) \tag{5.6.6}$$

Now, let us introduce:

$$\Delta\theta = \theta_{VN} - \theta_{LN} = \text{constant}$$

where θ_{VN} is the temperature of the vapor leaving the top tray. We find that:

$$L_N = R(k_1 + k_2(\theta_{VN} - \theta_R)) \tag{5.6.7}$$

where k_1 and k_2 are constants given by:

$$k_1 = 1 - k_2\Delta\theta \tag{5.6.8}$$

$$k_2 = c_p/\lambda \tag{5.6.9}$$

The two temperatures appearing in equation (5.6.7) can be measured, as can the external reflux flow. Therefore, L_N can be computed from equation (5.6.7). Figure 5.6.8 is a diagram of the solution for the problem where the internal reflux is calculated as L_N and fed to a controller with the set point $L_{N,\text{SET}}$ to control the reflux valve. This scheme corrects the external reflux flow (R) on the basis of the difference between the two stream temperatures.

The elementary control structures for a binary distillation column shown in Figures 5.6.1 through 5.6.6 can be developed further. McAvoy (1983) and Shinskey (1984) have studied different possibilities for controlling the setpoints of the boil-up and reflux controllers from measurements of the bottom product and distillate concentrations via a steady-state model of the column based on equation (3.12.9). This clearly indicates a long-awaited trend toward the use of model-based control systems.

Energy Integration (Buckley, 1982)

When several columns are connected in series to form a large system for the separation of a raw material into many different fractions, it is possible and desirable to couple the columns thermally so that the heat recovered in a condenser at one stage can be used for reboiling in another stage. This may create a control problem more difficult than existed for the simple case discussed above, because the great number of cross-couplings or interactions. An example of this is shown in Figure 5.6.9, where the bottom product from column 1 is the feed for column 2, and the condenser of column 1 is the reboiler for column 2. Figure 5.6.10 shows how the feed to a column can be heated by the bottom product of the same column for the purpose of increasing the enthalpy of the feed.

Figure 5.6.11 is a diagram of a comprehensive system for energy economy where some of the vapor from the top of the column is taken out at the accumulator and sent to a compressor where it takes on energy, and the temperature and pressure are increased to superheat the vapor (Shinskey, 1977). This vapor is used in the reboiler, and the condensate (distillate) goes to a second accumu-

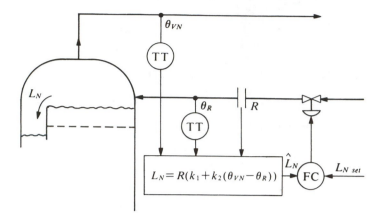

Figure 5.6.8. System for controlling the internal reflux flow.

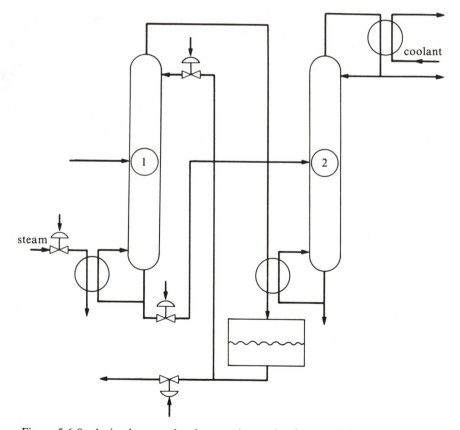

Figure 5.6.9. A simple example of energy integration for two distillation columns.

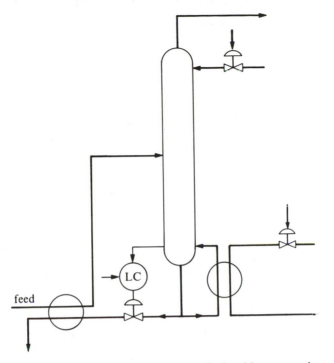

Figure 5.6.10. Energy integration of the feed and bottom product.

lator as shown at the bottom of the figure. From here, the distillate is sent under pressure partly to the next stage of the process and partly, under level control, back to the first accumulator. It thereby undergoes adiabatic expansion to decrease the temperature and pressure. This cycle represents a heat pump that gives good energy economy at the cost of adding the compressor.

Figure 5.6.12 is an enthalpy–pressure diagram for the case of a distillation column designed to separate propane and propylene, which have nearly the same boiling points (see Figure 3.12.18). Propylene will be the top product. The two horizontal lines in Figure 5.6.12 represent the states in the two accumulators, and the line slanting up to the right represents the enthalpy increase due to compression. This increase is slightly nonisentropic because of the losses in the compressor. We see that the vapor is superheated by about 10°C. The values given in Figure 5.6.11 are based on the upper accumulator having propylene at a pressure of 10.5 bar with saturated vapor at 21°C. If we assume that the pressure drop through the column is about 0.5 bar, the pressure at the bottom will be 11 bar. If we further assume that the bottom product is pure propane, the saturation temperature of the bottom product will be 30°C. A reasonable temperature drop between the primary and secondary side of the reboiler is at least 13°C, in which case the propylene must condense at 43°C,

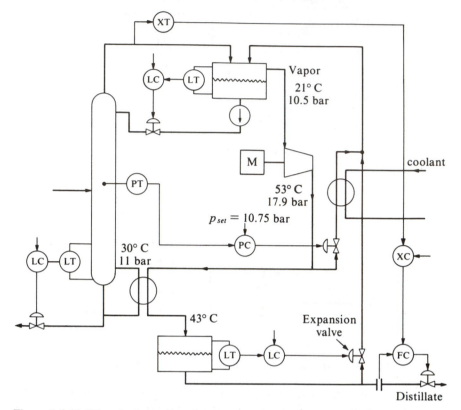

Figure 5.6.11. Use of a heat pump for energy-saving in a distillation column (based on Shinskey 1977).

corresponding to a pressure of 17.9 bar. Thus, both horizontal lines of Figure 5.6.12 have been determined, and the superheat temperature, given by the slanted line, is found to be about 10°C.

During evaporation of the propylene, which follows after the adiabatic expansion takes place along the lower horizontal line in the figure, there will be an enthalpy change of about 285 kJ/kg. This produces a cooling effect that leads to condensation of a corresponding amount of propylene from the distillation column. The enthalpy change experienced by the propylene passing from the compressor to the lower accumulator (represented by the upper horizontal line in the figure) is found to be about 315 kJ/kg and is available for evaporation in the bottom of the column.

The scheme shown in Figure 5.6.11 gives very strong energy coupling from top to bottom in the column. A number of alternative ways to control this system are possible, but all give more or less the same result. In any case, the following comments can be made:

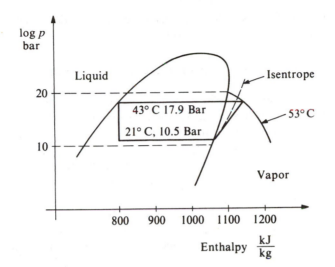

Figure 5.6.12. Enthalpy–pressure diagram for heat pump in Figure 5.6.11.

1. Because propylene and propane have a low relative volatility, the column must be operated with a high reflux ratio. This means that most of what enters the upper accumulator must be used for reflux.
2. Speed control of the compressor will not have much effect on the energy conditions in the system because it is the energy added to the compressor that is important. Therefore, the compressor can operate at constant speed.
3. In order to control the energy state, a means of manipulating the removal of energy from the system must be included.

Figure 5.6.11 shows that some of the propylene vapor is returned around the compressor and cooled. This vapor stream is used to control the pressure in the column. Also, the level at the bottom is controlled by the flow of bottom product, and the level in the upper accumulator is controlled by the reflux flow. The concentration of the top product can be controlled by means of the flow of distillate, as shown in the figure.

A general phenomenon that occurs when material or energy is recirculated in a process is that the holdup time increases. This leads to slower dynamics unless the recirculating stream is controlled so that the process "can't tell" whether the material or energy flow is recirculated or comes from an external source. In most cases this control requires some form of holdup capacity.

The many strong interactions that can occur in systems of distillation columns, when energy is to be used most efficiently, provide a strong argument for the use of some form of multivariable control strategy, perhaps based on the state space concept. This problem has not been treated extensively in the literature, but a great deal of effort is being devoted to it. Multivariable control of more conventional (freestanding) distillation columns is described extensively, and will be treated later in this chapter.

A Complex Distillation Control System (Bristol, 1980)

Figure 5.6.13 is an example of a relatively complex control system for the separation of butanes from gasoline and further separation of isobutane from normal butane. The most important elements in this system are summarized as follows:

1. The first column has feedforward from the measurement of the feed flow to the set point of the flow controller on the distillate, through a multiplier that corrects the strength of the feedforward on the basis of measurement of the concentration of the top product.
2. There is feedforward from the distillate flow to the set point of the reflux flow, which in turn is in cascade with the level controller for the accumulator.
3. For good energy efficiency, the cooler for the first column is the reboiler for the second column, but there is also a second reboiler where the primary medium is hot oil, which can be controlled.
4. Distillate from the first column passes through a surge tank with level controlled by the outflow of butane, which is also the feed for the second column.
5. The value of the feed flow to the second column is normally fed forward to the flow of isobutane distillate at the top [dynamic feedforward indicated as $g(t)$], where it contributes to the set point for the flow controller on the reflux. A number of correction factors are multiplied into the system to adjust the strength of the feedforward.
6. Feedforward to the isobutane flow is adjusted on the basis of analysis of the normal butane in the bottom section.
7. Feedforward to the reflux flow is adjusted on the basis of analysis of the isobutane at the top.
8. The isobutane at the top is completely condensed in an air cooler, and the pressure at the top is controlled with the help of a bypass. The reflux flow of isobutane is controlled by a flow controller whose set point is determined partly by the level controller for the accumulator and partly by feedforward from the feed stream.
9. The differential pressure across part of the second column normally acts on the addition of energy to the column's reboiler, but a minimal value selector can switch the energy addition to control based on the accumulator level.
10. The reboiler in column 1 has a system for calculating the energy addition on the basis of the flow measurement and the measured temperature difference. The control valve on the hot oil used for heating this reboiler

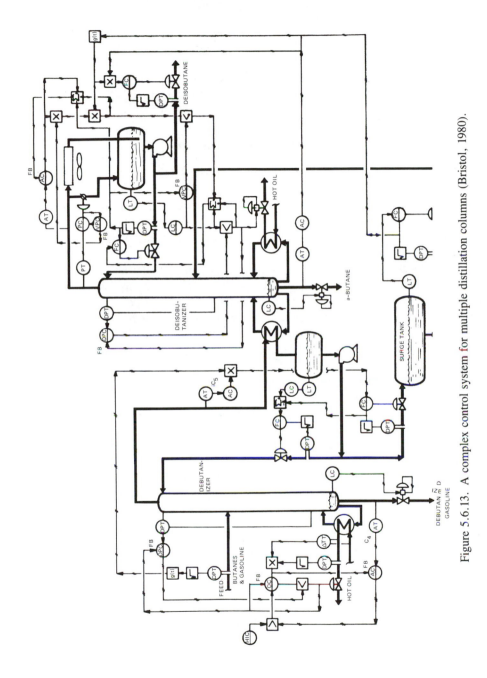

Figure 5.6.13. A complex control system for multiple distillation columns (Bristol, 1980).

can be controlled either from a heat flow controller based on the calculated heat flow or from a differential pressure controller that measures the differential pressure across the column, depending upon which gives the larger deviation.

11. The set point for the energy controller for the reboiler can be either entered manually or taken from the concentration controller for the bottom product, depending on which gives the larger deviation.

Multivariable Control of Distillation Columns

The structural solutions discussed in the previous section are based on physical considerations that lead to the natural pairing of variables (control variables and measured variables) in monovariable control systems. Knowledge of process dynamics characteristics has not been used to any great extent. Mass and energy balances at steady-state operating conditions were the dominant factors in determining the control structures.

Because distillation columns actually have strong interactions among the many different variables, many attempts have been made to use some of the tenets of modern control theory to get better operation, especially with respect to operating disturbances (such as feed variations), and when many columns are coupled together so that a disturbance can easily propagate through the system. Feedforward has already been mentioned as one means of improving control system characteristics without creating stability problems. Decoupling by means of dynamic cross-coupling terms has also been proposed and tried. Multivariable control based on estimated state variables and feedback from these estimated states, based on either optimal control theory or the theory of modal control, has also yielded good results in many cases. Adaptive monovariable controllers, especially those designed for minimum variance control, have also given good results in several systems.

Some of the most important results from the approaches mentioned above will be discussed in the following sections.

Characteristically, in the design of the conventional control structures described in the previous sections, there is no attempt to develop a precise description of the objectives of the control system. In optimal control theory, such a description is defined as the "objective functional" that one wants to minimize by generating the control variables properly. Even if one is not interested in a truly optimal solution (which can be very complicated and expensive), there is always at least some advantage in having a reasonable idea of how to weigh the effects of changes in the various system variables. This leads immediately to consideration of optimization, and as mentioned in section 2.16, one must decide whether to use dynamic optimization or static optimization. If static optimization is satisfactory, the objective of the control system will be to hold the variables at the "best possible" values near their set points that can be adjusted slowly in order to obtain static optimal conditions. The ability of the control system to respond to external disturbances and to those generated internally by interactions is of critical importance.

Decoupling

Several attempts have been made to achieve decoupling in control systems for distillation columns. One was reported by Luyben (1970), in which he treated a column described by a model that used the reflux flow and reboiler heat flow as control variables and concentration of the distillate and of the bottom product as the two outputs. Using the notation introduced in section 2.9, we can design a decoupling controller that eliminates, or at least reduces the effect of, the cross-couplings in the process when suitable pairings of the control and measurement variables in the process have been defined. The transfer function matrix $H_u(s)$ can be defined for the following variables:

u_1: reflux flow
u_2: reboiler heat flow
y_1: concentration of distillate
y_2: concentration of bottom product

The decoupling, given by equation (2.9.2), is:

$$H_{c2}(s) = H_u^{-1}(s) \; \mathrm{diag}(H_u(s)) \qquad (5.6.10)$$

Luyben (1970) found the elements of the frequency response matrix $H_{c2}(j\omega)$ as shown in Figure 5.6.14. The apparent resonance phenomena occurring around $\omega = 3$–4 (rad/min) are due to the fact that the process model includes transport delays. The equivalent phenomena are described in detail in section 3.8 (see Figure 3.8.10).

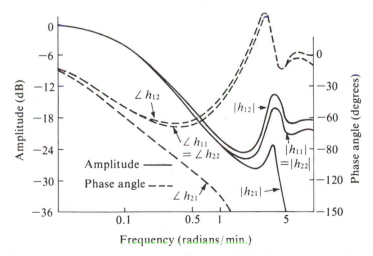

Figure 5.6.14. Frequency response functions for a distillation column (Luyben, 1970).

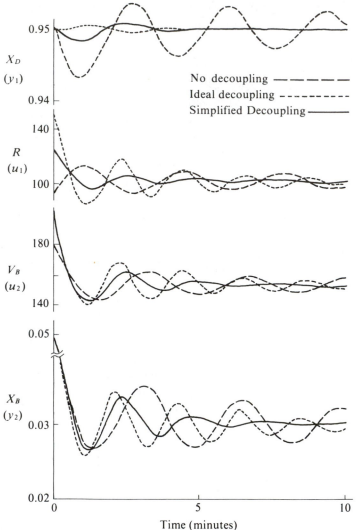

Figure 5.6.15. Results of use of decoupling in distillation column (Luyben, 1970).

Various approximations can be made on the basis of the ideal decoupler. Figure 5.6.15 shows the results obtained from an ideal decoupler and a simplified decoupler, compared with the system without decoupling. The upset to the system in each case was a step change in the set point for the controller on the concentration of the bottom product. As we see, the ideal decoupler gives nearly perfect decoupling with regard to concentration of the distillate. The simplified decoupler also gives a usable result. The oscillation of the bottom product concentration is due to the adjustment of the controller and is nearly the same for all three cases. It is important to note how the reflux flow and the reboiler power react. Since the step occurred in the set point for the bottom product concentration, it is most interesting to see how the reflux flow varies in the three cases. The ideal decoupler gives a response nearly opposite to that obtained from the system without decoupling!

Toijala and Fagervik (1972) reported on a simulation study comparing different methods of decoupling. They studied a binary distillation column with ten plates. The transfer functions between variations in the reflux flow (ΔR) and the vapor flow out of the reboiler (ΔV) and the concentration of the most volatile component at each plate (x_j) in the column were found as follows:

$$\frac{x_{11}}{\Delta R}(s) = \frac{0.542}{(1 + 14.2s)(1 + 0.14s)} \tag{5.6.11}$$

$$\frac{x_8}{\Delta R}(s) = \frac{0.673}{(1 + 12.2s)(1 + 0.083s)^2} \tag{5.6.12}$$

$$\frac{x_3}{\Delta R}(s) = \frac{0.462}{(1 + 12.2s)(1 + 0.083s)^7} \tag{5.6.13}$$

$$\frac{x_0}{\Delta R}(s) = \frac{0.516}{(1 + 18.2s)(1 + 0.083s)^{10}} \tag{5.6.14}$$

$$\frac{x_{11}}{\Delta V}(s) = \frac{-0.453 \cdot e^{-0.17s}}{(1 + 13.4s)(1 + 0.13s)} \tag{5.6.15}$$

$$\frac{x_8}{\Delta V}(s) = \frac{-0.575 \cdot e^{-0.12s}}{(1 + 11.5s)} \tag{5.6.16}$$

$$\frac{x_3}{\Delta V}(s) = \frac{-0.488 \cdot e^{-0.03s}}{(1 + 11.5s)} \tag{5.6.17}$$

$$\frac{x_0}{\Delta V}(s) = \frac{-0.579}{(1 + 17.4s)} \tag{5.6.18}$$

where the indices 0 and 11 refer to the reboiler and the accumulator, respectively. The time constants are given in minutes.

In addition to these transfer functions, transfer functions are given for the control valves, measuring elements, and the reboiler, respectively:

$$h_v(s) = \frac{1}{(1 + 0.05s)} \tag{5.6.19}$$

$$h_m(s) = \frac{1}{(1 + 0.083s)} \tag{5.6.20}$$

$$h_r(s) = \frac{1}{(1 + 0.167s)} \qquad (5.6.21)$$

Let us now look at the control structure shown in Figure 5.6.16, where the levels in the accumulator and column bottom are controlled by the distillate flow and bottom product flow, respectively, and the reflux flow and reboiler heat flow are controlled by the concentrations on plates 8 and 3, respectively. Using the notation of sections 2.8 and 2.9, we can find the elements of the process transfer function matrix.

$$h_{11}(s) = h_v(s) \cdot \frac{x_8}{\Delta R}(s) \cdot h_m(s) = \frac{0.673}{(1 + 0.05s)(1 + 12.2s)(1 + 0.083s)^3}$$
$$(5.6.22)$$

$$h_{12}(s) = h_v(s) \cdot h_r(s) \cdot \frac{x_8}{\Delta V}(s) \cdot h_m(s)$$
$$= \frac{-0.575 \cdot e^{-0.12s}}{(1 + 0.05s)(1 + 0.167s)(1 + 11.5s)(1 + 0.083s)} \qquad (5.6.23)$$

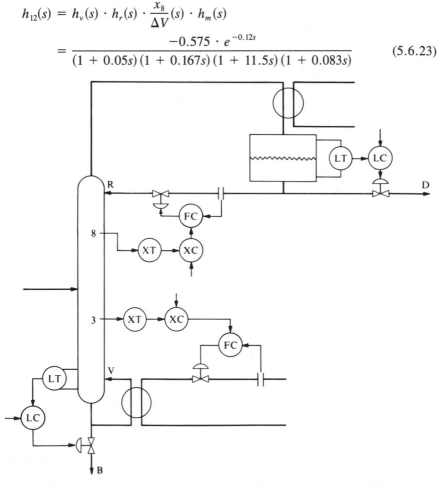

Figure 5.6.16. Control system studied by Toijala and Fagervik (1972).

$$h_{21}(s) = h_v(s) \cdot \frac{x_3}{\Delta R}(s) \cdot h_m(s) = \frac{0.462}{(1 + 0.05s)(1 + 12.2s)(1 + 0.083s)^8}$$

$$(5.6.24)$$

$$h_{22}(s) = h_v(s) \cdot h_r(s) \cdot \frac{x_3}{\Delta V} \cdot h_m(s)$$

$$= \frac{-0.488 \cdot e^{-0.03s}}{(1 + 0.05s)(1 + 0.167s)(1 + 11.5s)(1 + 0.083s)} \qquad (5.6.25)$$

Using the theory developed in sections 2.8, 2.9, and Appendix B, we find:

$$Y(s) = \frac{h_{12}(s)h_{21}(s)}{h_{11}(s)h_{22}(s)} = \frac{0.8089 \cdot e^{-0.09s}}{(1 + 0.083s)^5} \qquad (5.6.26)$$

The frequency responses $Y(j\omega)$ and $(1 - Y(j\omega))$ are shown in Figure 5.6.17. Figure 5.6.18 is the frequency response $h_{11}(j\omega)$ with the associated $M_1(j\omega)$ for the control system, based on measurement of the concentration on plate 8 and control of the reflux, where there is no control of the reboiler heat flow. Likewise, Figure 5.6.18 is the frequency response of $\Delta_2 = Y(j\omega)M_1(j\omega)$. Figure 5.6.19 gives the corresponding frequency responses for $h_{22}(j\omega)$, $M_2(j\omega)$, and $\Delta_1(j\omega) = Y(j\omega)M_2(j\omega)$ where $M_2(j\omega)$ is the frequency response of the closed control loop around the reboiler based on measurement of the concentration of plate 3 with constant reflux flow assumed.

Figure 5.6.20 shows the frequency responses $(1 - \Delta_1(j\omega))$ and $(1 - \Delta_2(j\omega))$. These frequency responses modify the frequency responses $h_{11}(j\omega)$ and $h_{22}(j\omega)$, respectively, and determine the response and stability conditions of the two closed loops when we do not use decoupling. It is important to note that we first adjust the upper loop so that it has a reasonable stability margin, and so that M_1 has good damping response. Then, keeping this loop fixed, we adjust the lower loop. The lower loop will be somewhat affected by the upper loop because the factor $(1 - Y(j\omega)M_1(j\omega))$ has a gain somewhat larger than one in the vicinity of the crossover frequency of the lower loop, and actually contributes a positive phase angle in this frequency range. This calls for a slight reduction in proportional gain.

The most important effect is that at low frequencies we get a reduction in the gain by a factor of $(1 - 0.8089) = 0.1911$, or -14.4 dB. If the loops contain PI controllers, the settling time will be about five times longer when there is cross-coupling unless the integral time is reduced by a factor of about five. This calls for a compromise in the tuning so that the system will tolerate being switched to manual.

Similar reasoning can be applied to the upper loop if we choose to adjust the lower loop first. Then the factor $(1 - Y(j\omega)M_2(j\omega))$, which corrects the frequency response $h_{11}(j\omega)$, will give essentially the same result as described above.

If we want to use decoupling, as described in section 2.9, and use the decoupler given by equation (2.9.3), namely:

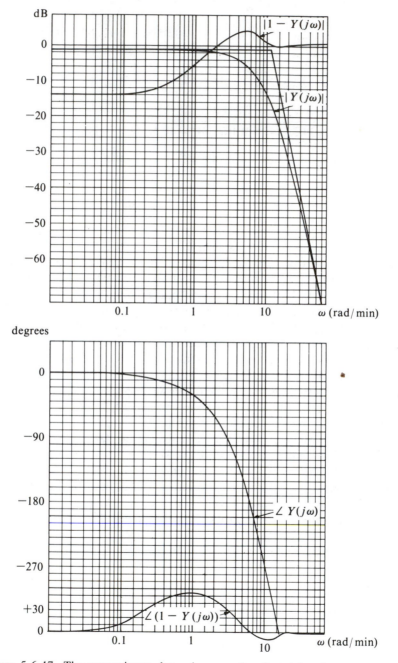

Figure 5.6.17. The approximate dynamic correction factor for the system of Figure 5.6.16.

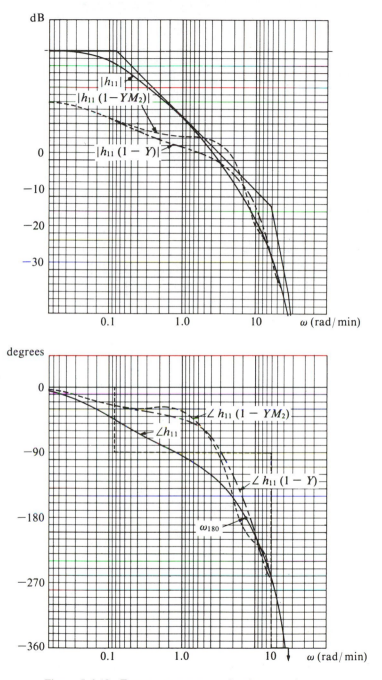

Figure 5.6.18. Frequency responses for the upper loop.

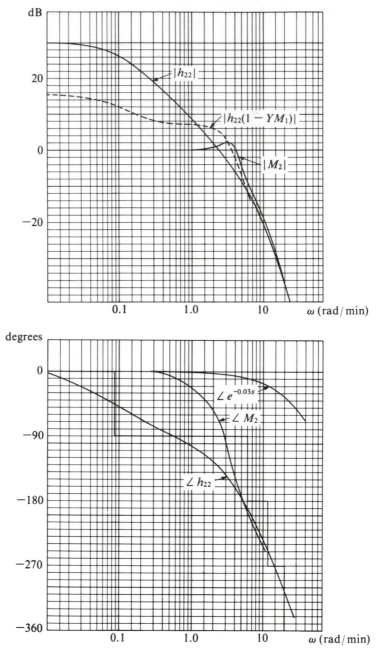

Figure 5.6.19. Frequency responses for the lower loop.

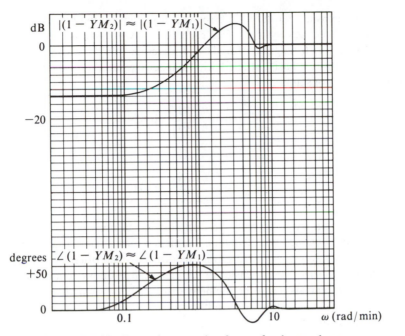

Figure 5.6.20. Dynamic correction factors for the two loops.

$$H_{C2}(s) = \begin{bmatrix} 1 & -\dfrac{h_{12}(s)}{h_{11}(s)} \\ -\dfrac{h_{21}(s)}{h_{22}(s)} & 1 \end{bmatrix} \tag{5.6.27}$$

then the decoupled process, in accordance with equation (2.9.4), will have the transfer function matrix

$$H_u(s)_m = \begin{bmatrix} h_{11}(s)(1 - Y(s)) & 0 \\ 0 & h_{22}(s)(1 - Y(s)) \end{bmatrix} \tag{5.6.28}$$

We see that the apparent transfer functions for the process are nearly the same as those that we had without decoupling. This is so because, as is evident from Figures 5.6.17 and 5.6.20:

$$1 - Y(s)M_1(s) \approx 1 - Y(s) \tag{5.6.29}$$

and:

$$1 - Y(s)M_2(s) \approx 1 - Y(s) \tag{5.6.30}$$

The decoupler in equation (5.6.27) contains transfer functions that are difficult to realize. Close analysis shows, however, that if these functions are replaced by constant terms, little error will result inside the possible frequency range of the feedback systems. For each of the two systems in this example, ω_{180} is about 4 to 6 rad/min. Toijala and Fagervik (1972), using a proposal of Luyben (1970) as a basis, studied what happens when the cross-couplings are replaced with constants:

$$C_{12} = -\frac{h_{12}(0)}{h_{11}(0)} = 0.854$$

and:

$$C_{21} = -\frac{h_{21}(0)}{h_{22}(0)} = 0.947$$

Simulation of the systems with these interaction terms gave nearly the same results as were obtained with dynamic interaction terms. This result is typical — complex, dynamic cross-couplings and feedforward can often be replaced with much simpler terms that are easier to use but change the system behavior very little. Further analysis using the methods discussed above, perhaps also including simulation, is nevertheless usually necessary in order to get a good picture of what can actually be accomplished.

State Feedback

Because a distillation column has many strong interactions (cross-couplings) between variables, and because there are many "inner" states that cannot be measured directly, this process is a good candidate for application of multivariable control based on state estimation. Many authors have described such control systems for experimental columns and reported convincing results. The practical implementation of these solutions, however, depends on many conditions, including, among other things, the operators' understanding of how the system works and affects the process, and how to make the system recover from possible shutdowns.

Pike and Thomas (1974) described a system based on a linear, discrete state space model with four state variables:

x_1: mass flow of distillate
x_2: mass flow of bottom product
x_3: concentration of top product
x_4: concentration of bottom product

The control variables are:

u_1: reflux flowrate
u_2: reboiler heat flow

The disturbances considered are:

v_1: feed concentration
v_2: feed flowrate

In this case all of the state variables can be measured, so no estimator is necessary. Because this model is very simple, it is reasonable to expect that significant improvements can be made by expanding the model and adding a state estimator.

The objective functional used is a discrete version of that given by equation (2.11.9). Matrices Q and P are chosen so that a unit change in each variable $(\Delta x_j, \Delta u_j)$ has the same cost effect as a unit change in vapor flow. This situation is known as the equal-cost case. The matrices Q and P are both diagonal. In another case, referred to as the low-cost case, matrix P is multiplied by the factor 0.1.

The control strategy is a multivariable, linear feedback from the states to the control variables through a constant matrix and feedforward from the current and future values of the disturbances (Anderson and Moore, 1971; Balchen, Solheim, and Fjeld, 1978).

This control strategy is compared with conventional control for both the low-cost case and the equal-cost case. In both cases the assumption is that the form of the disturbances is known, and/or the disturbances are assumed to be stochastic with known power spectra. The results are plotted in Figures 5.6.21a and 5.6.21b. From Figure 5.6.21a, we see that if the future values of the disturbances are assumed, the multivariable optimal control compared with conventional control is better by a factor of 2.5 to 3, while there is no significant improvement when the disturbances are assumed to be stochastic. The authors explain that conventional control will be about the same as optimal control when there is little cost assigned to the control variables; that is, they can be allowed to vary over wide ranges. In Figure 5.6.21b, where the costs for the control variables are high, we see that in each case the optimal control is better by a factor of about 2.5 than conventional control.

Another contribution to the development of multivariable optimal control for a distillation column was made by Gilles and Retzbach (1980), who studied distillation columns with pronounced temperature fronts along the column. This occurs in so-called extractive distillation where an extra component (extractant) is mixed into the feed to make distillation easier. Instead of a mathematical model based on state variables for each plate, Gilles and Retzbach chose a model with the following four states:

x_1: heat flow into the reboiler
x_2: vertical flow of vapor from the reboiler
x_3: position of the highest temperature knee (the propanol front)
x_4: position of the lowest temperature knee (extractant front)

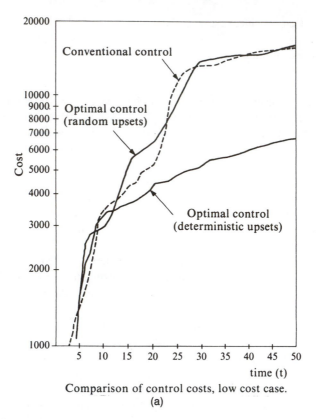

Comparison of control costs, low cost case.
(a)

Figure 5.6.21. Results of multivariable optimal control (Pike and Thomas, 1974). (Reprinted with permission from *Industrial & Engineering Chemistry Process Des. Dev., 13*, 2, 97–102. Copyright 1974, American Chemical Society.)

The control variables are:

u_1: control signal to the steam control valve in the reboiler
u_2: mass flow of the side stream removed near the bottom

The disturbances are:

v_1: feed flowrate
v_2: feed concentration

The measurements available are:

y_1: temperature at the highest temperature front
y_2: temperature at the lowest temperature front

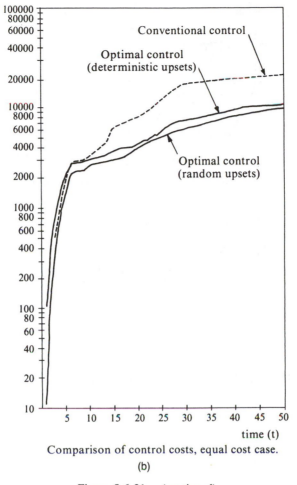

Comparison of control costs, equal cost case.

(b)

Figure 5.6.21.   (continued)

Figure 5.6.22 shows the profiles of the most important variables (propanol, water, extractant, temperature) along the column under steady-state conditions. The flow of extractant (F_E) is added, as shown in Figure 5.6.23, near the top of the column, and the feed stream (F_A) is added near the middle. The sidestream (S) is removed just above the bottom. It is assumed that the reflux stream (R) and the extractant stream (F_E) are equipped with ratio controllers so that they are proportional to the feed flowrate (F_A). The variation of the sidestream (ΔS) from its nominal value is, however, a control variable.

The mathematical model used for the state estimator has the following structure:

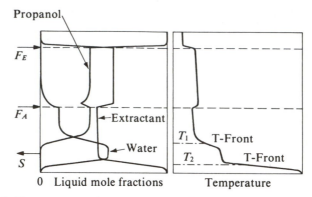

Figure 5.6.22. Temperature and concentration profiles in a column for extractive distillation (Gilles and Retzbach, 1980).

$$\dot{\mathbf{x}} = \begin{bmatrix} a_{11} & 0 & 0 & 0 \\ a_{21} & a_{22} & 0 & 0 \\ 0 & a_{32} & 0 & 0 \\ 0 & a_{42} & 0 & 0 \end{bmatrix} \mathbf{x} + \begin{bmatrix} b_{11} & 0 \\ 0 & 0 \\ 0 & b_{32} \\ 0 & b_{42} \end{bmatrix} \mathbf{u} + \begin{bmatrix} 0 & 0 \\ 0 & 0 \\ c_{31} & c_{32} \\ 0 & c_{42} \end{bmatrix} \mathbf{v} \quad (5.6.31)$$

$$y = \begin{bmatrix} 0 & 0 & d_{13} & 0 \\ 0 & 0 & 0 & d_{24} \end{bmatrix} x \quad (5.6.32)$$

In addition to the estimation of the states given by equation (5.6.31), the state estimator also includes estimation of the feed concentration (v_2). This is modeled as a constant so $\dot{v}_2 = 0$. Therefore, the state estimator will have five integrators.

The control strategy consists of feedback from the four estimated states and the estimated feed concentration, together with feedforward from the feed flow (v_1), which is assumed to be measurable.

The control system is shown in Figure 5.6.23. Figure 5.6.24 gives the results of a simulation of the system's behavior. Figure 5.6.24a shows that the position of the highest temperature knee moves very little (only a fraction of one plate) when the feed flowrate undergoes a step change. Figure 5.6.24b also shows this clearly in terms of the two measured temperatures. Figure 5.6.24c shows how the estimator tracks the step change, and that the control system acts to keep the changes in the two temperatures very small.

A critical analysis of this system leads to the conclusion that it is the two temperature measurements that make the system work so well, and the two first state variables are very strongly coupled to these measurements. Maybe it is not necessary to use two state variables to model the thermal dynamics of the reboiler. On the other hand, perhaps one state variable should be reserved for the dynamic phenomena in the top of the column so that feedforward from the

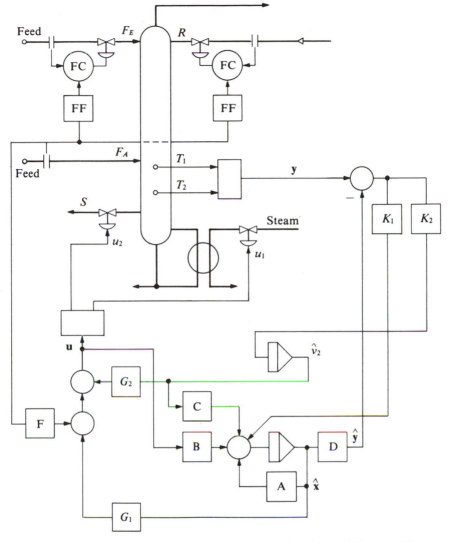

Figure 5.6.23. Structure of the control system for the column of Figure 5.6.22.

feed flowrate to the flows of extractant and reflux can be dynamic. In any case, this work is a good example of how a simple model can be very effective when used with state estimation.

Adaptive Control

There have been several attempts to use adaptive control systems on distillation columns. This is still very much in the active research stage and seems to be

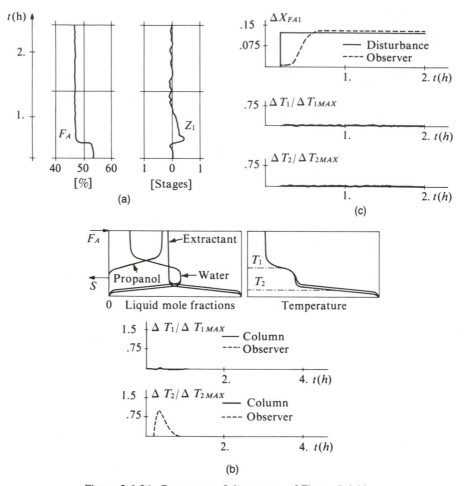

Figure 5.6.24. Responses of the system of Figure 5.6.23.

developing around single-variable controllers in relatively conventional structures, but ones in which the parameters adapt automatically (Sastry, Seborg,
and Wood, 1977; Dahlqvist, 1981).

Dahlqvist (1981) studied adaptive minimum-variance control of top product
concentration (at the top plate) by control of the reflux flow, and bottom
product concentration (at the bottom plate) by control of the reboiler heat flow.
At first Dahlqvist tried keeping these control loops independent of each other.
Then, after slight modification, they were used together. Dahlqvist also tried to
use the ratio between the reflux flow and the reboiler heat flow as a control
variable. Figure 5.6.25 gives the results of an experiment with an adaptive controller for top product concentration with reflux flow as a control variable. The
controller has eight parameters, which were determined by a previous experi-

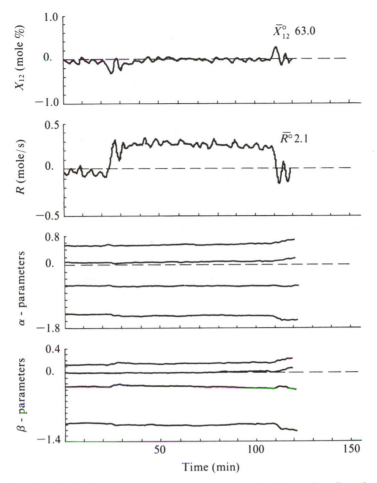

Figure 5.6.25. Distillation column with adaptive control of the reflux flow from top product concentration.

ment. The curves show the response of the system to a step change in reboiler heat flow. Figure 5.6.26 shows the result when the ratio of the reflux to the reboiler heat flow was used as a control variable. The system was disturbed by several small step changes in the reboiler steam flow.

Figure 5.6.27 shows the results obtained by controlling both the bottom and top product concentrations by the reboiler heat flow and the ratio of that flow to the reflux flow as control variables. In this case the system was upset by a step change in the feed rate. Dahlqvist points out that by choosing the right pair of variables, the adaptive controller need be no more complicated than two independent single-variable adaptive controllers. He makes no comparison with how other controllers will perform on this process, but it is clear that the con-

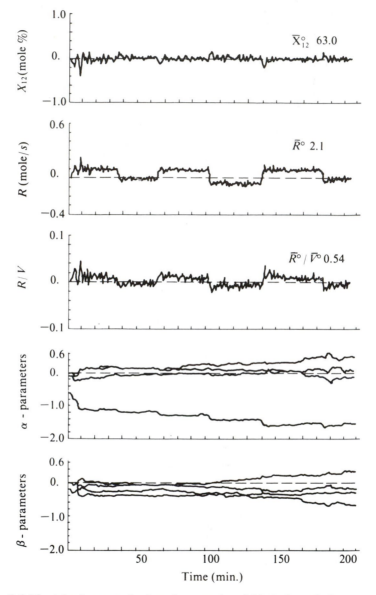

Figure 5.6.26. Adaptive control when the control variable is the ratio between reflux flow and reboiler steam flow.

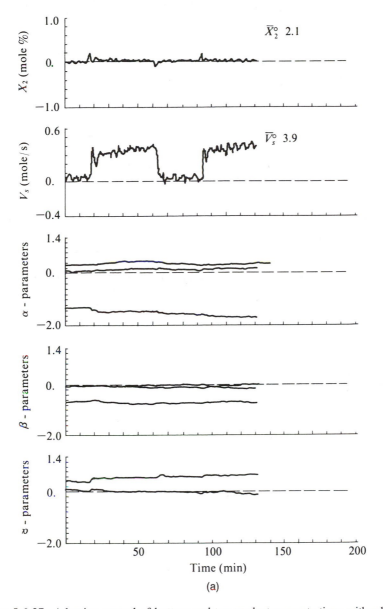

Figure 5.6.27. Adaptive control of bottom and top product concentrations with reboiler heat flow and the ratio between reflux and reboiler heat flow as control variables (Dahlqvist, 1981).

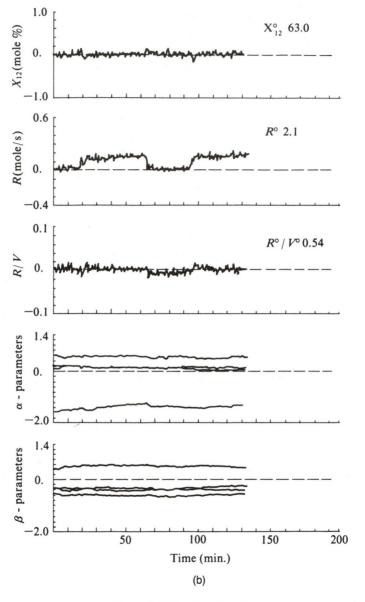

Figure 5.6.27.   (continued)

trollers that are used will need a significant degree of derivative action. A comparison with conventional controllers thus should include a full PID controller. A characteristic of the adaptive controllers just described is that they are very sensitive to the controller parameters that must be adapted. The question of what will happen if the adaptive action is removed and the controller operates with fixed parameters has yet to be studied.

Multiple Distillation Columns

Most of the previous examples have considered binary distillation in isolated, or single, columns. However, the raw materials or products of many industrial processes consist of many components that must be separated by either a series of distillation columns in cascade or a single large column with many different feeds and/or sidestream products. Multicomponent distillation is more complicated than binary distillation, and it is becoming more and more commonly used in industry. Multicomponent distillation in a single column will not be considered further here; we will instead turn our attention to columns in cascade.

A typical example of the use of cascade coupled distillation columns is the distillation of a mixture of hydrocarbons resulting from the pyrolysis (cracking) of ethane or an ethane–propane mixture. Ethane is one of the most important products of the separation of "rich gas" from an oil field. Ethane (C_2H_6) is split by pyrolysis into hydrogen (H_2) at about 3.7% by weight, methane (CH_4) at about 3.3%, acetylene (C_2H_2) at about 0.2%, ethylene (C_2H_4) at about 49%, ethane (C_2H_6) at 39%, propylene (C_3H_6) at 1.1%, propane (C_3H_8) at about 0.2%, butane (C_4H_{10}) at 0.3%, pentanes (C_5 +) at 1.6%, and other hydrocarbons to give the full 100%. If the feed to the cracker is a mixture of ethane and propane, the product composition will be different. Figure 5.6.28 is an example of the percentage weight distribution of products for a feed composition of 25% ethane and 75% propane by weight. We see that production of ethylene, which is the main objective, increases approximately linearly with conversion, and that the production of pentanes increases rapidly with increasing conversion while the production of propylene decreases.

In order to get an idea of how difficult or easy it is to separate different components by distillation, the boiling points of some of the most important components are listed in Table 5.6.1. Figure 3.12.17a contains similar information.

It is clear from Table 5.6.1 that it is relatively easy to separate methane and hydrogen from ethylene and heavier components if one can just cool them sufficiently. Figure 5.6.29 illustrates how the first stage in this separation takes place. To the left is the demethanizer (column 1), where methane and hydrogen leave from the top, and ethylene and the heavier components leave at the bottom. The pressure in this column is about 32 bar, which gives a temperature of about −92°C at the top and about 4°C at the bottom. From Figure 3.12.17a we see that this last temperature corresponds to the boiling point of ethylene. The

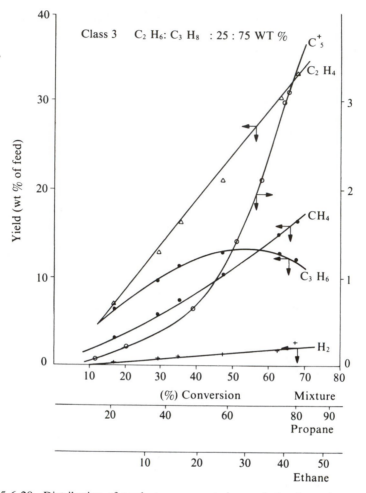

Figure 5.6.28. Distribution of product components in pyrolysis of an ethane–propane mixture (Froment et al., 1976). (Reprinted with permission from *Industrial & Engineering Chemistry Process Des. Dev.*, *15*, 4, 495–504. Copyright 1976, American Chemical Society.)

bottom product from the first column is sent to the de-ethanizer (column 2), where the propylene and heavy components are removed from the bottom, and ethylene, ethane, and acetylene leave as top product along with any possible residual methane. The pressure in this column is about 28 bar, which gives a temperature at the top of −10°C and about +80°C at the bottom, this being the boiling point of propylene.

The acetylene is removed in a special reactor, not by distillation, and the ethylene–ethane mixture remaining is sent to column 3. The de-ethanizer column provides very sharp separation, with the amount of propylene and

Table 5.6.1 Boiling points of common hydrocarbons.

MATERIAL	BOILING POINT AT ATMOS. PRESSURE (°C)
Hydrogen	−252.87
Methane	−164.0
Ethylene	−103.71
Ethane	− 88.63
Acetylene	− 84.2
Propylene	− 47.4
Propane	− 42.07
Isobutane	− 11.72
n-butane	− 0.56
Isopentane	+ 27.94
n-pentane	+ 36.28
Isohexane	+ 60.22

heavier components leaving with the top product being less than 1%. Similarly, C_2-hydrocarbons do not appear in the propylene product in the bottom.

Depending on how much methane gets through the demethanizer (column 1), it may be necessary to include additional columns for demethanization before separating the ethylene from the ethane. There are no additional columns for this purpose in Figure 5.6.29.

In column 3, which separates the C_2-hydrocarbons, the ethane is the bottom product, and ethylene is the top product. The pressure in this column is about 9.7 bar, giving top and bottom temperatures of −56°C and −36°C, respectively. The latter is the boiling point for ethane. The ethane that comes from the bottom of this column is returned to the storage tank upstream of the pyrolysis process.

From the top of column 3 the ethylene is sent to a compressor, using the principles shown in Figure 5.6.11. The compressed ethylene is condensed in the column's reboiler stage. Part of the ethylene also passes through equipment that cools the process gas earlier in the process.

The bottom product from column 2, consisting of propylene and heavier components, is sent to column 4 (the depropanizer) where propane and propylene are the top products and butane and heavier components the bottom products. The pressure in this column is about 11 bar, giving top and bottom temperatures of 24.3°C and 76°C, respectively, the latter being the boiling point of the estimated butane mixture.

The top product of column 4 is sent to column 5 for separation of propylene from propane. This occurs at about 12.4 bar. The top temperature is about 23.5°C, and the bottom temperature, the boiling point of propane, is about 31°C. This column is one of the most difficult operations in the system because propylene and propane have a low relative volatility (see Figure 3.12.17a). At 12.4 bar, the boiling point difference between these components is only about

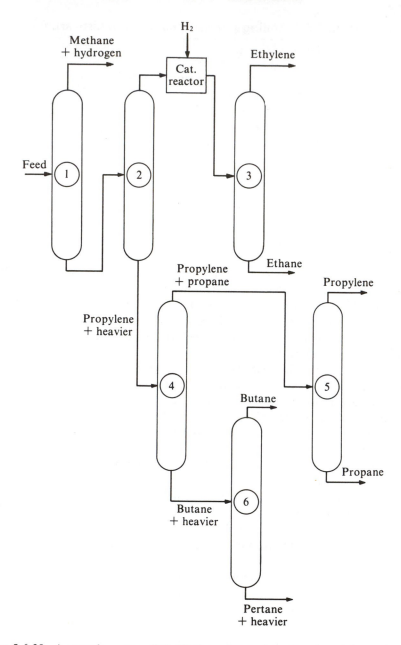

Figure 5.6.29. A cascade system of distillation columns for separation of the products after pyrolysis of ethane (ethane–propane).

10°C. Therefore, this column must contain many plates, perhaps 180 or more. The propylene is a valuable product used for the production of polypropylene. The propane is used partly as a fuel, and part is returned to the pyrolysis unit. The bottom product of column 4 contains butane and heavier components. These are sent to column 6, where the top product is a mixture of normal butane and isobutane. The bottom product consists of pentanes and heavier components. The pressure in column 6 is about 5 bar, and the temperatures at top and bottom are 42°C and 104°C, respectively. The bottom temperature is the boiling point of isopentane.

If desired, the butane from the top can be treated further in another column to split the isobutane at the top and n-butane at the bottom. The unseparated mixture can also be used as is. Similarly, the pentanes in the bottom product of column 6 can be distilled further to split the isopentane (top product) from the n-pentane and heavier components (bottom product). Alternatively, this mixture can also be used unseparated.

The purpose of this discussion of Figure 5.6.29 is to show a process with six cascade coupled distillation columns, all equipped with control systems of the type described earlier. We see that it is not satisfactory that each column simply operate in a stable manner; the system must also operate so that the disturbances created by one column have the least possible effect on the other columns. This is especially true, for example, in the case of column 4 (the C_3–C_4 separator), where the top product must not experience large changes in flowrate. This is so because this flow is also the feed for column 5, which separates propylene from propane and is the most difficult column to operate. Therefore, the accumulator for column 4 should have level control via the reflux flow, which is two to three times greater than the distillate flow. Because the distillate flow from column 4 is the feed for column 5, it is advantageous to provide flow control on this stream, with the flow controller being updated slowly on the basis of temperature control (concentration control) from the top of column 4.

Optimization of Systems of Distillation Columns

The many control systems involved in a large process for the separation of the components of crude oil or the separation of hydrocarbons after pyrolysis (cracking) provide an obvious milieu for some sort of overall optimization structure. The simplest way to do this is to adjust the set points of the individual controllers and perhaps the ratio settings of the ratio controllers.

The fundamental principles for this were discussed in section 2.16. It often happens that the optimal operating conditions (combination of set points) is determined by a combination of boundary conditions or limitations due to equipment or product specifications, as shown in Figure 2.16.1. An example of an optimization concept for a relatively small part of a distillation system will now be discussed (Baxley, 1969).

The problem is, how does one operate a column for maximum economy when it has varying limitations and disturbances, especially in the feed? Consider the column of Figure 5.6.3 where the cooling power is controlled relative to the removal of distillate, and the reflux flow is controlled by the accumulator level. This gives, indirectly, a constant reflux ratio. The ratio between the cooling power and the flow of distillate can be changed on the basis of analysis of the top product.

The following limitations apply to such a column:

- Each plate has a limited capacity in terms of vapor and liquid flow, depending on the design of the plate.

- The reboiler capacity is limited by the area of the heat exchanger and the heat capacity of the primary medium.

- The capacity of the condenser is determined by the heat transfer surface and availability of the cooling medium.

It is useful to express the capacity of a column in terms of the maximum allowable flow of vapor from the top. Since the reboiler and the condenser together produce the material that becomes vapor at the top, the vapor flow at the top can be used as a parameter to characterize the capacity of the column and its associated equipment.

In practice the flow of vapor is increased until either optimal conditions or the capacity of the column is reached (see Figure 2.16.2). Optimal operating conditions can be defined as occurring when the incremental operating cost is the same as the incremental product value. This optimal point can fall either within or outside the possible physical operating boundaries set by the equipment.

Figure 5.6.30 is a flowchart of the procedure for optimization of such a column. The following steps make up this logical scheme:

1. First, determine the vapor flow at the top of the column (V_N), by either direct measurement or calculation on the basis of mass and energy balances around the top of the column and the condenser.
2. Next, determine the maximum top vapor flow to the condenser, which requires that the valve controlling the addition of coolant be "full open." If a linear relationship between valve opening and vapor flow can be assumed, then the maximum vapor flow from the top (V_{N1}) can be determined by linear extrapolation from measurements of the vapor flow and the valve stem position.
3. Similarly, determine the maximum top vapor flow (V_{N2}) given by the limitations of the reboiler. If the reboiler is heated by steam at a measurable pressure and a fixed pressure limit, the maximum (V_{N2}) can be found by extrapolation from the existing vapor flow and the existing pressure in the

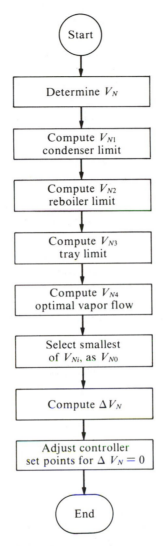

Figure 5.6.30. Calculation scheme for optimization of a distillation column (Baxley, 1969).

supply to the reboiler, together with the temperature at the bottom of the column.

4. Find the maximum allowable top vapor flow (V_{N3}) as determined by the column plates from measurement of the pressure drop over the column and linear extrapolation to a previously determined pressure limit.

5. Find the optimal top vapor flow (V_{N4}) given by the incremental operating cost, which is equal to the incremental product value, by assuming a simple static model of the economics of the column that includes the run-

ning costs of energy and the products and parameters that characterize the separation performance of the column.

6. Find the minimum value of the constrained vapor flows (V_{Ni}, $i = 1, 2, 3, 4$), and use this value (V_{NO}) as the set point for adjustment of the top vapor flow. The difference between this calculated value and the actual measured value of the top vapor flow ($\Delta V_N = V_{NO} - V_N$) is a measure of how much the column deviates from the optimum.

7. Calculate ΔV_N.

8. Adjust one or more controller set points by means of a standard control algorithm until $\Delta V_N = 0$. If more than one set point is adjusted, this should be done at some fixed relationship.

There are many other ways to achieve optimization of the conventional control systems' set points in a distillation column, but all are based on some sort of quantitative formulation of an optimal criterion and a satisfactory mathematical model of the column. In processes that handle large quantities of raw materials, it is often very profitable to make even small improvements in production or quality so optimization routines can pay for themselves in a very short time. The considerations that apply to a single column can also apply to systems of columns, so similar optimization schemes can be used for entire processing systems and plants.

5.7. REFRIGERATION CYCLE CONTROL

Refrigeration is another thermal process that needs extensive automatic control. The basic technology of this process was discussed in section 3.13.

A simple refrigeration cycle is shown in Figure 5.7.1. It consists of a compressor that compresses a vaporized refrigerant such that both the pressure and temperature of the gas rise significantly. There are many special refrigerants, such as R12 (freon), but in principle any gas can be used if the pressure and temperature are suitable. In the condenser, the gas is cooled under constant pressure until it condenses. The refrigerant must be chosen with proper reference to the refrigeration temperature. At the expansion valve, the pressure decreases with constant enthalpy (adiabatic expansion); that is, energy is neither gained nor lost. Vaporization occurs in the evaporator where the heat required is obtained from the primary medium, which in turn cools as it gives up this heat. The pressure, and therefore the temperature, in the evaporator are chosen in accordance with the temperature one wants in the primary medium (the temperature of the coolant). There will be a little superheating of the vapor after the evaporator before it returns to the compressor. Compression can be assumed to be isentropic; that is, it takes place at constant entropy. The cycle is shown in Figure 5.7.2. The enthalpy change that gives the cooling power is given by the value of Δh_1 in Figure 5.7.2, where the units are J/kg refrigerant. The enthalpy added during compression is Δh_2.

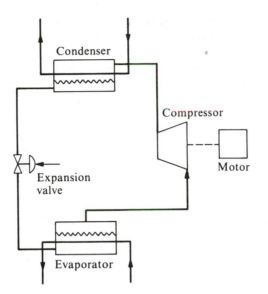

Figure 5.7.1. A simple refrigeration system.

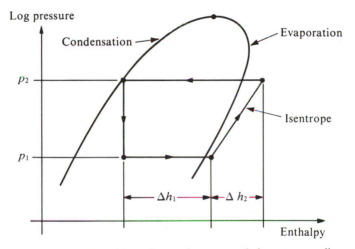

Figure 5.7.2. A simple refrigeration cycle on an enthalpy–pressure diagram.

The sum of the enthalpy change that takes place in the condenser and the enthalpy that is removed from the system through heating the secondary medium in the condenser is given by $\Delta h_1 + \Delta h_2$.

Because the transport of heat from the medium to be cooled (the process medium) is determined by the temperature difference between this medium and the cooling medium (the refrigerant in the evaporator), a change in heat transport requires control of the temperature (or pressure) in the evaporator. The actual energy transport through the evaporator is seen to be, according to Figure 5.7.2, equal to Δh_1 multiplied by the mass flow of the refrigerant. This mass flow is controlled by the expansion valve. There is some storage capacity of the refrigerant ahead of the expansion valve, and the amount allowed to pass through the valve is just sufficient to obtain the desired pressure in the evaporator. This can be achieved by means of a pressure control system, as shown in Figure 5.7.3. The set point for the pressure control system can be used to change the temperature in the evaporator.

In large refrigeration systems the compressor will be equipped with some means of capacity control in order to maintain the desired pressure p_2 in the condenser. This can be done either by controlling the speed of the compressor as shown in Figure 5.7.3 or, in the case of a constant-speed compressor, by a controlled return valve from the high pressure side to the low pressure side. If the condenser is oversized enough that all the vapor is condensed, there is no need for a control system on the secondary side of the condenser. If, however, the secondary medium in the condenser is a process liquid whose energy is used elsewhere for heating, then control might be needed. This will not be discussed further here.

In the system shown in Figure 5.7.2, the temperature of the input to the compressor can be slightly above the boiling point of the refrigerant; that is, the refrigerant vapor can be superheated. Usually, this temperature difference will be about 5 to 10°C. In small systems such as home refrigerators and air conditioners, the expansion valve and control system consist of a relatively simple, combined unit known as a thermostatic expansion valve. This unit has a measuring element for pressure in the evaporator and a small temperature-regulating

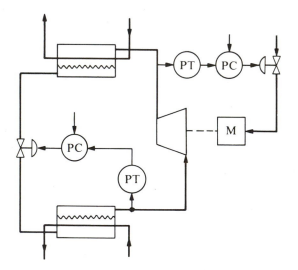

Figure 5.7.3. Simple refrigeration cycle with two control systems.

system that consists of a closed measuring bulb filled with an equilibrium mixture of gas and liquid phases of the refrigerant. This chamber is placed in the superheating zone and therefore has a pressure somewhat higher than the evaporator pressure. The degree of superheating given by the difference between the two measured pressures is used to control the expansion valve.

Several other control structures can be applied to the refrigeration system shown in Figure 5.7.1. Figure 5.7.4 shows a system in which the evaporator is a reservoir containing liquid refrigerant. Evaporation occurs at the surface of the reservoir, and the primary medium, the medium that is to be cooled, passes through a pipe completely submerged in the liquid refrigerant. The point of this arrangement is to get the best possible heat transfer to the refrigerant side. In this arrangement it is the level control in the evaporator that operates the expansion valve by passing just enough refrigerant through the valve to keep the liquid level constant. The power added to the compressor motor is controlled by the evaporator pressure. The pressure (and temperature) in the condenser will be floating, in the sense that they will be determined by the amount of cooling power taken out from the secondary side of the condenser. If one wishes to maintain a constant temperature in the output of the primary side of the evaporator, the temperature control system can be arranged to control the set point of the compressor pressure controller. Then, if the flow of the primary medium in the evaporator is increased, and it is desired to hold the temperature of the outflow of the primary constant, the pressure and temperature *in* the evaporator must decrease. As a result, the compressor must run faster in order to process more vaporized refrigerant. The motor speed will increase if the pressure set point for the evaporator is decreased, as indicated in Figure 5.7.4.

A multivariable control structure, as shown in Figure 2.13.3, can be derived from the model shown in Figure 3.13.6. Here, θ_1, θ_2, W_1, and W_2 may be defined as state variables. Both pressures (p_1 and p_2) and all enthalpies are computed from known algebraic relationships. The control variables in this system may be the opening of the expansion valve (u_1), the speed of the compressor (n), and the secondary temperature of the condenser (θ_{20}). This type of structure will be particularly useful in multiple-stage refrigeration systems with strong interactions and large load disturbances.

The control literature contains surprisingly little information regarding control structures for refrigeration systems, perhaps because detailed dynamic analyses of these systems did not begin to appear until relatively recently, (Danig, 1963; Marshall and James, 1975).

5.8. CONTROL OF CHEMICAL REACTORS

Section 3.14 provided an overview of some of the theoretical problems concerning chemical reactors and illustrations of some of the control problems these unit operations present.

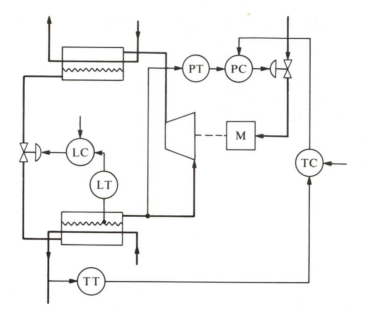

Figure 5.7.4. Simple refrigeration system with three control systems.

The behavior of most chemical reactions is very dependent on temperature. When a process is exothermic, that is, the process generates heat, it can be difficult to keep the reaction at a temperature that gives satisfactory results (usually measured as maximum yield within the product specifications) because the process operating under this objective can be inherently unstable (see section 3.14). Use of feedback to stabilize the process by means of dynamic cooling (such that the eigenvalues of the system are moved to the left half-plane) is therefore a very important but often difficult control problem.

Another factor in the design of chemical processes, including chemical reactors, is that it is desirable to have the best possible energy utilization and economy. One way to do this is to use heat generated by the reaction in a heat exchanger to preheat cold raw materials prior to reaction. Such forms of energy economy represent positive feedback that makes control more difficult and also more necessary.

The most common problem in controlling a chemical reactor is to maintain a constant (or some other desired) temperature profile as determined by steady-state optimization considerations. Many investigators have studied the development of various control structures to achieve such control. Conventional structures consisting of multiple single-variable control loops and multivariable structures with and without state estimators have been proposed and examined (Balchen et al. 1973; Michelsen, Vakil, and Foss, 1973; Vakil, Michelson, and Foss, 1973; Silva, Wallman, and Foss, 1979; Wallman, Silva, and Foss, 1979;

Viswanadham, Patnaik, and Sarma, 1979; Foss, Edmunds, and Kouvaitakis, 1980; Ray, 1981; Shinnar, 1981; Wallman and Foss, 1981).

One of the simplest problem situations occurs in the case of a chemical reactor consisting of a well-mixed tank with continuous addition of the reactants, as in the system shown in Figure 5.8.1. This system is known as a continuous stirred tank reactor (CSTR). The individual reactant flows are controlled by flow controllers FC-1 and FC-2, whose ratios are such that reactant B is metered in the proper relationship to reactant A.

The temperature of the reactor is controlled by a cooling medium that is added through a control valve. A pump (P) maintains good circulation of the coolant through the jacket surrounding the reactor. The flow of coolant is controlled by its temperature, measured by temperature controller TC-1, as it leaves the cooling jacket. The temperature in the reactor itself is controlled by

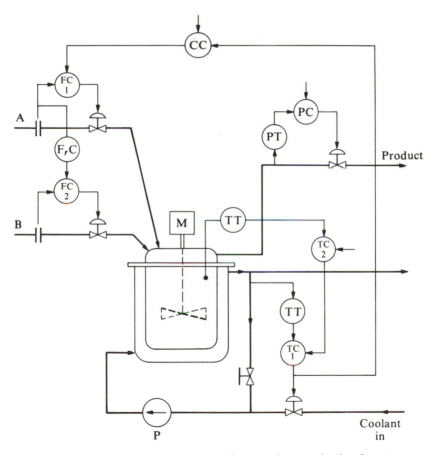

Figure 5.8.1. A typical control system for a continuous stirred tank reactor.

temperature controller TC-2, which is in cascade with TC-1. Possible changes due to changing coolant characteristics will be eliminated by the first temperature control loop. If it is necessary to maintain a given pressure in the reactor, this is done with a pressure controller that operates a control valve on the product outflow from the reactor.

Because the capacity of the cooling system can be a limiting factor on the production rate of the reactor, measurement of the opening of the control valve in the cooling loop can be used as an indication of how far from its maximum capacity the reactor is being operated. This measurement can be used through controller CC to act on the set point of the flow controller in the reactant stream, and can be set up to ensure, for example, that the cooling valve cannot exceed 90% full opening.

Another type of reactor widely used in industry is one for the synthesis of ammonia from hydrogen and nitrogen. A schematic diagram of the various major units that make up a typical ammonia plant (see section 6.4) is shown in Figure 5.8.2. Figure 5.8.3 gives a detailed cross section of the synthesis reactor itself. A cold mixture of hydrogen and nitrogen at the proper stoichiometric ratio ($N_2 + 3H_2 \rightarrow 2NH_3$) from a multistage compressor upstream of the reac-

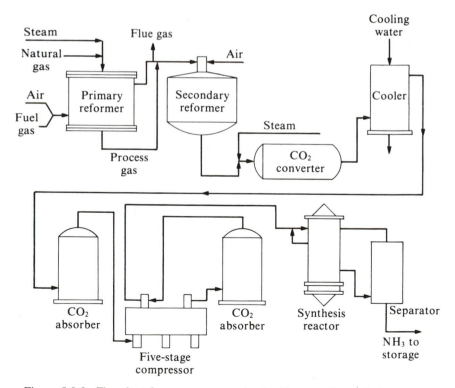

Figure 5.8.2. Flowchart for processes associated with ammonia synthesis reactor.

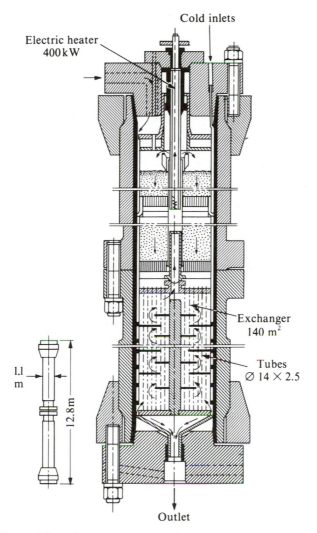

Figure 5.8.3. Schematic diagram of ammonia synthesis reactor.

tor enters at a pressure of about 270 bar and a temperature of 100°C. The first part of the reactor contains an iron oxide catalyst, and in order to have uniform operation it is important that its temperature be held at approximately 450°C. The incoming gas exchanges heat with the outgoing gas in the heat exchanger tubes located near the outlet of the reactor. In order to maintain the required temperature pattern (profile), the reactor is equipped with one or more parallel paths for the cold gas, as shown in Figure 5.8.4. Here we see that the incoming gas is split into three paths, each with a control valve, to make it possible to control the temperature profile.

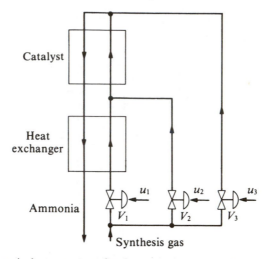

Figure 5.8.4. A typical arrangement for the temperature control variables in an ammonia synthesis reactor.

A conventional control system for this process controls the three gas flow valves (possibly two) depending on the three (possibly two) temperature measurements along the reactor, with special attention to the section containing the catalyst. Figure 5.8.5 is a solution where the valve in the direct path (V_1) is controlled on the basis of the temperature measured in the synthesis gas after the heat exchanger. The valve in the first parallel path (V_2) is controlled on the basis of the temperature of the synthesis gas after heat exchange in the catalytic section but before it actually goes into the catalytic section. The valve in the second parallel path (V_3) is controlled by a temperature controller acting on measurement of the reacted gas near the exit of the catalytic section. An alternative to the solution of Figure 5.8.5 is to control the three parallel flows by ratio control based on temperature measurements. The solution of Figure 5.8.5 has proved to function satisfactorily under steady-state conditions, and stability of the reactor usually is not a problem.

More complicated control systems including multivariable state feedback and feedforward systems have been studied with the objective of managing different operating conditions such as varying flow, temperature, or composition of the synthesis gas (Viswanadham, Patnaik, and Sarma, 1979). A comparison was made between the performance of a complete multivariable control system with control variables u_2 and u_3 (see Figure 5.8.5) and feedback taken either from nine state variables, all of which are temperatures, distributed along the reactor or from only two temperatures measured at the inlet and outlet of the catalytic section. Feedforward was taken from measurement of the gas temperature in the reactor inlet and from the synthesis gas composition. The gas flow was assumed constant. The results of this study showed that it is satisfactory to use two temperature measurements in the catalytic section, and that the improve-

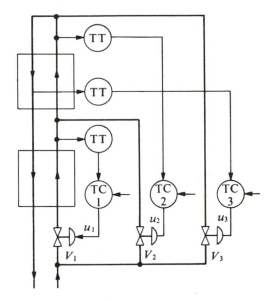

Figure 5.8.5. A possible structure for temperature control of an ammonia synthesis reactor.

ments obtained by using feedforward from the disturbances cannot justify additional instrumentation.

Another industrial process that has been the subject of intense efforts to develop a better control strategy is the so-called fluid catalytic cracking process (Gould, Evans, and Kurihara, 1970; Lee and Weekman, 1976). A simplified flowchart for this process is given in Figure 5.8.6. The process consists mainly of two parts: a reactor, where heavy oil is exposed to a catalyst, and a regenerator, for regeneration of the catalyst used in the reactor. The feed to the unit is a heavy fraction of distilled crude oil that reacts endothermically to form various lighter hydrocarbons as well as carbon, which collects on the catalyst. The product gas leaves at the top of the reactor, while catalyst is continuously removed from the bottom and, together with air, blown into the bottom of the regenerator. In the regenerator the carbon on the catalyst is burned off to form H_2O, CO, and CO_2. The combustion gases leave the regenerator at the top.

The regenerated catalyst, which is heated by the exothermic combustion process, is returned to the reactor and mixed with the feed, thus providing the heat necessary to evaporate the feed and promote the cracking reaction. The feed passes through a gas-fired preheating chamber before it contacts the catalyst and goes into the reactor. Both the reactor and the regenerator have the characteristic of a fluidized bed, in that the catalyst is kept fluidized in the rising gas or air in each unit.

There are several possible control variables in the process shown in Figure 5.8.6. To simplify the situation somewhat, assume that the amount of cata-

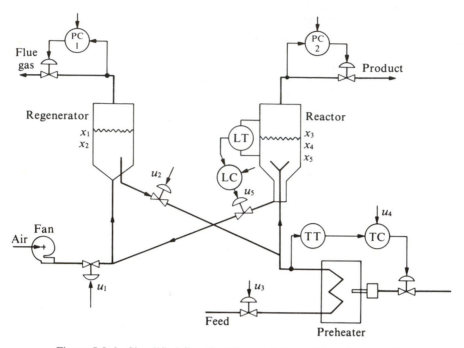

Figure 5.8.6. Simplified flowchart for catalytic cracking of heavy oil.

lyst in the reactor is controlled by means of the flow of catalyst (u_5) leaving the bottom of the reactor. There are then four control variables.

u_1: air flow in the bottom of the regenerator
u_2: flow of regenerated catalyst back to the reactor
u_3: flow of feed
u_4: temperature of the feed after the preheater (actually the set point of the temperature controller)

If more detail is needed, an extra state variable can be added to characterize the preheating process, but in the following we will assume relatively fast dynamics.

A set of state variables that describes this process with sufficiently rich detail is (Gould, Evans, and Kurihara, 1970):

$x_1 = \theta_{rg}$: temperature in the regenerator ($\approx 627°C$)
$x_2 = c_{rc}$: carbon on the regenerated catalyst (≈ 0.6 wt%)
$x_3 = \theta_{ra}$: temperature in the reactor ($\approx 500°C$)
$x_4 = c_{tot}$: total carbon on the spent catalyst (≈ 1.5 wt%)
$x_5 = c_{cat}$: catalytic carbon on the spent catalyst (≈ 0.9 wt%)

It should be pointed out that one must distinguish between the two types of carbon on the catalyst in the reactor (catalytic carbon and noncatalytic carbon).

The dominant disturbance in this system is a factor, v_1, that gives the rate of formation of carbon for the actual oil feed.

Using the state variables listed above, the following structure containing five coupled nonlinear differential equations can be formulated:

$$\dot{x}_1 = f_1(x_1, x_2, x_3, u_1, u_2) \qquad (5.8.1)$$

$$\dot{x}_2 = f_2(x_1, x_2, x_4, u_1, u_2) \qquad (5.8.2)$$

$$\dot{x}_3 = f_3(x_1, x_2, x_3, x_5, u_3, u_4) \qquad (5.8.3)$$

$$\dot{x}_4 = f_4(x_2, x_3, x_4, x_5, v_1) \qquad (5.8.4)$$

$$\dot{x}_5 = f_5(x_2, x_3, x_5, u_2) \qquad (5.8.5)$$

Usually, the following three measurements are available in the system:

$y_1 = x_1$: temperature in the regenerator
$y_2 = x_3$: temperature in the reactor
y_3:   oxygen concentration in the combustion gas

The following three values are used to evaluate the process yield:

z_1: flow of light fractions from the top of the reactor
z_2: flow of gasoline from the top of the reactor
z_3: flow of heavy components from the top of the reactor (which are recirculated to the feed after distillation)

The quantities $\mathbf{y}$ and $\mathbf{z}$ can be modeled as functions of $\mathbf{x}$ and $\mathbf{u}$ as follows:

$$y_3 = g_3(x_1, x_2, u_1) \qquad (5.8.6)$$

$$z_1 = h_1(x_2, x_3, u_3) \qquad (5.8.7)$$

$$z_2 = h_2(x_2, x_3, x_5, u_3) \qquad (5.8.8)$$

$$z_3 = h_3(x_2, x_3, x_5, u_3) \qquad (5.8.9)$$

Lee and Kugelman (1973) also developed a model for this process where they chose to use four state variables instead of five as in the example above. They did not differentiate between the catalytic carbon (x_5) and the total carbon (x_4) on the catalyst in the reactor.

The conventional way to control this process is shown in Figure 5.8.7. Two control systems are added to the system consisting of temperature control

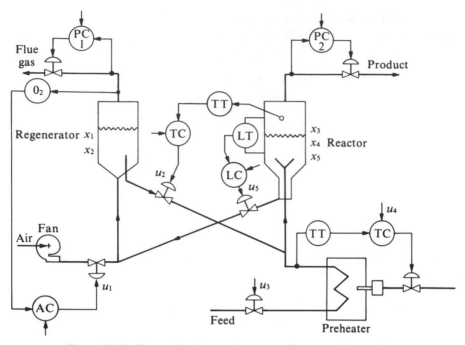

Figure 5.8.7. Conventional control structure for a catalytic cracker.

around the preheater, level control in the reactor by means of flow of spent catalyst (u_5), and pressure control in the tops of both the reactor and the regenerator. One loop controls the air supplied to the regenerator (u_1) on the basis of the measured oxygen in the combustion gases. This is a natural approach. Alternatively, carbon monoxide could be measured and used instead of oxygen. The other added loop is a system to control the flow of regenerated catalyst (u_2) depending on the temperature in the reactor. This is logical because the temperature is highly dependent on the addition of catalyst, which not only promotes the reaction but provides heat for evaporation of the oil.

However, this simple control system does not perform particularly well, because of the many interactions in the process. These can be seen in equations (5.8.1) through (5.8.5). A simulation showed that even with the best possible choice of controller parameters the system is relatively oscillatory. These results are shown in Figure 5.8.8a.

On the basis of optimal control theory where the objective function expressed the profit obtained with production of various products from the top of the reactor, Gould, Evans, and Kurihara (1970) arrived at a different strategy, as shown in Figure 5.8.9. Here, control is concentrated on the regenerator so that the air flow to the bottom is controlled by the temperature in the regenerator, and the output of regenerated catalyst is controlled by an oxygen analysis of the combustion gases. No measurement of the reactor temperature is used in this system.

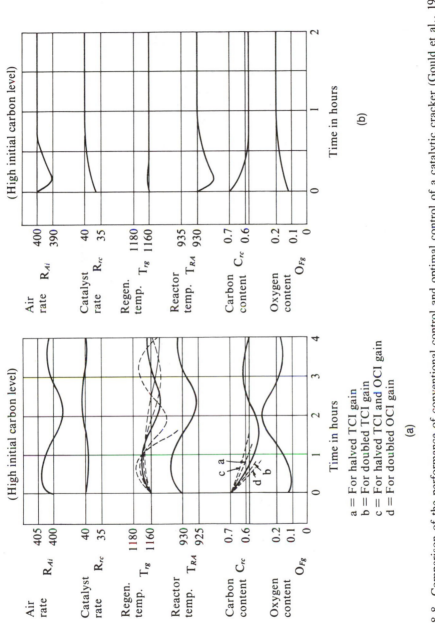

Figure 5.8.8. Comparison of the performance of conventional control and optimal control of a catalytic cracker (Gould et al., 1970). (Reprinted with permission from *Automatica, 6,* L. A. Gould, L. B. Evans, and H. Kurihara. Copyright 1970, Pergamon Press, Ltd.)

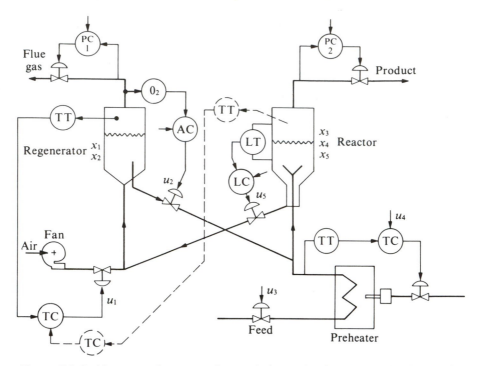

Figure 5.8.9. New control structure for catalytic cracker based on optimal control (Gould et al., 1970). (Reprinted with permission from *Automatica, 6,* L. A. Gould, L. B. Evans, and H. Kurihara. Copyright 1970, Pergamon Press, Ltd.)

Lee and Weekman (1976) thought that the above solution was too risky and proposed an extension, as shown by the dashed line in Figure 5.8.9. The extension added cascade control of the temperature in the reactor via temperature control in the regenerator. The result of the solution of Gould et al. (1970) compared to the conventional solution is essentially decoupling of the process because control is concentrated on the regenerator. The extension of the cascade controller from the reactor temperature, which can be slow, can maintain the energy balance (see Figure 5.8.8b).

Schuldt and Smith (1971) studied use of linear-quadratic optimal control based on a model with six state variables (including the level and mass in the reactor) and four control variables (u_1, u_2, u_4, and u_5 in Figure 5.8.6). Using measurements of the level in the reactor, oxygen concentration in the combustion gas, reactor temperature, and temperature in the regenerator, they developed a multivariable feedback matrix with constant elements (a multivariable P controller), which gave the favorable response shown in Figure 5.8.10.

A lesson to be learned from the above examples is that a multivariable control strategy based on a state space description of quite low order will usually lead to a proper control structure. However, it is necessary to make a detailed

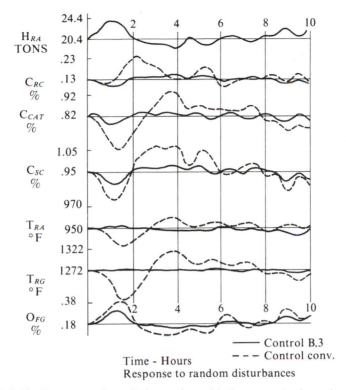

Figure 5.8.10. Response of catalytic cracker with linear-quadratic optimal control (Schuldt and Smith, 1971).

investigation of the relative significance of each control contribution in order to determine which simplifications can be used.

5.9. CONTROL OF COMPRESSORS

Compressors are used for many different purposes in a wide variety of modern industrial processes. They are especially important in refrigeration systems where vapor at low pressure undergoes a pressure and temperature increase followed by expansion. This is used extensively in low-temperature distillation to provide suitable operating conditions with respect to temperature and pressure. Furthermore, compressors play an important role in the transportation of gases through pipelines. The physical phenomena relating to compression and the transport of gases and vapors were described in section 3.4. In this section we will consider the most important aspects regarding control of compressors.

Of the many types of compressors used in industry, the most important are the piston (reciprocating) compressor and the centrifugal compressor. We will limit our discussion to centrifugal compressors, but most of what can be said about them also applies to piston compressors.

The pressure increase given to a gas or vapor by a compressor depends on two quantities: the mass flow through the compressor and the compressor speed. The mass flow can be controlled by valves either in series with the compressor or in parallel, in a recirculating bypass around the unit. The compressor speed can be controlled through the drive motor.

A special phenomenon takes place in a centrifugal compressor designed for high efficiency, an effect known as surging. When the gas flowrate into the intake of the compressor falls below a certain limit, the slope of the pressure–flow characteristic changes sign. An interpretation of this is that the inner resistance of the compressor becomes negative. The gas flow intermittently changes direction, causing serious pressure changes that can damage the compressor and its drive equipment. This is illustrated in Figure 5.9.1 with pressure–flow characteristics for different compressor speeds, where we can see that the pressure increase for a given speed reaches a maximum at a given volumetric flow. Also shown is a typical process pressure characteristic at an acceptable operating point. The surge limit is given by the dashed curve. This curve can be shown to be approximately parabolic when the compressor is operated adiabatically (Magliozzi, 1967; White, 1972). We find:

$$p_2 - p_1 = k_1 q_1^2 \frac{p_1}{\theta_1} \tag{5.9.1}$$

where:

p_2 = compressor outlet pressure (bar)
p_1 = compressor intake pressure (bar)

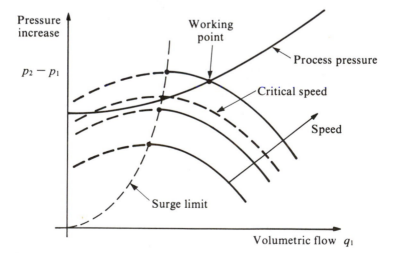

Figure 5.9.1. Pressure–flow characteristics for a centrifugal compressor.

q_1 = volume flow on intake side (m³/s)
θ_1 = absolute temperature on intake side (°K)
k_1 = constant factor

The pressure increase is a quadratic function of the volume flow on the intake side, as shown in Figure 5.9.1. The volumetric flow of the gas, measured by means of the differential pressure across an orifice, is:

$$q_1 = k_2 \sqrt{\frac{\Delta p \cdot \theta_1}{p_1}} \qquad (5.9.2)$$

where Δp is the differential pressure across the orifice (bar), and k_2 is a constant factor. Substituting equation (5.9.2) into equation (5.9.1), we get:

$$p_2 - p_1 = k_1 k_2^2 \Delta p = k_3 \Delta p \qquad (5.9.3)$$

where k_3 is a constant factor. This means that the surge curve is given by a linear relationship between the pressure drop across the orifice on the intake (suction) side (Δp) and the pressure rise over the compressor ($p_2 - p_1$), as shown by the dashed line in Figure 5.9.2. If we want to maintain the operating point a certain distance away from the critical line, we can use a bypass around the compressor (see Figure 5.9.3) to assure that the volumetric flow on the suction side is always a little higher than the critical value. An example of this is indicated by the straight line labeled "control curve" in Figure 5.9.2. The flow on the suction side is now controlled by the valve in the feedback path such that we have:

$$p_2 - p_1 = k_4 \cdot \Delta p \qquad (5.9.4)$$

where $k_4 < k_3$.

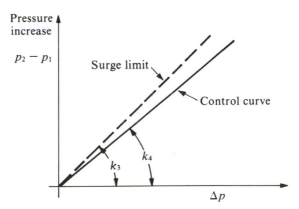

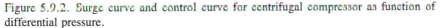

Figure 5.9.2. Surge curve and control curve for centrifugal compressor as function of differential pressure.

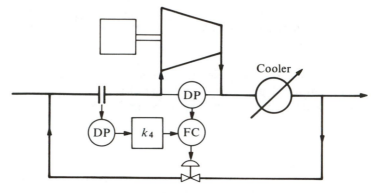

Figure 5.9.3. A typical anti-surge control for a centrifugal compressor.

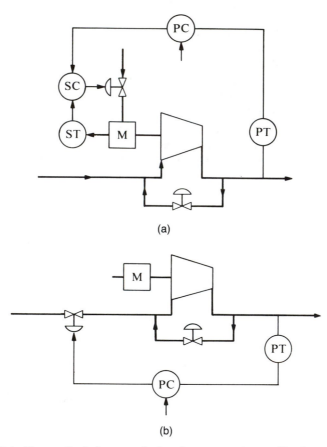

(a)

(b)

Figure 5.9.4. Two methods for control of outlet pressure in centrifugal compressor.

The instrumentation required to solve equation (5.9.4) is shown in Figure 5.9.3. The differential pressure from the orifice is multiplied by the constant k_4 and is then sent to flow controller FC, whose set point is determined by the pressure drop over the compressor ($p_2 - p_1$). The valve in the feedback path will be open only when the pressure drop over the orifice falls below the control curve. The gas cooler seen in Figure 5.9.3 is found in many types of systems.

Control of the pressure on the high pressure side of the compressor can be achieved in a number of ways. Possibilities include either speed control of the drive motor on the compressor, as shown in Figure 5.9.4a, or a valve in the suction line, as shown in Figure 5.9.4b. An advantage to the solution in Figure 5.9.4a is that it involves less energy loss, but on the other hand the capital cost is much higher than that of Figure 5.9.4b.

Chapter 6
PROCESS CONTROL IN LARGE
INDUSTRIAL COMPLEXES

In the preceding chapters, we looked at many of the commonly used control schemes for elementary processes, unit operations, and limited process segments. In modern industry, many process complexes include perhaps hundreds of elementary processes and unit operations, which are interconnected to perform complicated functions. These large processes must combine raw materials and energy to form useful products in a manner that is economically, societally, and environmentally acceptable.

Among the industries often referred to as the chemical process industries are the pulp and paper industry, the chemical and petrochemical industries, oil refining, the fertilizer industry, food· processing, large parts of the metallurgical industry, and the cement industry. All are characterized and dominated by large, highly integrated and highly automated installations.

It is typical in these industries that a process section designed to make certain changes of states of the incoming raw materials will often produce one or more by-products in addition to the main product. Frequently, new processes must be designed and built to handle these by-products. The result is a network of integrated processes in which the material and energy flows are very closely interwoven. Today, energy integration of many of these complex plants is nearly total.

A major goal is that all energy and raw materials that enter the process should leave the process in some useful form without loss. This creates problems not only with equipment, but also with respect to process dynamic behavior and process control. Highly efficient energy and raw material utilization leads, as discussed earlier, to positive feedback in one form or another, which can make automatic control more difficult.

In this chapter we will consider the basic control problems in some of the chemical process industries. In those cases where the elementary processes and unit operations have been discussed in detail in earlier chapters, we will give only a general overview, but we will present detailed discussions of a few industry-specific problems. Although it is impossible to consider all industries, many of the solution schemes developed for one type of process can be applied in other industries.

6.1. PROCESS CONTROL IN THE PULP AND PAPER INDUSTRY

Pulp and/or paper plants can range from large, very complex integrated processes to relatively simple units, depending on the raw materials used and the products.

Although paper can be made from many different materials, most of it is made from wood pulp, and most wood pulp is produced either mechanically (groundwood) or chemically (sulfite or sulfate, the latter also commonly known as kraft). There are several other pulping processes (thermal-mechanical pulp, semichemical pulp, etc.), as well as pulps made from cotton, bamboo, bagasse, and other grassy plants, but in our discussion we will generally consider only the two major types listed above. Because each pulp has its own physical and chemical characteristics, most paper is formulated from a mixture of various types. In general the chemical pulps have longer fibers than the mechanical pulps and give better strength and printing qualities, but they are much more expensive than the mechanical pulps. The mixture chosen for a product depends not only on the desired product characteristics or quality, but also on the cost.

A typical groundwood pulp process is shown in Figure 6.1.1. Conceptually, the grinding process is a matter of forcing the debarked wood "stick" against a rotating carborundum-type wheel. This separates the fibers from each other, forming a pulp. It is important to note that this pulp contains all of the components (notably, lignin) that existed in the original wood. Two major characteristics of this type of pulp are that the fibers have comparatively short length, and, because of the lignin remaining, the pulp is difficult to bleach and tends to darken when exposed to light.

There are two major processes for producing chemical pulps. The objective of each is to dissolve the lignin from the wood, leaving individual fibers. Pulp that is cooked with a sulfite-based acid cooking liquor (usually sodium or magnesium sulfite) is known as sulfite pulp. Pulp produced with an alkaline cooking liquor is known as sulfate, or kraft, pulp. Either process can be of batch or continuous form, but most pulp mills built within the last twenty years or so have been continuous kraft processes.

In a batch pulping process, cooking takes place in a batch digester (a pressurized vessel), to which wood chips and cooking chemicals are added. The cook takes place under a predetermined temperature–time profile. Most mills have more than one of these digesters so that, even though each digester is a batch unit, the total production of pulp is relatively smooth.

In continuous digesters, chips and cooking liquor are continuously supplied to the digester to produce pulp continuously. As the chips flow through the digester, they experience a time–temperature profile similar to that used in the batch digester, although the cooking time may not necessarily be the same in both cases.

We will discuss digesters in some detail in sections 6.1.1 and 6.1.2.

In order to get an overview of the pulp and paper-making process, consider the fully integrated pulp and paper mill shown in Figure 6.1.2. Examination of such a mill will provide the perspective needed to appreciate later detailed discussions of control of individual processes within the system.

The flow of the major raw material, wood, begins in the wood yard at the upper left corner of the figure. The wood is cut to length and sent to drum bark-

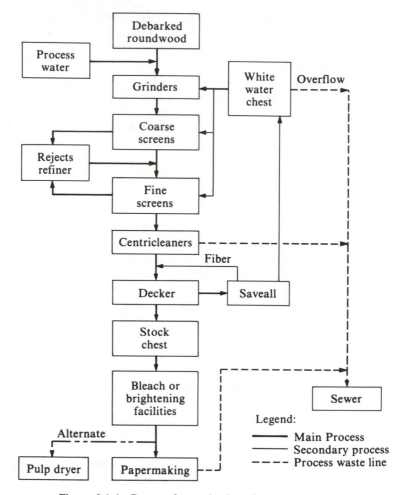

Figure 6.1.1. Process for production of groundwood pulp.

ers for bark removal. It then goes to the pulp mill, for either grinding (mechanical pulp) or preparation for cooking (chemical pulp). As this is a chemical pulp mill, the next process is chipping, where the wood is cut into chips measuring about $3 \times 5 \times 0.7$ cm. The chips are then classified for size uniformity, and the accepts move on to the digester.

In the digester, the cooking liquor dissolves the lignin in the wood, allowing the fibers to separate when they are blown from the digester. As the cooked pulp leaves the digester, it is separated from most of the cooking liquor, which is sent to the recovery system where it is concentrated by evaporation. The organic components are then burned to provide steam, and the inorganic components are used in the makeup of new liquor. The recovery process, which is crucial in the economics of any chemical pulp mill, contains many unit operations and is currently the object of considerable research in terms of both process and process control.

After leaving the digester, the pulp is washed, screened, and cleaned to remove any remaining liquor or foreign materials, and then it is bleached to the desired whiteness (known as brightness in the paper industry). The cleaning–screening–bleaching operation uses many kinds of separation devices such as cyclones, rotary and/or flat screens, vacuum drum washers, and so on. A four-stage bleach plant is shown in Figure 6.1.2. Usually, the stages use different bleaching and extraction agents, depending on the type of pulp, ultimate product, cost, and philosophy of mill management.

Next, the bleached pulp is refined by conical or disk refiners. These devices condition the pulp by adjusting fiber length and surface area in order to develop the required physical properties of the pulp.

At this point, the pulp is usually mixed with other pulp types according to the recipe, or furnish, required to make the desired product on the paper machine. The properly mixed furnish is diluted to less than 0.5% fiber and sent to the paper machine headbox. Here it is uniformly distributed and delivered to the moving Fourdrinier wire. This "wire" is made of a synthetic material with a carefully designed pattern of woof and warp. It passes over various configurations of solid rolls, suction boxes, and foils—all designed to remove water from the stock (the fiber–water suspension), which has been delivered to the wire from the headbox.

Following the wire, the newly formed paper web enters a press section where water is removed by mechanical pressing. The next method of water removal is drying on steam-heated rotating drying cylinders. At this stage, machines that make tissue products (paper towels, napkins, facial tissues, etc.) differ, in that they have a single, large (6 to 8 meters)-diameter dryer. Virtually all other machines have multiple sections of smaller-diameter (1 to 2 meters) drying cylinders.

Paper machines also differ in terms of auxiliary processes. Some have coaters to apply coatings for printing papers, some have size presses to apply special resins for waterproofing and similar characteristics, some have systems for applying plastic surfaces, and so forth.

There are also several approaches to the wire section, including systems that have two wires rather than one. Board machines often have multiple Fourdriniers in order to make laminated products. Details of the actual systems in use are many and complex, but the control problems differ in detail rather than concept. Excellent descriptions of the many processes in the industry can be found in the literature (Britt, 1970; Saltman, 1979; Casey, 1980; Kline, 1982; Mummé, 1983).

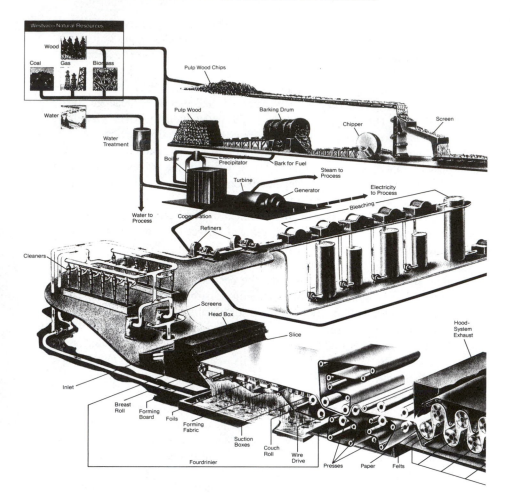

Figure 6.1.2. A fully integrated pulp and paper mill. (Reprinted with permission of WESTVACO Corporation, 1986.)

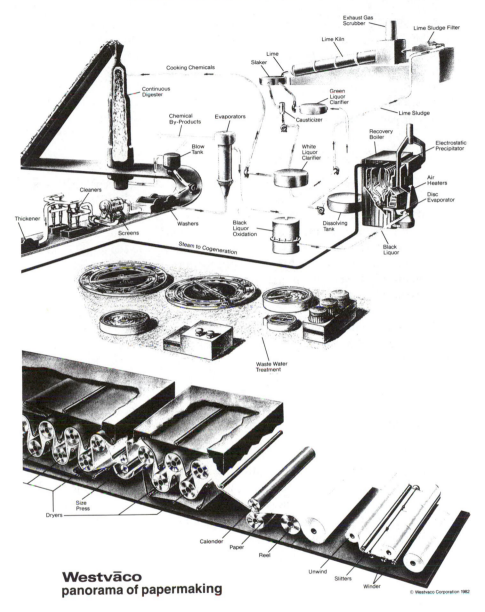

Westvāco
panorama of papermaking

© Westvaco Corporation 1982

Figure 6.1.2. (continued)

Because it is impossible to discuss all the processes in a pulp and paper mill in a book such as this, we concentrate our discussion on the basic control techniques for the following processes:

- Batch digesters
- A continuous kraft digester
- The paper machine

Control of a Batch Digester

Sulfite pulp is produced in an acidic cooking liquor. The chief cooking agent is sulfur dioxide (SO_2) dissolved in water at a concentration between 4% and 9%. About 2 to 3% of the sulfur dioxide is in the form of a bisulfite. The SO_2 is absorbed in water, and the pH of the liquor is controlled by adding a base, usually calcium, magnesium, ammonia, or sodium. The ratio between the SO_2 and the base is such that most of the SO_2 occurs in the form of sulfurous acid (H_2SO_3). Depending on which base is used, one gets calcium bisulfite [$Ca(HSO_3)_2$], magnesium bisulfite [$Mg(HSO_3)_2$], sodium bisulfite [$NaHSO_3$], or ammonium bisulfite [NH_4HSO_3].

The main reaction in the sulfite process takes place between the bisulfite ions and the lignin in the wood. The chemical reactions are very complex and not completely understood, but extensive research is in progress to learn more about the reaction so that the process can be developed further. (See, for example, Casey, 1980.)

There are many differences between kraft and sulfite pulps, and many products contain a mixture of them in order to capitalize on the advantages of each. Kraft is also used to give strength to paper made chiefly from groundwood pulps.

Figure 6.1.3 is a sketch of a typical batch sulfite digester, a large pressure vessel equipped with pumps and valves for handling the chips, liquor, steam, and pulp. Usually there are several identical digesters on the site. This makes it possible to schedule the cooking periods, usually about 8 hours, to maintain a more or less steady flow of pulp from the pulping process as a whole. Operation includes the following steps:

1. Filling the digester with wood chips and cooking liquor.
2. Bringing the digester and contents up to temperature by the addition of steam.
3. Cooking under prescribed time, pressure, and temperature conditions.
4. Blowing the digester (removing the pulp from the digester).

A conventional instrumentation system for a batch kraft digester is shown in Figure 6.1.4. This system works as follows: At the start of each cook (batch), steam is added to the bottom of the digester through control valve FV-1. The steam flow is measured by an orifice that sends a flow signal to flow controller

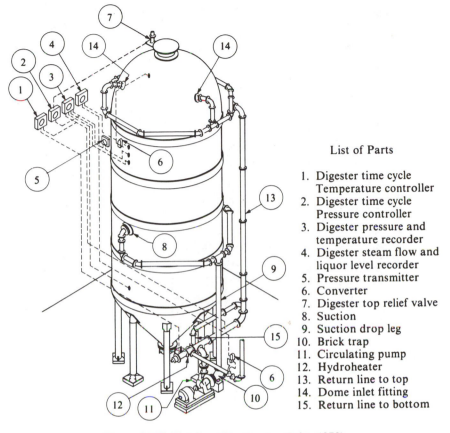

List of Parts

1. Digester time cycle
   Temperature controller
2. Digester time cycle
   Pressure controller
3. Digester pressure and
   temperature recorder
4. Digester steam flow and
   liquor level recorder
5. Pressure transmitter
6. Converter
7. Digester top relief valve
8. Suction
9. Suction drop leg
10. Brick trap
11. Circulating pump
12. Hydroheater
13. Return line to top
14. Dome inlet fitting
15. Return line to bottom

Figure 6.1.3. Batch sulfite digester (Britt, 1970).

FRC-1, which, by means of a selector (LSR-1), chooses the smaller of two signals to operate the valve FV-1. A pressure sensor (PT-2) at the top of the digester sends a signal to pressure controller CPC-2, which has a preset time–pressure set point program. The output from the controller also goes to the minimum signal selector mentioned earlier. As a result, once the cook is in progress, the pressure controller takes over control of the steam valve FV-1. In this way, the pressure, and therefore the temperature, will follow the prescribed time–pressure program.

Two temperature elements are located inside the digester. The outputs of these elements are sent to a recorder (TPR-2), which records the two temperatures as well as the pressure. This provides a measure of the temperature uniformity (circulation) in the digester during the cook, and the relation between temperature and pressure indicates the amount of inert gases in the digester.

Because the digester is heated by steam, an equivalent amount of water (condensate) or vapor must be removed from it. This is done with valve FV-3,

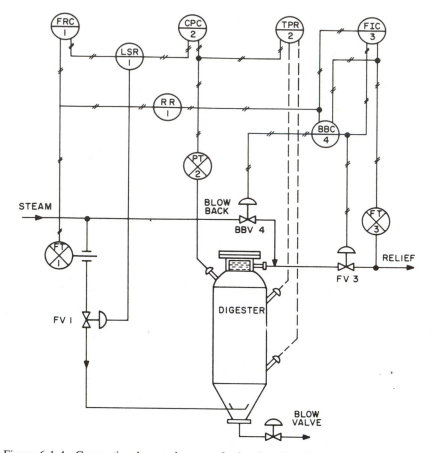

Figure 6.1.4. Conventional control system for batch sulfite digester (Lavigne, 1977).

which is controlled by flow controller FIC-3. The set point of this controller is operated by a reverse relay whose signal is taken from a measurement of the steam flow. A large amount of water or vapor is removed when the steam flow is low, and vice versa.

There is a screen, or filter, at the top of the digester to prevent solids from leaving through the relief line with the water or vapor. In order to prevent buildup or plugging of this screen by solids, the screen is backflushed with steam by closing the relief valve FV-3 and opening valve BBV-4. This is controlled by controller BBC-4, which compares the set point of FIC-3 with the signal to the relief valve FV-3. When the filter clogs, the relief valve must open completely. In that situation, BBC-4 takes control, closing FV-3 and opening BBV-4 for a short period.

When cooking is completed, the new pulp is literally blown out the bottom of the digester through the blow valve. The blow forces the pulp out of the digester, and the sudden pressure release causes the cooked chips to explode, thereby separating the individual fibers from each other. It is this phenomenon that changes a soggy, cooked wood chip into an equivalent amount of separate fibers.

The control system indicated by Figure 6.1.4 can be implemented either by traditional, individual controllers or by a computer.

Even though the chemical reactions that take place in a digester are not completely understood, it is possible to develop useful mathematical models of the process. These models make it possible to estimate unmeasurable states and to manipulate the most important control variables in order to obtain a pulp product that meets the desired quality and degree of lignin removal.

Many pulp mills now use sophisticated computer systems to control the scheduling of the loading, operation, and blowing of multiple digesters at a single site, in order to optimize production and to minimize steam consumption. Extensive research led to the use of these systems, which radically reduce the energy requirements of batch digesters, so that they become economically attractive once again.

The sulfate and sulfite processes have many features in common. Chari (1973) developed a model that was used by Wells, Johns, and Chapman (1975) to develop a control system. The model has four state variables:

$x_1 = L_\alpha$: concentration of α-lignin
$x_2 = L_\beta$: concentration of β-lignin
$x_3 = (OH)^-$: concentration of hydroxyl ions
$x_4 = (HS)^-$: concentration of hydrosulfide ions

In practice, there is only one control variable, temperature:

$u = \theta$: temperature in the digester

Measurements of several process variables can be made during the cook in order to maintain control of the process, but most of them concern elementary variables that do not appear directly in the model. The most important measurement that gives information about the state of the process is the kappa number, which is a measure of the degree of delignification of the wood. Because the kappa number is the nearly universally accepted measure of the degree of cooking, and is the most important measurement of the quality of the cook, we are interested in calculating it on the basis of the state variables. Then we have:

$$y_1 = \kappa: \text{kappa number}$$

and:

$$y_1 = g_1(\mathbf{x})$$

The differential equations that describe the changes of the state variables have the form:

$$\dot{L}_\alpha = -R_{11} - R_{31} + R_{12} + R_{32} \tag{6.1.1}$$

$$\dot{L}_\beta = -R_{21} - R_{41} + R_{22} + R_{32} \tag{6.1.2}$$

where:

$$R_{ij} = L_l^{m_l} C_k^{n_k} \exp\left(A_i - \frac{E_i}{R\theta}\right) \tag{6.1.3}$$

and:

L_l = concentration of lignin type l, either α or β
m_l = power of corresponding lignin type
C_k = concentration of ion type k, either $(OH)^-$ or $(HS)^-$
n_k = power of corresponding ion type k
A_i = Arrhenius frequency factor
E_i = activation energy
R = gas constant
θ = absolute temperature ($°K$)

Equations (6.1.1) and (6.1.2) express the way the two types of lignin change as functions of their concentrations, the ion concentration, and the temperature. Similar equations can be written for the stoichiometry for the concentrations of the two ions. The result is a state space equation of the form:

$$\dot{x} = f(x, u, p) \tag{6.1.4}$$

where p is the vector of parameters that must be estimated.

Wells et al. (1975) designed a system that calculated predicted values of the state variables of the state variables and generated a temperature pattern that ultimately led to the desired kappa number on the basis of the model given by equation (6.1.4). This is shown in Figure 6.1.5. The approach has many possibilities that should be developed further.

An older, simpler approach based on the so-called H factor was developed by Vroom (1957). The H factor is defined as follows:

$$H = \frac{1}{60} \cdot \int_0^t \exp\left(43.18 - \frac{1.611 \cdot 10^4}{\theta(\tau)}\right) \cdot d\tau \tag{6.1.5}$$

where $\theta(\tau)$ is the temperature function and τ is the running time.

The H factor is an approximation of the integral of the reaction rate [see equations (6.1.1 through 6.1.3)] and is a measure of the lignin content if the exponent m_l in equation (6.1.3) is equal to 1. Different cooking programs have

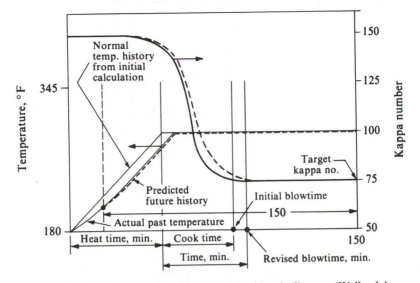

Figure 6.1.5. Predictive control based on model of batch digester (Wells, Johns, and Chapman, 1975).

been based on computation of the H factor and manipulation of the temperature and cooking time to reach a specific value of the H factor. Even though this method is a very rough approximation of the actual conditions, operators have used it successfully to produce a more uniform pulp product than is possible (or at least usual) with a time–temperature program alone.

A more recent approach uses the S factor. Applied in the same manner as the H factor (Haegglund, Wallin, and Karlsson, 1980), it is defined as:

$$S = \int_0^t (p_{SO_2})^\alpha \exp\left(C - \frac{D}{\theta(\tau)}\right) d\tau \qquad (6.1.6)$$

where p_{SO_2} is the partial pressure of SO_2, and α, C, and D are constants dependent on the wood species.

Control of a Continuous Sulfate (Kraft) Digester

The modern continuous kraft digester has caused a dramatic change in the economics and production capacity of the pulp industry. Probably the most widely used digester of this type is the Kamyr digester, originally designed and built cooperatively by the Karlstad Mekaniska Verkstad in Sweden and Myrens Verksted in Norway.

The most important elements of a Kamyr digester appear schematically in Figure 6.1.6. The flow of chips to the digester is controlled by a chip feeder that has a controllable, variable-speed drive on a five-pocket rotor. From the chip meter the chips pass into the low-pressure feeder and presteaming vessel.

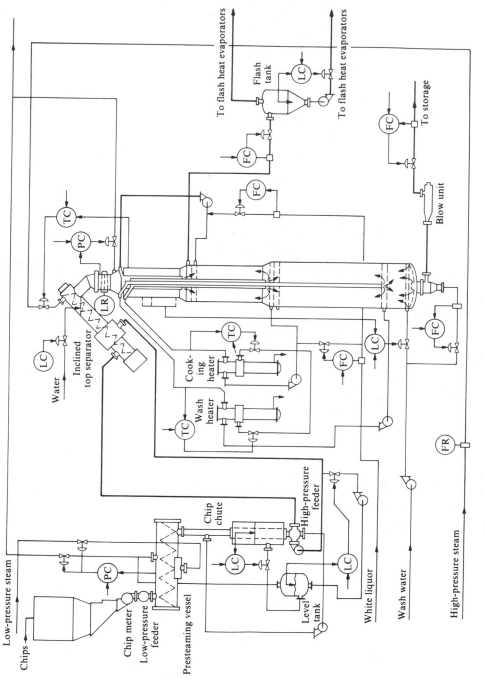

Figure 6.1.6. Kamyr kraft digester with conventional control (Britt, 1970).

This vessel, a cylindrical tank, operates at a pressure of about 1 to 1.5 bar. The chips are exposed to steam as they are transported through the cylinder by a slow-speed auger. This process drives the noncondensable gases and turpentines from the chips. These products are vented at this point so that they will not interfere with later heating and diffusion processes in the chips. The condensed steam is absorbed by the chips, increasing the density of the chips sufficiently that they will sink in the cooking liquor in the digester, a characteristic absolutely necessary for the process to work. The presteamer helps to make the chip temperature and moisture more uniform prior to entry into the digester, and the moisture and heat also tend to swell the chips, making penetration of the cooking liquor easier.

After presteaming, the chips drop into a level-controlled receiver partly filled with white liquor. At this point, the active alkali begins to penetrate the chips. The high-pressure chip feeder is located at the outlet of this tank. This feeder has four pockets, each of which goes across the rotor so that the chips and liquor can enter the pockets freely and leave at high pressure. The pressure on the high side of this feeder is about 11 to 12 bar.

The chips and liquor are then pumped to a separator at the top of the digester. This separator has a cylindrical screen containing a transport screw, or auger. The liquor is separated from the chips and is returned to the high-pressure chip feeder while the screw takes the chips to the top of the digester itself.

As we have mentioned, pretreatment makes the chips heavy enough to sink in the liquor in the impregnating zone, which is the first zone of the digester. It takes about 45 minutes for the chips to pass through this zone. During this time, they are exposed to a temperature that varies between 100°C and 127°C. The most important function of this zone is to provide the time and conditions for the liquor to impregnate the chips by diffusion. The temperature is too low for significant chemical reaction to occur.

The zones and various inlet–outlet lines of the digester are clearly shown in Figure 6.1.7.

The next zone is the heating zone, where the temperature of the chips is raised to about 150°C. The heating medium is hot cooking liquor that is pumped into the zone through a set of concentric perforated tubes or nozzles and is removed from the zone through a heat exchanger that controls the liquor temperature.

The diameter of the digester increases slightly from top to bottom to help inhibit vertical motion and mixing of the chips, as it is desirable to maintain plug flow of the chips and liquor as they pass through the digester. This requires special attention because the chip density changes as cooking progresses.

In the lowest part of the heating zone the temperature is held at about 165°C. Here, the heating system is separate from that in the upper zone.

The hot chips next move into the cooking zone. The temperature in this zone increases from top to bottom because the chemical reactions are exothermic, and there is little heat lost to the surroundings. The time required for the chips

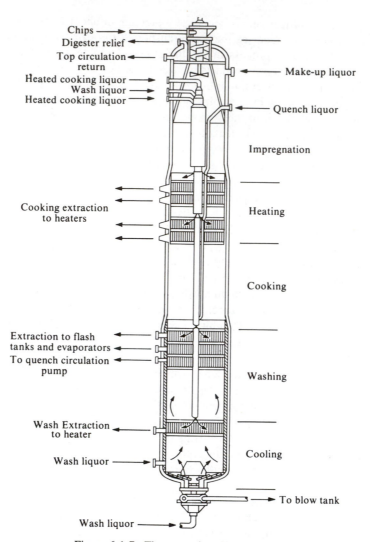

Chips
Digester relief
Top circulation return
Heated cooking liquor
Wash liquor
Heated cooking liquor
Make-up liquor
Quench liquor
Impregnation
Cooking extraction to heaters
Heating
Cooking
Extraction to flash tanks and evaporators
To quench circulation pump
Washing
Wash Extraction to heater
Wash liquor
Cooling
To blow tank
Wash liquor

Figure 6.1.7. The zones in a Kamyr digester

to pass through the cooking zone varies, depending on the wood species and the desired product qualities and characteristics, but a typical time would be about 90 minutes.

The reactions are quenched at the bottom of the cooking zone by addition of cool liquor. The cool liquor forces some of the used cooking liquor (black liquor) out of the chips and cools the total mass to a temperature insufficient to support the chemical reactions. The major importance of proper quenching is that it stops the reaction and maintains the strength of the pulp.

The cooled mass continues downward into a washing zone where dilute liquor and warm water are forced through it to remove liquor and dissolved solids from the pulp. The wash liquor enters the washing zone at about 135°C, but the temperature at the bottom of the zone is about 75 to 80°C. The washing can be very sophisticated, with most modern washing stages being counter-current diffusion washers.

The cooled mass is removed from the bottom of the digester by slowly rotating wings, or scrapers, on a central axis. This device leads the pulp to the blow valve. The pressure at this point is about 14 bar, but after the blow valve it is reduced to atmospheric.

In principle, the control problem for a continuous kraft digester is the same as for the batch system. The most important difference is that the chips move continuously through the digester as they undergo the different phases of the process. Therefore, the temperature profiles and the concentrations of the cooking liquor in the different zones of the digester are critical to proper operation of the system.

Beyond the usual basic functions, it is possible to design control functions based on the use of a mathematical model to estimate the unmeasurable states. In the following paragraphs we will first look at conventional control solutions and then at more sophisticated possibilities.

Figure 6.1.6 shows a number of control loops that affect the treatment of the chips as they pass through the digester. The steam added to the presteaming unit is controlled by a pressure controller. However, the diagram shows one control valve in the steam supply line and another in the line to the condenser. This requires that the movement and characteristics of the two valves be matched to each other

The next system is level control of the liquid in the tank just before the high-pressure chip feeder. This is accomplished with a valve in the liquor line to the level tank. The level in this tank is also controlled by a pump in the high-pressure system. Water is added at high pressure at the separator at the top of the digester to hold the level constant.

The high-pressure steam added at the top of the digester is controlled on the basis of the temperature slightly farther down in the digester. A pressure controller controls the removal of steam to the condenser. These two controllers are strongly coupled in the sense that the temperature–pressure relationship is set for saturated steam. Although the system can function well if the two control loops have different bandwidths, only one of the controllers can have integral action.

It is necessary to manipulate the addition of chips to precisely control the chip flowrate through the digester. This system is not shown in Figure 6.1.6, but is described in some detail in Figure 6.1.8, which represents part of a comprehensive computer-based digester control system. We see in this figure that control of the chip feeder system velocity and withdrawal of pulp from the

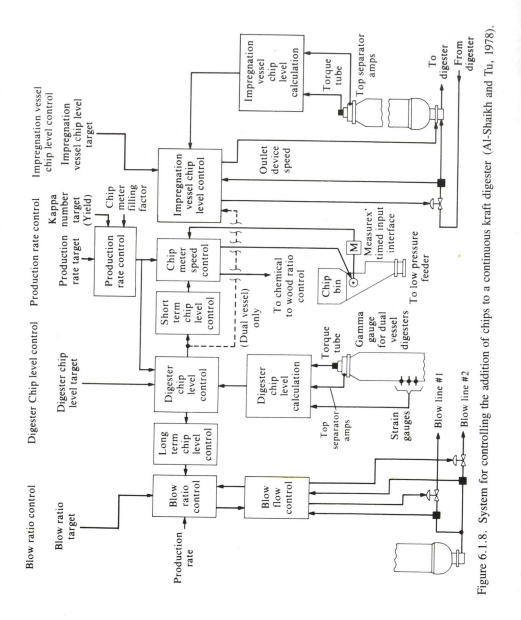

Figure 6.1.8. System for controlling the addition of chips to a continuous kraft digester (Al-Shaikh and Tu, 1978).

bottom of the digester are the two major control variables with respect to chip flow control.

The most important control functions in the digester itself are those that control the addition of cooking liquor and the temperature profile along the digester. In Figure 6.1.6 we see that the white liquor is added at two places: the recirculation system before the impregnating zone and the heating zone, where the liquor is circulated through heat exchangers for temperature control. In this scheme the white liquor flow is held constant in the upper section, and the total flow is held constant by a control valve on the line into the heating zone. There is no control on the strength of the liquor in the digester in Figure 6.1.6. In Figure 6.1.9 we see a system for controlling the set points of the two flow controllers under a supervisory control system including measurement of the alkalinity of the cooking liquor and calculation of the required amount of white liquor.

Similarly, the liquor in the cooking zone is pumped out of the bottom of the zone, circulated through a heat exchanger, and returned to the center of the cooking zone.

Temperature control, as shown in Figure 6.1.6, is relatively simple. The heating zone of the digester is divided into an upper and a lower temperature region. Each has its own temperature control system so that a profile can be maintained. Because temperature is a critical factor in cooking, it is reasonable and possible to control the temperature set points on the basis of model calculations of the type discussed in section 6.1.1. This is shown in Figure 6.1.10. We see that two models are used here, one for calculation of the temperature pattern and one for calculation of the reaction process, as described in section 6.1.1. The latter model makes it possible to change the temperature profile in order to obtain the desired kappa number for the pulp.

Wash liquor is added to the washing section at the bottom of the digester, and in Figure 6.1.6 we see that this is controlled by the level at the top of the washing zone. This system must be viewed in combination with the chip level controller described earlier. The removal of pulp from the bottom is also flow-controlled.

The technology and details of operation of continuous digesters change very rapidly. Therefore, the brief discussion in this section should be considered as representative of the process, rather than a definitive description of any particular unit.

Control Problems on the Paper Machine

We will limit our discussion to the Fourdrinier machine, as it is the most widely used type of paper machine in the industry today. As was the case for the continuous digester, there is no truly typical paper machine, but the machine we will discuss could very well exist in practice.

The Fourdrinier machine, pictured in Figure 6.1.2, is usually treated as consisting of two major sections—the wet end and the dry end. We define them

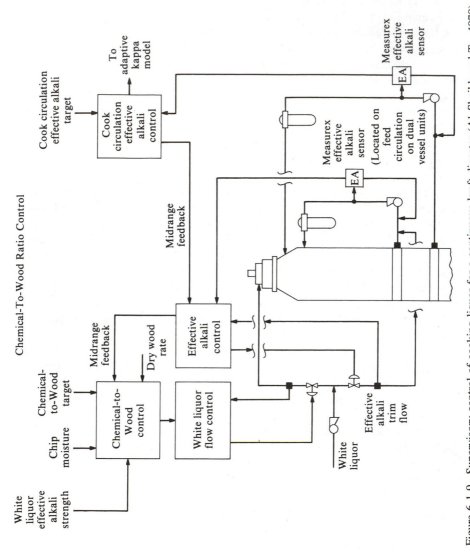

Figure 6.1.9. Supervisory control of cooking liquor for a continuous kraft digester (Al-Shaikh and Tu, 1978).

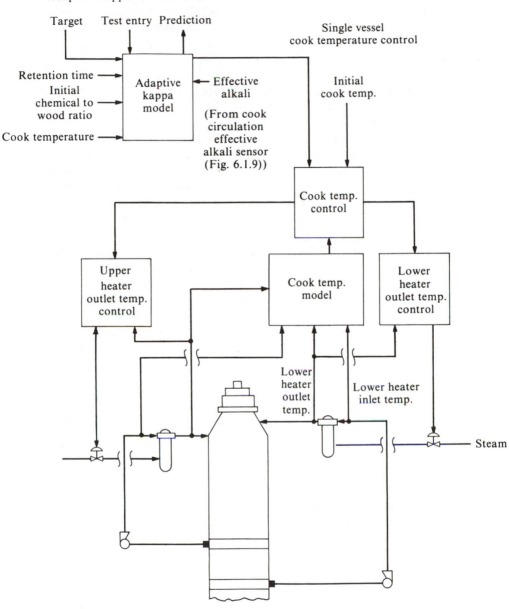

Figure 6.1.10. Model-based system for control of temperature profile in a continuous digester (Al-Shaikh and Tu, 1978).

here as follows: the wet end consists of all the equipment up to the point where the paper (sheet) leaves the press section, and the dry end includes all subsequent processing up to the final reel (windup of the continuous sheet) at the end of the paper machine.

The Wet End. Figure 6.1.11 shows the unit processes and instrumentation systems for the wet end of a representative paper machine. This diagram contains not only the headbox and Fourdrinier, but also much of the stock supply and recovery systems.

The final stock, mixed and proportioned for use on the paper machine, is stored in the machine chest shown at the lower left corner of Figure 6.1.11. The level in the machine chest is held constant by controlling the line between it and the mixed stock chest directly upstream. It is important that this level be constant, as it provides a hydraulic head that affects the pumping rate on the outlet side of the chest. Whitewater (water that contains very little fiber) is added to the flow at the inlet of the pump on the outlet side of the machine chest. The addition of whitewater is controlled by a *consistency* controller, which has the function of maintaining a uniform, desired consistency in the stock line downstream of the machine chest. Consistency control is extremely important, as it is the means by which the amount of fiber supplied to the paper machine is controlled. The primary control mechanism is flow, but excellent flow control will not assure delivery of the proper amount of fiber unless the consistency is known and constant. The amount of fiber delivered determines the weight of the final paper produced. Measurement of consistency is difficult and provides only an approximation of the amount of fiber in the suspension. Therefore, it is necessary to update the set point of the consistency controller on the basis of measurements of paper quality at a later stage in the process.

The stock is next pumped to the mixed stock refiner for final adjustment of fiber length and surface area. There are several popular refiner designs, but all of them, in one way or another, expose the pulp to both cutting and brushing actions. In conical refiners, this action is created by passing the pulp through the space between concentric cones, one fixed and one rotating. Each cone has longitudinally mounted cutting bars. Decreasing the distance between the cones and increasing the power causes more refining action to take place. The extent of refining is measured (inferentially) by a property known as freeness. Freeness is actually a measure of the rate at which water drains from a pulp under a standard set of conditions, and this rate is a function of the surface area of the fibers. Control of the relative position of the refiner cones, which is measured by the power used by the refiner, is usually based on so-called freeness controllers, but, as shown in the figure, it can also be based on measurement of the vacuum in a suction box near the end of the Fourdrinier section. The principle of the latter is that as it becomes more difficult to drain the water from the sheet on the wire, the vacuum in the suction box will increase. In terms of refining, this means that the surface area of the fibers has increased, causing the water to

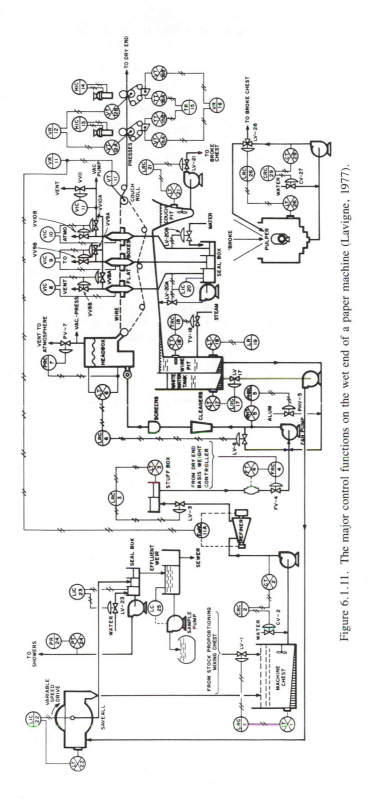

Figure 6.1.11. The major control functions on the wet end of a paper machine (Lavigne, 1977).

bond more tightly to the fibers as a whole. If such an increase is undesirable, the control system must cause the refiner cones to separate slightly in order to reduce the power drawn by the refiner, thus reducing the energy transferred to the fibers and decreasing the cutting and brushing action as desired. From the diagram, we see that there is a rather long transport delay between the refiner and the measurement point on the wire. Therefore, this freeness (refiner) control system is quite slow and often unreliable. On the other hand, freeness controllers can provide reliable service and can usually be installed in shorter control loops.

Following the refiner, the stock passes through a control valve to a small, level-controlled surge tank known as the stuff box. The series position of the control valve means that the refiner operates at the line pressure supplied by the pump. Flow from the stuff box is controlled as precisely as possible by a magnetic flowmeter and a control valve. The set point of this flow control system is determined on the basis of measurement of the basis weight (total mass/area, where the area used in the definition depends on the product) of the finished paper at the dry end of the machine. This valve is the basis weight valve. Clearly, the level control on the stuff box holds the pressure upstream of the basis weight valve constant, enabling the valve to provide precise flow control of stock to the paper machine. Precise flow control and precise consistency control are both necessary if basis weight is to be uniform.

The refined, controlled stock flow is diluted with whitewater taken from the wire pit, which is a collector below the wire section. The whitewater, more specifically known as tray collector water in this application, is mixed into the main flow at the inlet of a large pump known as the fan pump. Flow of the diluted stock from the fan pump is controlled by a valve and recirculation line. The flow to the machine passes through a final screening and/or cleaning process just prior to the headbox. The control valve in the recirculating line is controlled on the basis of the liquid level in the headbox.

Control of headbox consistency by means of controlling the thick stock consistency (at the outlet of the machine chest), flow control of the thick stock, and dilution with a relatively constant flow of whitewater at the fan pump, as shown in Figure 6.1.11, is not always a satisfactory means of obtaining good basis weight uniformity of the finished product. Ideally, one would like to measure the consistency of stock in the headbox; but because of the low consistency (0.1–0.5%), conventional consistency meters cannot be used. There have been many attempts to develop new measuring instruments, but none has been completely satisfactory. However, the consistency set point can be adjusted on the basis of a signal from measurement of the basis weight at the dry end of the paper machine, but with limited flexibility (see discussion in the next section).

The headbox, which has a very complex design on a modern machine, has a full-width, manually (usually) controlled opening (slice) through which the thin (very low consistency), uniformly distributed stock is forced out onto the wire at a precisely determined velocity. The ratio of the jet velocity to the wire speed depends on the machine and the product. The wire speed can be quite high—over 1000 m/min on a modern paper or tissue machine. To get a jet velocity this high, the headbox is pressurized by an air pad over the liquid (pond) level in the headbox. The total pressure in the headbox is measured and controlled by a pressure controller that adds or vents air on the air pad side. Control of the jet velocity is critical, and there have been many attempts to improve on the system just described (Fjeld, 1978).

Once the stock is on the wire, the process of water removal begins. Some water is forced through the wire by gravity, but most is removed by suction created by foils, table rolls, and/or suction boxes located beneath the wire. On the machine shown in Figure 6.1.11, there are no foils, only three suction boxes, each having its own vacuum control system.

Whitewater from the suction boxes goes to the seal pit for each box and then on to the wire pit and either the fan pump or the saveall—a filter device that removes fiber from the whitewater, or at least thickens the whitewater so that it can be recycled through the stock system.

The continuous web of paper passes from the wire to a press section consisting of one to three press nips, where the sheet is mechanically pressed between a hard roll and a continuous wool felt. A critical problem in pressing is that water removal must be maximized without damaging the surface of the sheet. Figure 6.1.11 shows no automatic control function in the press section, but there is often control on the vacuum and always instrumentation for measuring vacuum and temperature. Major research in pressing has been directed toward the mass transfer mechanisms of water removal, but so far there appears to be little that advanced control techniques can do to improve the process.

Figure 6.1.12 is a plot of the solids profile through the entire system. Typical consistencies appearing at various points in the process are as follows:

Machine chest	6–10%
Thick stock	2–5%
Headbox	0.1–0.5%
Dry end of the wire	15–25%
Out of wet presses	30–45%
Dry end reel	94–98%

These ranges represent a wide variety of products and paper machines.

A major research goal is to develop a method of making a high-quality sheet by some type of economically feasible dry process, but research in this direction has not yet been successful. At the present time, the only feasible method of making paper in large quantities is the method of dilution and formation followed by expensive drying.

The Dry End. The dry end of a paper machine is shown in Figure 6.1.13. A major feature of this part of paper, board, and newsprint (not tissue) machines

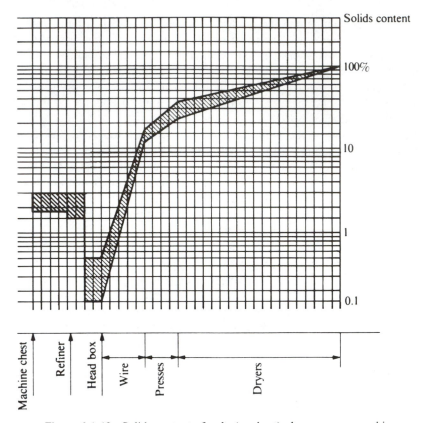

Figure 6.1.12. Solids content of pulp (or sheet) along a paper machine.

is the large number of steam-heated cylindrical dryers. These dryers are usually about 1 to 1.5 meters in diameter and the width of the machine (up to 8 or 9 meters). There may be from 40 to 70 of these dryers, grouped into as many as ten or more controllable groups divided into top and bottom sections. Each section has a continuous dryer felt covering that portion of the surface of each dryer in the section on which the paper contacts the dryer. The felt has two functions: to hold the sheet in direct contact with the dryer surface to provide good heat transfer; and to help promote drying by enhancing favorable heat and mass transfer conditions needed to get the moisture or vapor away from the sheet quickly. Modern high-speed machines have complex systems for ventilation of the dryer area. Hoods and other devices over the dryer sections help to force the vapor out of this area. These ventilation systems are economic necessities from the energy standpoint, and they also provide a means of recovering some of the energy used in drying.

Many of the control functions on the dry end are shown in Figure 6.1.13. First, let us look at the steam flow through the dryers, starting at the dry end and proceeding backward to the first dryer section. Steam is supplied to

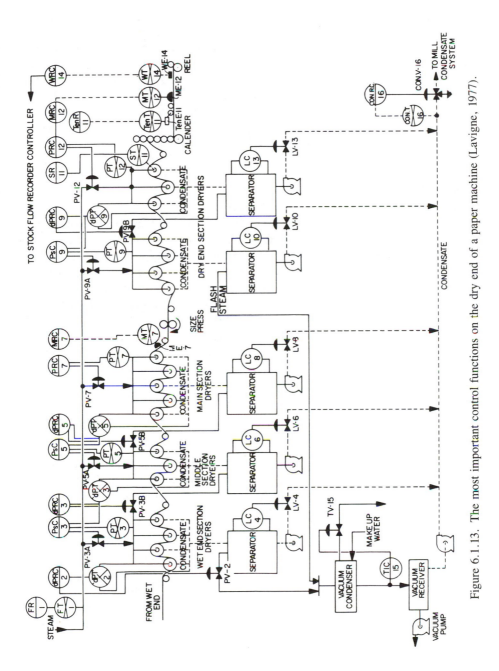

Figure 6.1.13. The most important control functions on the dry end of a paper machine (Lavigne, 1977).

each dryer section from two sources: through a control valve from the main steam plant system and from a condensate-separator system from the previous dryer section (except for the section at the dry end). The purpose of this arrangement is to extract as much heat as possible from the steam supplied to the paper machine.

The machine in the figure has five drying sections. The pressure in the first section (starting at the dry end) is controlled by pressure controller PRC-12. It also controls the temperature because the steam is saturated. The set point of this controller is determined by measurement of the moisture content of the sheet at the dry end. This is done by means of moisture controller MRC-12, usually an infrared, microwave, or capacitance type device. Because this last dryer section in the machine is very close to the point of moisture measurement, the transport delay between the measurement and the point of heat application is quite short, allowing a rapid response.

The steam and condensate removed from the first section go to a separator where the water and vapor are separated in a level-controlled separator tank. The steam fraction is sent through a control valve to the next dryer section in line. The driving force is a pressure differential between the pressures of the first and second dryer sections. In the second section, fresh steam is added to the recovered vapor, the new steam being manipulated by a valve controlled by a pressure controller that receives its measurement signal either from pressure element PT-9 or from the differential pressure from the previous section. The flow is determined such that the desired differential pressure between sections is obtained. As a result of this arrangement, there is a gradual decrease in steam pressure as one moves from the dry end backward to the wet end dryer section, and the shape and magnitude of the pressure curve along the machine depend upon the product being made, the behavioral characteristics of the machine, the desired machine direction moisture profile, and the final moisture as measured by MRC-12 and controlled by PRC-12.

In the machine shown in Figure 6.1.13, any flash steam produced by the second dryer section separator is sent to a condenser. This probably does not represent common practice, where all flash steam is used by upstream sections until the section at the wet end has been supplied.

The third dryer section also gets some fresh steam from the steam network. This flow is controlled by pressure controller PRC-7. The set point of this controller can be controlled by a moisture controller based on a moisture measurement in the center of the drying operation on the machine (MRC-7) if this measurement is available. The pressure and differential pressure practices described above are repeated for each section all the way back to the wet end dryers. Figure 6.1.13 shows how the steam from the third stage is used in the remaining two upstream stages.

Many machines operate with only the countercurrent steam flow. These machines add raw steam only to the dry end section and rely on flashing the condensate all the way back to the wet end. All the steam in each dryer section is condensed. As a result, the wet end section may actually operate at a pressure slightly less than atmospheric. A machine with four dryer sections might operate those sections at pressures of 7, 4.8, 1.4, and 0.5 bar, respectively. This practice is also good from a product quality standpoint—it is usually desirable to heat the sheet slowly and progressively, so the use of flash steam is ideal.

Temperature control of the drying section of a paper machine must be a compromise between efficient energy use and maintenance of the proper temperature (moisture) profile through the machine. As we have said, the profile is responsible for many of the final sheet characteristics, including smoothness, wrinkling, curl, moisture uniformity, porosity, and so on. The profile also affects certain operating characteristics such as sheet break frequency and sheet tension on the machine.

Good control of the basis weight is also critical to maintaining a low sheet break frequency. Figure 6.1.13 shows a basis weight meter (WT-14) at the dry end of the machine. It is connected to a basis weight controller (WRC-14) that controls the set point of the thick stock flow controller at the wet end (see Figure 6.1.12). The basis weight meter is an expensive, complex device that uses beta-ray absorption as its measuring principle.

Supervisory Control Systems. The conventional systems described above form a basis for the realization of more advanced control schemes that can improve the quality of the product. Many researchers have attempted to develop adaptive control schemes to reduce the variance of moisture and basis weight in paper (Åstrom, 1964; Åstrom and Bohlin, 1966; Cegrell and Hedqvist, 1973; Borisson and Wittenmark, 1974; Fjeld and Grimnes, 1974; Al-Shaikh, 1978). Adaptive controllers have provided significant improvements in control of moisture (through the set point of the pressure controller in the last dryer section) and in basis weight control (through control of the set point for the thick stock flow controller). For the machine they were investigating, Åstrom and Haggman (1974) found a reasonable transfer function for the wet end, from the flow of the thick stock to the headbox consistency, to be:

$$h(s) = \frac{K(1 + 191s)e^{-30s}}{(1 + 27.6s)(1 + 231s)} \qquad (6.1.7)$$

The corresponding transfer function with respect to the basis weight at the dry end of the machine had a transport delay of 70 seconds instead of 30 seconds. The zero and pole given respectively by the factors 191 and 231 are due to the recirculation of fiber in the whitewater used to dilute the thick stock. If we ignore these two factors, which give a gain of about 0.83 at the middle frequencies in the range of interest, we see that the system is dominated by one time constant and a transport delay. According to the theory presented in section 2.7, the most significant thing that can happen with feedback from the basis weight measurement and the use of a model based on equation

(6.1.7), including the 70-second transport delay, will be an improvement in the ability of the system to react quickly and with good damping to changes in the basis weight set point. Moisture and basis weight control have also been studied with respect to feedforward and decoupling between these loops. If the basis weight measurement is adjusted for moisture, then cross-coupling will be small.

Figure 6.1.14 shows graphically how the moisture variance on a paper machine was reduced by adaptive control compared with traditional control (Fjeld, 1981).

Virtually all of the discussion above has applied to conditions of the sheet in the machine direction. Nothing explicit has been said about conditions in the cross-machine direction. Most of the cross-machine characteristics of the paper are determined on the very first portion of the wire, ahead of the so-called dry line, which is the point on the wire where enough water has been removed that the fibers are no longer in suspension in the fiber–water mixture. At this point the appearance of the stock literally changes from liquid to dry. Several adjustable spindles across the slice make it possible to shape the lip, which is the upper edge of the slice, to give a variable opening across the slice. This is commonly done to adjust the amount of stock in the cross-machine direction to obtain a uniform cross-machine basis weight. The adjustment of these spindles is based on measurements made by the basis weight and moisture meters, which scan across the machine. These devices can provide cross-machine information, they can be held at a fixed cross-machine location to give only a machine direction measurement, or they can scan and integrate cross-machine and machine direction information as desired. At present, few machines use cross-machine information to adjust the slice spindle settings automatically. The various values available from the basis weight and moisture meters can be used, as seen above, to control steam and thick stock set points, but they can also be used as monitors and alarms. A survey of some cross-direction control algorithms is given in Wilhelm and Fjeld (1983).

Cross-machine control is also being applied to moisture control. Sophisticated control systems based on scanning moisture meters are being marketed by several manufacturers. Control of the cross-machine moisture profile cannot be obtained from the drying cylinders. It is implemented with auxiliary heaters, usually infrared radiators, that are sectionalized in the cross-machine direction. Even on machines equipped with these cross-machine trimming devices, the major drying is still done by the traditional steam drying cylinders.

An important control problem on a paper machine is control of the drive systems of the many different rotating shafts on the machine. Of special importance is control of the machine drive itself—the series of motors that drive all of the rotating elements from the wire turning mechanism, through the presses and the dryers, to the reel. Proper operation requires that each rotating shaft have a certain relationship to the others. They are *not* operated at the same speed; a certain amount of stretch is necessary to maintain the quality of the sheet surface, which dries during its journey through the machine.

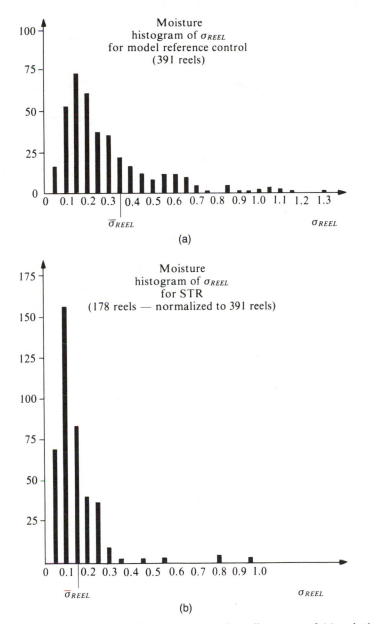

Figure 6.1.14. Comparison of moisture variance under ordinary control (a) and adaptive control (b) (Fjeld and Wilhelm, 1981).

Strictly speaking, the machine drive is not a process control system, but it is a critical part of a paper machine. A schematic diagram of a machine drive is presented in Figure 6.1.15.

Each dryer within a section may have its own drive motor, or the section may be geared together and driven by a common motor. The motors are usually dc motors with thyristor control. The speed control system ensures (a) that each section operates at a speed that has the proper ratio to the speed of each other

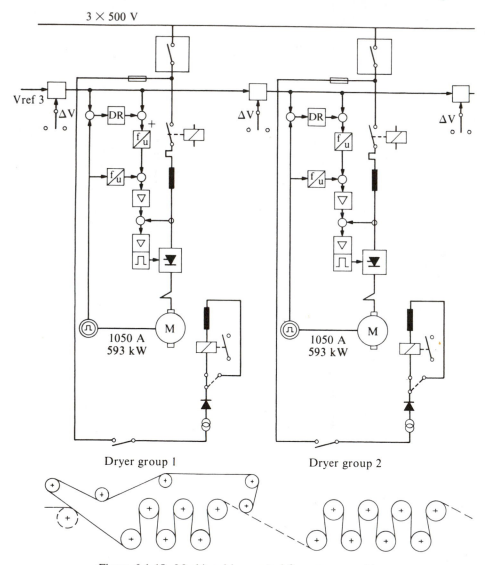

Figure 6.1.15. Machine drive control for a paper machine.

section in the machine; and (b) that when the machine speeds up or slows down, each section changes speed appropriately and precisely. Figure 6.1.15 provides an overview of how this system is arranged. The set point for the speed control system of one section comes from the previous sections. It is absolutely necessary that the speed control be precise. This is done by measuring the rotation speed of the motor shaft by means of a pickup that emits a digital pulse frequency proportional to the speed. The integral of this frequency is a

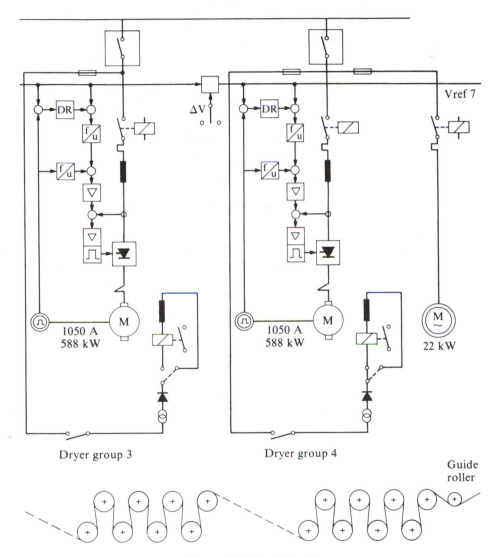

Figure 6.1.15 (continued)

measure of the angular displacement of the shaft. The stretch of the sheet along the paper machine is determined by the difference in velocities between the sections, and, as we have mentioned, the amount of stretch desired depends on a host of factors, most relating to product quality and machine runability. As we see on the figure, the drive motors for the wet end sections must also be precisely controlled at the appropriate ratio to the speeds of the dryer sections.

6.2. CONTROL OF PROCESSES FOR OIL EXTRACTION AND REFINING

Instrumentation and automatic control play a major role in the complex technology of oil production processes (such as offshore oil platforms) and systems that process crude oil into its many useful products. We will give an overview of the most important control problems and their solutions in the following specific applications:

- Oil–gas separation on an offshore oil platform
- Processing of wet gas
- Refining of crude oil

Process Control of Oil–Gas Separation on a Production Platform

Figure 6.2.1 is a flowchart of one of the most important processes that take place on an offshore production platform. A mixture of oil and gas comes up from the various well heads through control valves. The problem is to separate gas, oil, water, and mud from each other by a gradual pressure reduction distributed over four stages.

Of the five horizontal separator tanks, the one at the lower left of Figure 6.2.1 is a testing tank. A pipe from the well head carries a mixture of oil, gas, water, and mud into the inlet separator (Figure 6.2.2a). The pressure at this point is 67.5 bar, and the temperature is 89°C. The four separators are large horizontal tanks that can be pressure-controlled in order to promote separation of the various gas fractions by vaporization in the individual tanks. The fraction condensed at any given pressure goes with the oil and water to the bottom of the tank. Removal of oil and condensate is controlled by means of a level controller operating a control valve on the outlet stream. The vapor flow is controlled by a pressure controller operating on the basis of the pressure in the separator. Each separator has a similar system, but the pressure at each stage is different, typically 67.5 bar at the first stage, 21.7 bar in the first crude oil separator, 5.9 bar in the second crude oil separator, and 1.6 bar in the third-stage crude oil separator (Figure 6.2.2b). The temperatures in the separators are held at approximately 80 to 90°C. The vapor from each separator stage goes through a scrubber to remove any remaining liquid particles and on to a centrifugal compressor. The compressor raises the gas pressure so that the gas can be reinjected at a higher pressure into an earlier stage of the process. This is a typical countercurrent process where the vapor from the third separator goes through the first compressor back to separator number two, the vapor from separator number two goes through compressor number two back to separator number one, and the vapor from separator number one goes through compressor number three either to the gas network or back into the well head. Each compressor is equipped with an anti-surge control system. Figure 6.2.1 does not show how the three compressor motors operate or how their speeds are controlled. If each compressor has its own motor, often a turbine, then each must have a speed controller whose set point can be controlled from a pressure controller on the suction side of the compressor. It is also possible to drive all three compressors from a single shaft driven by a common motor. In this case it is necessary to choose one of the compressors as the object of the suction pressure control so that the required pressure difference is maintained across the outlet valve of each of the separators.

The system given in Figure 6.2.1 describes the only major control system on an oil platform. Usually an oil platform is highly instrumented, but there are relatively few control loops.

Control of Wet–Gas Processing

Figure 6.2.3 is a schematic flowchart of the process units used for processing wet gas, which is the gas sent by pipeline directly from the separation process at the oil field to a processing plant. The purpose of the processing shown in the figure is to remove the heavy hydrocarbons and send the light components out to a gas network system where they are usually distributed for use as a fuel. The heavy hydrocarbons are usually separated by stepwise fractionation into ethane, propane, isobutane, n-butane, and gasoline. Which fractions are chosen for use as fuel depends upon the market situation. (It is interesting to note that at one time in the United States only the "casing head" gasoline was separated, and the remainder of the wet gas was burned. This practice was finally banned by law!) The fractionated products are widely used by the petrochemical industry, usually cooled or compressed into a liquid form. Figure 6.2.4 is a more detailed drawing of the separation process used after the lightest fractions (mostly methane) have been removed. The plant consists of five distillation columns. The process control principles used here are the same as those described and applied in section 5.6. In the present section we will summarize the control solutions and explain why the various techniques are used.

Distillation column 1 in Figure 6.2.4 is used to separate ethane and lighter components from propane and the heavier components. The level in the accumulator is controlled by the cooling power to the condenser, so some of the vapor at the top of the column must also leave the condenser in the form of vapor, as either ethane or some lighter components. The pressure at the top of the column is controlled by a valve in the condenser vapor outlet stream. The

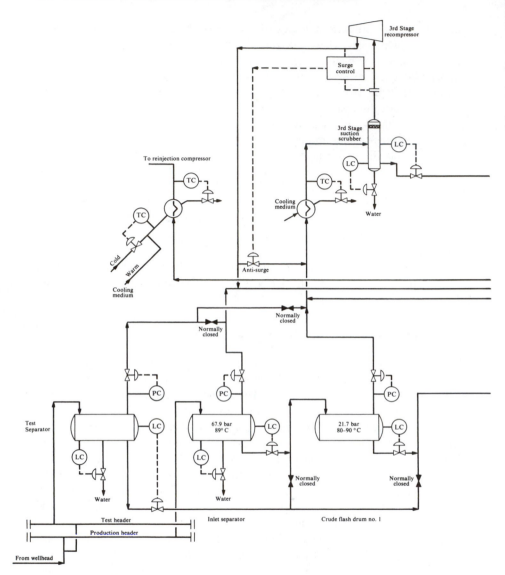

Figure 6.2.1. Flowchart for oil-gas separation and recompression of the gas on an off-shore oil platform.

reflux and distillate flows in this simplified flowsheet are both held at constant values by controllers. One problem with this control arrangement is that when the cooling power to the condenser varies, the temperature in the accumulator will vary. This causes the internal reflux to vary even if the external reflux is held constant. A system of the type described in Figure 5.6.8 could be used to maintain a constant internal reflux value.

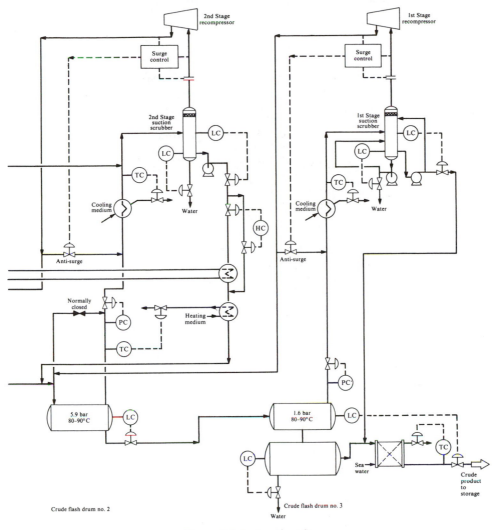

Figure 6.2.1. (continued)

The power added to the reboiler at the bottom of column 1 is controlled on the basis of the temperature somewhat higher in the column. This temperature is an indirect measurement of concentration of the bottom product. The steam flow to the reboiler is controlled by changing the liquid level of the condensate. As discussed earlier, this control functions by changing the surface area of the reboiler, but any other reboiler control principle could be used. The mass balance is maintained by level control in the bottom section, and is realized by controlling the flow of the bottom product leaving the column.

The distillate from column 1 is used as the feed for column 2. It contains ethane and some lighter components. Pressure control at the top of the second

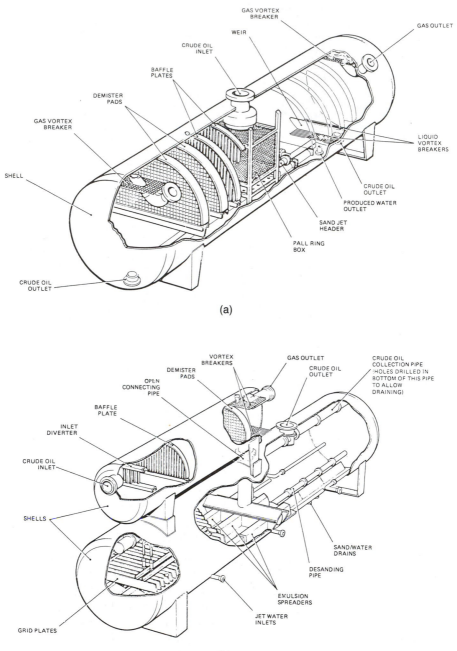

Figure 6.2.2. Cross-section of (a) inlet separator and (b) crude flash drum and coalescer.

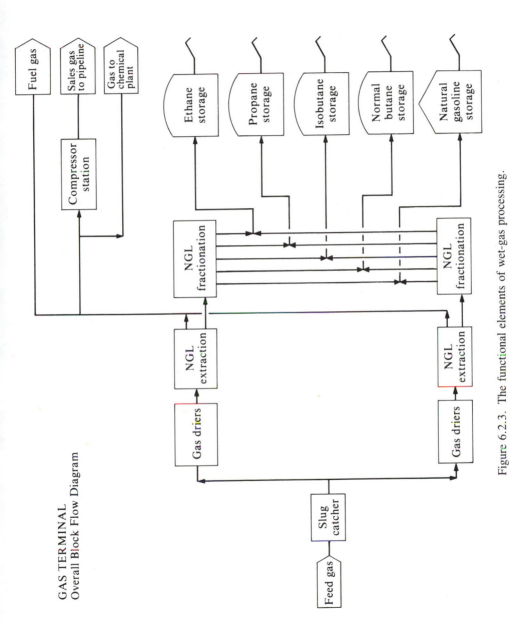

GAS TERMINAL
Overall Block Flow Diagram

Figure 6.2.3. The functional elements of wet-gas processing.

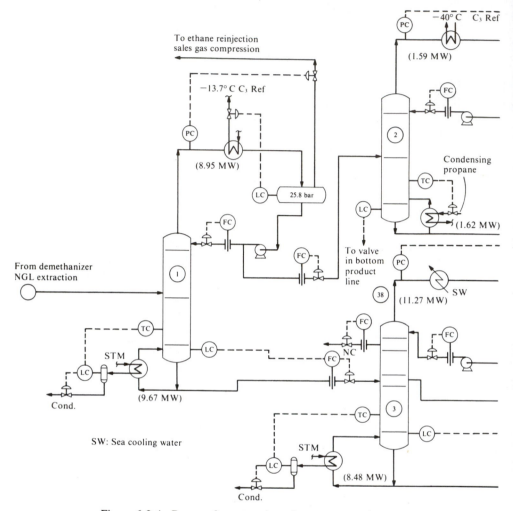

Figure 6.2.4. Process for separation of components of wet-gas.

column is obtained by relieving vapor and gas from the accumulator, and the reflux is held constant, cascaded from the level control of the accumulator. The power supplied to the reboiler of this column is controlled from the temperature at the bottom of the column, and thus the quality of the ethane bottom product is controlled. Columns 3 and 4 have essentially the same control systems as the first two columns.

The pressure profile through the process is very important. The profile is determined by the pressure of the feed, which is 26.7 bar. This requires top and bottom pressures of 26.9 and 26.5 bar, respectively. The pressures in columns 2 and 3 must be lower than those in column 1. Column 2 is operated at about 20 bar and column 3 at 12 bar. This makes it possible to use condensing

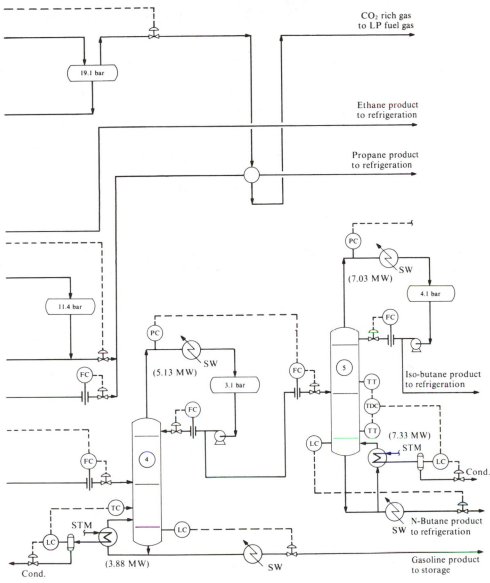

Figure 6.2.4. (continued)

propane in the reboiler in column 2 at a temperature of about 0°C to −6°C. Steam can be used in the reboiler for column 3 at a temperature of about 100°C.

The pressure in column 4 of Figure 6.2.4, which must be lower than that in column 3, is held at 4 bar. However, the pressure in column 5 is 5 bar, higher than that in column 4. This means that the pump after the condenser in column 4 must produce a pressure increase of at least 3 bar so that the flow controller on the feed for column 5 can work. The temperature in the reboiler of column 5, because of the pressure used there, will be 55°C. Therefore, this reboiler can be heated with steam.

These comments describe a typical example in which the conditions of the process operation and technology influence, directly and indirectly, the control system design.

An installation such as that shown in Figure 6.2.3 is an ideal candidate for a supervisory, steady-state optimization system designed to adjust the set points in the control systems so as to obtain an economic optimum. The factors that directly affect this optimum are energy costs, the values of the various products (often a function of market conditions and time), and product quality. Since the product flows in a system of distillation columns can be adjusted or modified to some extent, there will always be some set of operating conditions that give an optimum. In section 2.16 we discussed how an optimum can be determined either in economic terms or with respect to process limitations (see Figure 2.16.5). In section 5.6 we also described an optimization problem with respect to a particular distillation column (see Figure 5.6.30). Similar systems have been developed for complete factory complexes (Latour, 1979).

Control Tasks in Oil Refining

After the oil, gas, and water are separated at the oil field, the crude oil is transported, either by ship or pipeline, to some type of oil refinery. Large, modern oil refineries are among those plants at the forefront of chemical processing technology, and have a wide variety of important, interesting automatic control problems.

An oil refinery will have one of many different process designs, depending on which type of crude oil is to be refined and which products are to be produced. Figure 6.2.5 is a general schematic diagram for one type of oil refinery. The crude oil enters the process at the left side of the diagram and immediately goes through a separation process to separate the heaviest component from the other components. The bottom product is sent to a fluid coking unit to remove coke from the heavy bottom product; to a catalytic cracker, which cracks the heavy hydrocarbons into lighter forms; to a unit process for removing sulfur; to a catalytic re-former (reactor), which re-forms the various components into molecules of product having higher commercial value; and to a liquid–liquid extraction process (Udex) for further refining. In the upper part of the diagram the lighter fractions of the crude oil are separated into the major fractions by distillation, which produces various grades of gasoline. Some of this distillate is also polymerized to form butane, propane, high octane gasoline, and various forms of liquid gas.

Figure 6.2.6 illustrates the temperature distribution under which some of the important fractions are obtained by distillation, and in Figure 6.2.7 a simplified arrangement of control systems around a multiproduct crude oil distillation

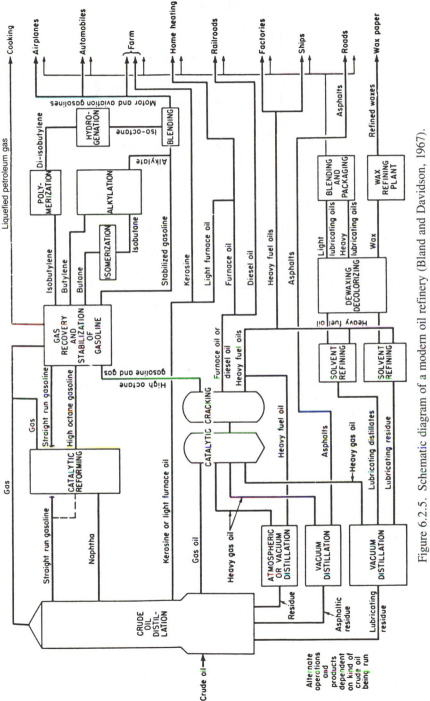

Figure 6.2.5. Schematic diagram of a modern oil refinery (Bland and Davidson, 1967).

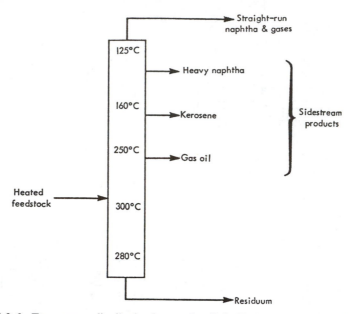

Figure 6.2.6. Temperature distribution in a crude oil distillation column (Speight, 1980).

column is shown. Figure 6.2.8 shows how the distribution of products can be obtained in a system consisting of three distillation columns.

A conventional control system for a catalytic cracker is shown in Figure 6.2.9 (also, see Figure 5.8.7). We see that most of the instrumentation is for monitoring rather than automatic control. We also see that the temperature in the reactor controls the set point for the flow controller on the air supply line to the regenerator. The return flow of regenerated catalyst is not controlled; it is only monitored by instrumentation. The feed flowrate is controlled, and the feed temperature is also controlled by passing the feed through a heat exchanger (see the lower right portion of the figure). The differential pressure between the regenerator and the reactor is controlled by a valve in the flue gas discharge line from the regenerator. The regenerator should operate at a slightly lower pressure than the reactor.

As we mentioned in section 5.8, the catalytic cracker is a good example of a process where different control structures offer many possibilities for improved operation. Developments toward this objective continue to appear, and certainly these developments will expand the application of multivariable control. In processes of this complexity, substantial gains and benefits can be expected.

A conventional control system for a polymerization reactor is shown in Figure 6.2.10. The purpose of this reaction is to combine two or more molecules to create a single, larger molecule having its elements in the same proportions as the feed. Polymerization of light olefins into hydrocarbons of high molecular weight is important in the production of high octane gasoline from the various

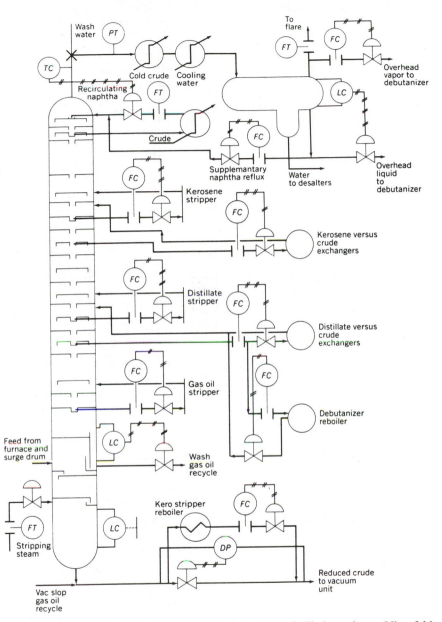

Figure 6.2.7. Control systems for multiproduct crude oil distillation column (Nisenfeld and Seeman, 1981).

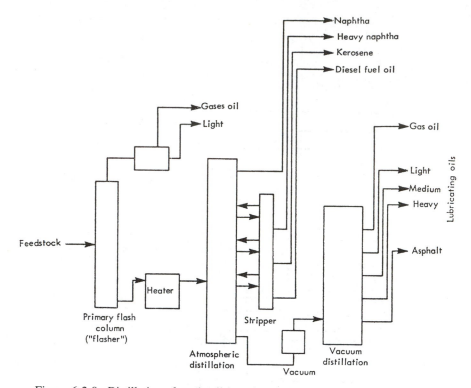

Figure 6.2.8. Distillation of crude oil in a three-column process (Speight, 1980).

gases produced by the catalytic cracker. The polymerization process usually takes place in a catalytic reactor where the reaction occurs at temperatures between 150 and 220°C and pressures between 10 and 85 bar, depending on the product specifications. The reaction is exothermic, and the temperature is controlled by a bypass valve around a heat exchanger that uses heat produced by the reaction to heat the feed reactants. This is shown at the left side of Figure 6.2.10. The polymerized reaction product is sent to a distillation column where propane is taken off the top, and butane is the bottom product. This column uses temperature control based on a plate near the top via the flow of distillate (propane). The flow of distillate is measured, and this measurement is used to provide ratio control to the reflux flow, making it possible to operate the column at the desired external reflux ratio. Propane leaves the top of the column as both a vapor and a liquid, and both forms are returned and mixed with the feed to the polymerization reactor. The liquid flow is the more important of the two. The gas flow, however, is used to maintain constant pressure in the reactor system. This is accomplished by a pressure controller that manipulates the gas flowrate.

The last column in the system separates butane from the polymer gasoline. From the control point of view, we note especially that we have a constant re-

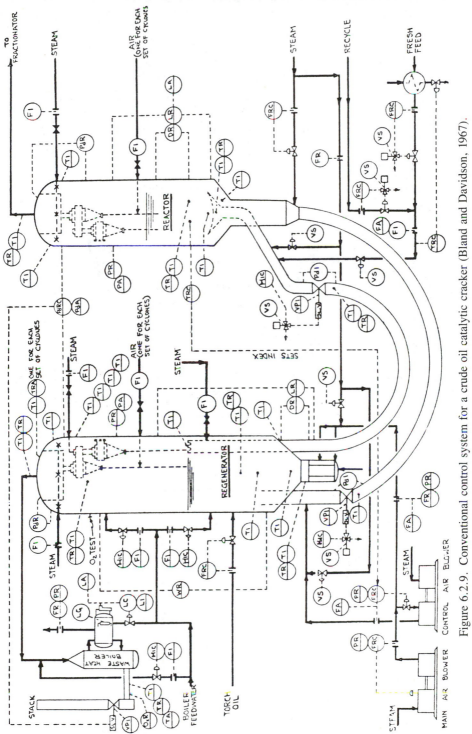

Figure 6.2.9. Conventional control system for a crude oil catalytic cracker (Bland and Davidson, 1967).

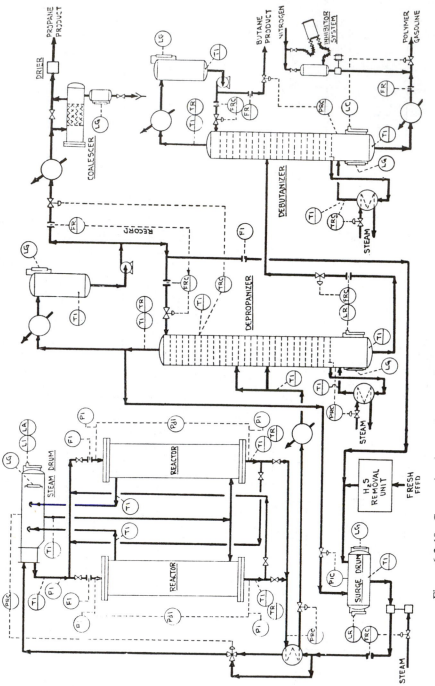

Figure 6.2.10. Conventional control system for a polymerization reactor (Bland and Davidson, 1967).

flux ratio and pressure control based on a measurement in the bottom of the column and manipulation of the butane flow. The reason why the pressure measurement is made at the bottom of the column in this case instead of at the top, which is more common, is that addition of nitrogen to the bottom product is a pressure-sensitive process.

6.3. PROCESS CONTROL IN THE PETROCHEMICAL INDUSTRY

In section 6.2 we discussed how oil and gas are extracted from an oil field, how the oil and gas are separated for further processing, and how they are prepared for use as products or still further processing (example: the separation of wet gas into several liquid products including ethane, propane, and butane). These products are the raw materials for the petrochemical industry.

The science of organic chemistry and petrochemical technology have made it possible to synthesize nearly any organic molecule by various combination processes (making large molecules from small ones) and splitting (cracking) processes (making small molecules from large ones). There are a very large number of these processes, and for just about every individual product there are several alternative process paths and starting points. The choice of each individual industrial process design must take into account the cost and availability of raw materials, the product market, and the problems associated with the process technology. Once a choice has been made and a plant has been built, changes in any of the conditions on which the original decision was based may require adjustments and/or changes in the plant.

Ethylene (C_2H_4), one of the most important intermediate materials for the petrochemical industry is actually a manufactured raw material, as it is a product of the refining systems we discussed earlier. It can be produced in many ways, but the most important is pyrolysis (cracking) of ethane (C_2H_6) or propane (C_3H_6). Ethylene is principally used as follows:

- For production of polyethylene, which is one of the most widely used plastics in the form of films and tubes.
- As an intermediate material for the production of EDC (ethylene dichloride), which in turn is an important intermediate material for production of VCM (vinyl chloride monomer) and PVC (polyvinyl chloride).
- For production of ethylene oxide, which is an intermediate material for several petrochemical processes.

First we will discuss the production of ethylene; then, the use of ethylene for the production of polyethylene and VCM, respectively; and finally, production of PVC from VCM.

As usual, we will provide only the basic process technology required to describe process control problems that play an important role in all these industrial processes. None of these processes can function completely or properly without instrumentation and automatic process control.

Ethylene Production. The Tasks of Process Control

Figure 6.3.1 is a simplified flowsheet of a system for producing ethylene from ethane and propane. The most important steps in this very complex system are as follows:

1. Pyrolysis (cracking) occurs first, in which ethane (C_2H_6) and/or propane (C_3H_8) is split into lighter hydrocarbons at high temperatures (350–850°C).
2. After quenching, washing, compression, precooling (chilling), drying, and refrigeration, the condensed gas goes to the demethanizer.
3. In the demethanizer, methane (CH_4) and hydrogen (H_2) are removed.
4. The next step is de-ethanization, where ethane is separated from propane and other heavier components.
5. Then comes the acetylene hydrogenation stage, a catalytic reaction where acetylene (C_2H_2) is converted to ethane by the addition of hydrogen (H_2). This is done because the presence of acetylene in ethylene is undesirable. After another demethanization, the gas goes to the next stage.
6. Residual ethane is now separated from the ethylene. The ethane is returned to the cracker, while the ethylene is refrigerated and sent to storage.
7. The heaviest components in the de-ethanizer (see item 4 above) are separated in a depropanizing process where propane and lighter components are separated from butane and the heavier components.
8. In the C_3-separation process, propane is separated from propylene, and both are sent to storage.
9. Finally, a debutanizer separates butane from the heavier gasolines.

A flowchart for the pyrolysis plant (cracker) is shown in Figure 6.3.2. A mixture of ethane and propane from storage tanks is heated in parallel tubes by gas-fired radiant heaters according to a certain specified temperature profile. Steam is also mixed into the flow to manipulate the composition of the products of the pyrolysis (see Figure 5.6.23).

The major control variables in this process are:

1. The flow of ethane and propane.
2. The flow of steam.
3. The heat added by the burners to maintain the temperature profile in the heaters.

In Figure 6.3.3 we see a flow controller that can hold the ethane/propane flow constant at the desired value. There is also a flow controller on the steam

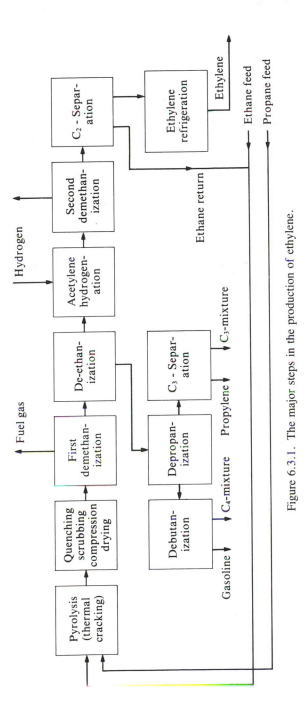

Figure 6.3.1. The major steps in the production of ethylene.

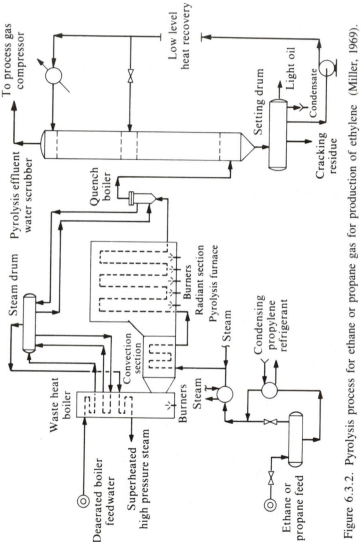

Figure 6.3.2. Pyrolysis process for ethane or propane gas for production of ethylene (Miller, 1969).

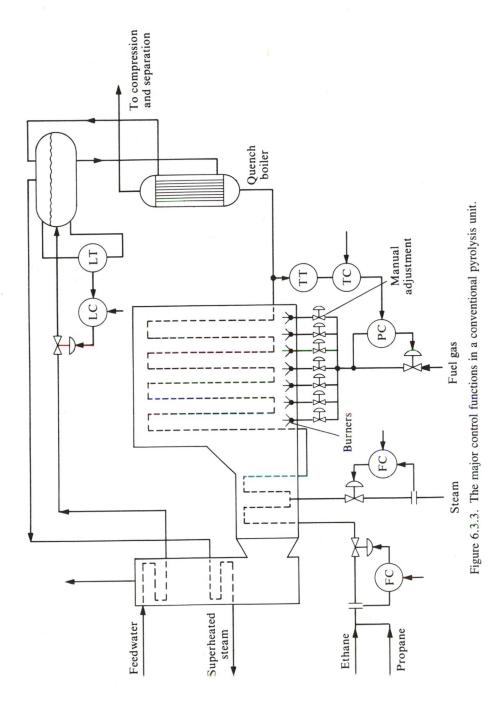

Figure 6.3.3. The major control functions in a conventional pyrolysis unit.

line. In the cracker there are about 96 burners, each of which receives its fuel gas through a manually adjusted valve. These manual valves trim the individual flows, but the total flow is controlled by a pressure controller on the manifold that supplies the fuel to the burners. The set point for the pressure controller is manipulated by the signal from a temperature controller, operating on the basis of a measurement of the temperature of the product gases leaving the cracker. This temperature is in the neighborhood of 820°C. As the holdup time for the gases in the furnace is only a few seconds, it is impossible to use any form of control other than that which holds the temperature profile and holdup time at their proper values. The holdup time is determined by the flowrate of the incoming gas. In the system of Figure 6.3.3 there is no device for analyzing the light hydrocarbon products of pyrolysis, so there is no means of directly controlling the set points of the control loops. In order to accomplish this, one could use a gas chromatograph to analyze the product composition and, on the basis of that composition, adjust the temperature profile, the holdup time, and the steam flow.

It is especially important that the temperature profile in this process be controlled precisely because of its strong influence on the product. Variations in the heat value of the fuel gas for the burners can lead to changes in the heating power. Even though the fuel gas system of Figure 6.3.3 has pressure control, it is only through the actions of the temperature control of the product gas in cascade with the pressure control that these changes can be reduced. In installations where the heat content of the gas can change rapidly, the heat content should be measured and the pressure set point corrected rapidly by a feedforward action.

A mathematical model of the product composition's dependence on the process variables, complete with updating of model parameters by gas chromatographic analysis, is a possible further extension to the conventional solution. Such a technique is presently under development.

Wilson and Belanger (1971) used dynamic optimization (Pontryagin's maximum principle) to develop a control strategy for the pyrolysis process. They used the following control variables: u_1, feed rate of hydrocarbons; u_2, steam/hydrocarbon ratio in the feed; u_3, ratio between the maximum reactor temperature and the maximum allowable temperature; u_4, pressure at the reactor outlet. The object of the optimization concerned the production of coke, which fouls the cracker. They hoped to find a control variable pattern over the operating cycle that would give the maximum total profit with respect to the values of raw materials and product, and to the cost of stopping the process in order to remove the coke. Simulation studies have shown that dynamic optimal control should be able to improve profits by perhaps 10 to 20%.

The optimum control variable pattern will be the same from period to period (between each coke removal operation), so a simple parameterization and static optimization of the control variable pattern should give quite satisfactory results.

Most of the operations following pyrolysis are distillation processes operating in the temperature range between $-90°C$ and $+104°C$, and in the pressure range between 10 bar and 32 bar.

The most difficult of these distillation operations is the one that separates propylene form propane because the relative volatility between these substances is very low. At a pressure of 11 bar, propylene boils at about 20°C and propane at about 32°C (see Figure 3.12.17). Using Figure 6.3.4, let us look further into the control problems associated with this operation.

Figure 6.3.4 is a proposed control system for this column. Operation of the improved system is as follows, beginning at the top of the column and working toward the bottom:

1. The pressure in the column is controlled via the flow of primary medium to the condensers.
2. The distillate is flow-controlled, and the set point for the flow controller is determined by a combination of feedforward from the feed stream flow and analysis of the concentration of propylene in the top product.
3. The flow of distillate controls the set point for the reflux flow, which is also adjusted from the level in the accumulator. In the plant illustrated by Figure 6.3.4, the reflux flowrate is about 120,000 kg/hr, the distillate flow about 9000 kg/hr, and the feed rate about 13,000 kg/hr. The purpose of the dynamic feedforward is, as described in section 5.6, to ensure that the reflux flow follows changes in the feed rate at the proper proportion.
4. The bottom of the column has three reboilers, two heated by water and one by low pressure steam. Variations in the water temperature are compensated by changing the flow of steam to the third reboiler. We see that the water flow in the two reboilers is controlled at a constant value by means of a bypass valve. The heat power supplied to these two reboilers is the product of the temperature difference across the reboiler and the flow of heating water (see section 2.7). The flow of low pressure steam is measured, and its heat content is determined and added to that of the heating water to generate a value for the total power added to the reboilers. This is then compared with the desired power addition, and the set point of the steam flow is adjusted to eliminate the deviation.
5. The set point for the reboiler power controller of step four (above) is controlled by the concentration of the bottom product and a feedforward from the feed flow.
6. Flow of the bottom product is controlled on the basis of the level in the bottom.

This system provides substantially improved performance compared with the far simpler scheme that has long been used but has given relatively poor performance.

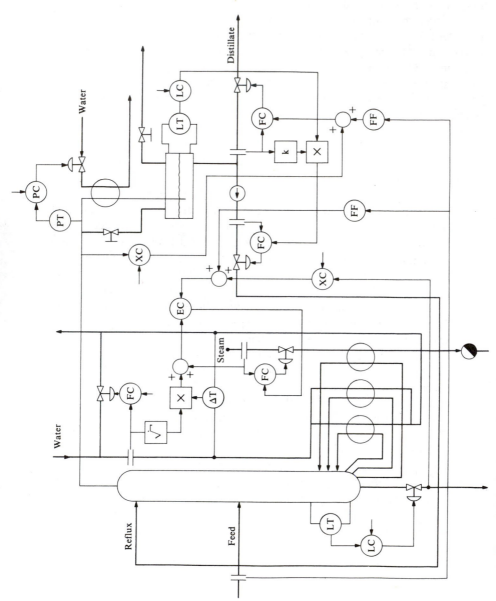

Figure 6.3.4. An improved control system for separation of propylene and propane by distillation.

Automatic Control and Polyethylene Production

The polymers that can be made from ethylene comprise the largest group of today's modern industrial polymers. Of these, polyethylene is the most important because of its very wide range of applications and relatively low cost.

Polyethylene exists in two forms, depending on whether it is made by a high pressure process or a low pressure process. It was first produced in about 1933 in England by Imperial Chemicals Ltd. This early process operated at a pressure greater than 1000 bar and a temperature between 100°C and 300°C. Because of these very difficult operating conditions, researchers were interested in developing a different commercial process. In 1953 a German chemist, K. Ziegler, developed a method for producing polyethylene by a catalytic process operated at atmospheric pressure and close to room temperature. The method had actually been discovered by accident about ten years earlier by another German chemist, but he did nothing with it. Since 1953, catalytic production of polyethylene has been developed further by companies in the United States such as Phillips Petroleum Co. and Standard Oil Co. of Indiana.

The various catalytic processes differ from each other primarily in the catalyst used. In the Phillips process, shown in Figure 6.3.5a, the catalyst consists of chromium oxide deposited on aluminum silicate and suspended in a solvent. Polymerization takes place at 130 to 170°C and 35 to 100 bar. If the temperature is above 130°C, the polymer will dissolve in the solvent; otherwise, it will develop as a suspension.

The polymer forms in the continuous stirred tank reactor and is removed continuously through a cooler. It then goes to a liquid–gas separator where any unreacted ethylene is separated and recycled to the reactor feed. The product stream moves on to a cleaning and dilution process and then to a unit that separates the catalyst. The next stage in the process removes the solvent from the product stream. The solvent is recycled and the polymer goes to a drying unit.

The major control problems connected with the polyethylene reactor system occur at the reactor and the liquid–gas separator. A typical control system for the process is shown in Figure 6.3.6. The ethylene feed rate is flow-controlled and proportioned to the catalyst's flowrate. The set point for the ethylene flow controller is controlled on the basis of measurement of the gas flow out of the liquid–gas separator. The flow of solvent is held constant. A water-cooled jacket surrounds the reactor; the water is circulated through a heat exchanger to remove the heat absorbed from the reactor, and its temperature as it leaves the heat exchanger is controlled by manipulation of the flow of the primary medium in the exchanger. This temperature controller is cascaded with a temperature controller operating on the basis of the temperature measured inside the reactor.

The pressure in the reactor is controlled by a valve on the line between the reactor and the separator. The pressure in the separator is controlled by a similar valve in the outlet gas line, and the level of liquid in the separator is controlled

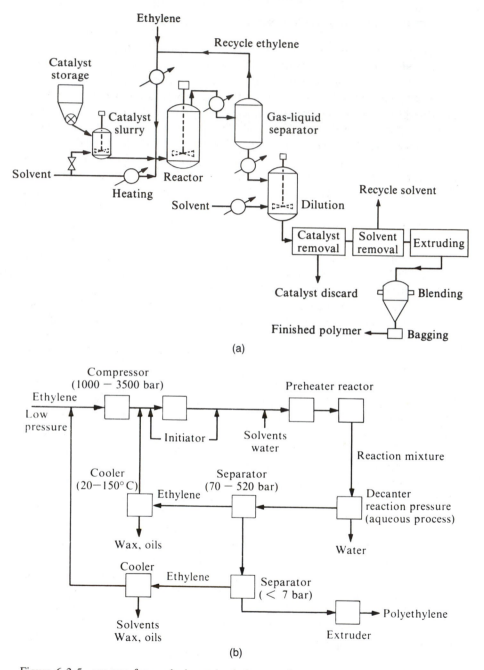

Figure 6.3.5. process for producing polyethylene. a. Low pressure process (the Phillips process). b. High pressure process (Miller, 1969).

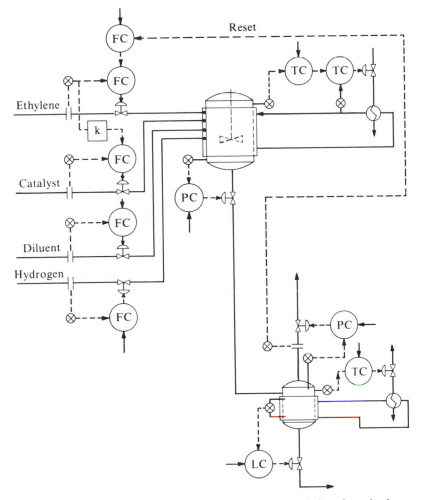

Figure 6.3.6. A typical control scheme for a low pressure catalytic polymerization process (based on Sittig, 1976).

by a valve in the liquid outlet line. The separator is also cooled by a cooling jacket and water–heat exchanger system similar to that found on the reactor.

Control of Vinyl Chloride Monomer (VCM) Production

Ethylene is also important as the raw material for production of VCM. Several processes can be used; the one chosen depends on the cost and availability of the raw materials. An early VCM plant used acetylene (C_2H_2), produced from calcium carbide in an electric furnace, and hydrogen chloride (HCl), produced from chlorine, but this process is no longer economically competitive with processes that use ethylene as the basic raw material.

Today's most common industrial process for VCM production is diagrammed in Figure 6.3.7. We see that the ethylene undergoes two different processes. In one it is chlorinated, and in the other it is oxychlorinated with oxygen and hydrochloric acid. The product is ethylene dichloride (EDC), more properly called 1,2-dichlorethane (CH_2Cl–CH_2Cl). EDC is thermally cracked (pyrolyzed) to produce vinyl chloride and hydrogen chloride:

$$CH_2Cl\text{–}CH_2Cl \rightarrow CH_2\!=\!CHCl + HCl$$

A separation process then separates the vinyl chloride monomer from the HCl. The monomer is the product, and the HCl is returned to the oxychlorination stage.

Vinyl chloride monomer (VCM) is a gas with a sweetish, slightly pungent odor. At atmospheric pressure its boiling point is $-13.4°C$, and its vapor pressure at $20°C$ is 3.5 bar. This means that VCM can be liquefied by water cooling at relatively low pressure. VCM is stored and transported in liquid form. The major use for VCM is as a raw material for production of polyvinyl chloride (PVC). The designation "monomer" clearly distinguishes between the raw material vinyl chloride and the product after polymerization. Production of PVC is discussed in the next section.

A simplified schematic diagram of a process for direct chlorination of ethylene is given in Figure 6.3.8. Controlled flows of ethylene and chlorine, both in the gas phase, are fed to the reactor. The set point of the ethylene flow is ratio-controlled from the measured chlorine flow. There is a separately controlled line in parallel with the control valve on the main ethylene stream that trims the total ethylene flow on the basis of the amount of free chlorine measured in the EDC product.

The reactor is built as a heat exchanger because the reaction between ethylene and chlorine is highly exothermic. The reaction takes place under temperature control at about 90 to 130°C and a pressure of 7 to 10 bar. Because the reaction is strongly exothermic, the temperature controller must have strong feedback. The cooling action is provided by water.

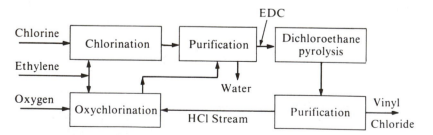

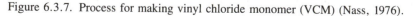

Figure 6.3.7. Process for making vinyl chloride monomer (VCM) (Nass, 1976).

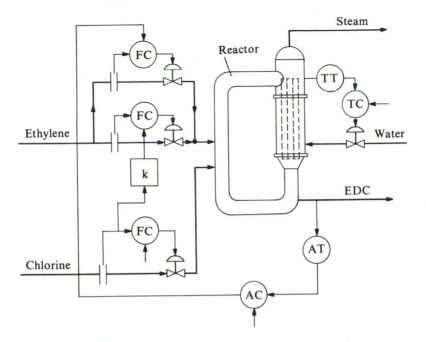

Figure 6.3.8. Control of process for direct chlorination of ethylene.

Figure 6.3.9 is a simplified flow diagram for the oxyhydrochlorination reactor, which is the second branch of the process in which ethylene is converted to EDC. This is the most difficult of all the processes in the entire VCM plant and takes place in a fluidized bed reactor with a catalyst consisting mostly of copper chloride.

The ethylene used in the oxyhydrochlorination reactor must be purer than that used in the direct chlorination process, and this need determines the performance requirements of the ethylene refining processing prior to the reactor. Three streams feed the reactor: ethylene, hydrogen chloride, and oxygen (in the form of air). All of the streams are flow-controlled, and the ethylene flow, as in the case of direct chlorination, consists of two branches, one for trimming on the basis of product analysis. This reaction is so strongly exothermic that it produces steam in the heat exchanger built into the reactor. Temperature control is based on a temperature measurement made near the middle of the reactor, and is in cascade with the pressure controller on the steam. There are also several temperature indicators along the reactor. The operating conditions for the reactor are a temperature of 200 to 350°C and a pressure of 2 to 10 bar.

After the two streams of EDC have been mixed, the product is washed with acid, lye, and water. Then the gases are separated in a series of three distillation columns before the EDC is pyrolyzed (cracked). This process is shown in Figure 6.3.10.

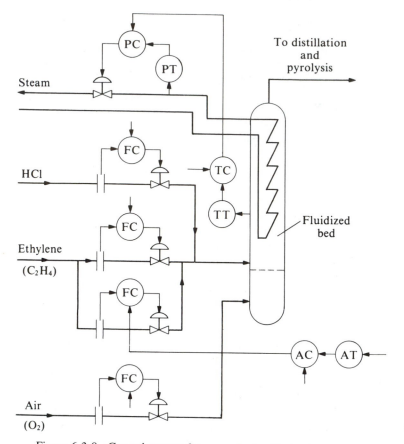

Figure 6.3.9. Control system for an oxyhydrochlorination reactor.

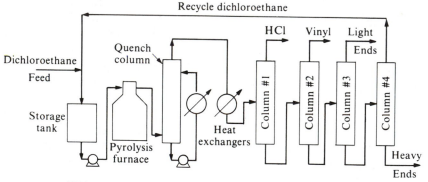

High-pressure pyrolysis unit for dichloroethane.

Figure 6.3.10. Process for pyrolysis of EDC (Albright, 1974).

The cracker has much in common with the one described earlier (Figure 6.3.2). The control problems are also about the same, in that the flow of EDC must be precisely controlled in order to obtain the correct final product. The product is a mixture of VCM, HCl, and EDC. Obviously, it is desirable to get as much VCM and as little EDC as possible in the product, as the EDC must be returned to the inlet of the cracker for reprocessing.

After pyrolysis (Figure 6.3.10) the product is cooled dramatically (from 500°C to 50°C) and then distilled to remove HCl as the top product for return to the oxychlorination stage. The bottom product is fed to a second column that separates the VCM and the uncracked EDC. The EDC is returned to the cracker, and the VCM is sent to a storage tank. A simplified diagram of the column used for separating HCl from VCM and EDC is shown in Figure 6.3.11. An important characteristic of this column is that there are five feed streams, consisting of three liquids, one gas–liquid flow, and one gas flow, which enter the column between plates 28 and 34. The flows are dominated by two streams, the top stream of HCl gas and the lowest stream which contains liquid VCM and EDC. The column also has two reflux flows, one being the liquid HCl returned to plate 80 and the other being the liquid VCM returned to plate 18.

The control structure for this column is quite ordinary. A controlled flow of steam is added to the reboiler where the boiling point is about 100°C, and the set point for the flow is determined by a temperature controller operating from the temperature at plate 15. The reflux flow to plate 80 is flow-controlled on the basis of the temperature at plate 38.

Control of Polyvinyl Chloride (PVC) Production

Although polyethylene is the plastic being produced in the greatest tonnage today, polymerized vinyl chloride once had that distinction. This product was first produced in 1835!

There are four different processes available for the polymerization of vinyl chloride. The reactions can take place in pure vinyl chloride, in an aqueous suspension, in an aqueous emulsion, or in a solution. The reactions are highly exothermic, and if they are allowed to generate high temperatures, the product will have poor stability. The main problem in this polymerization, then, is temperature control. The polymer product does not dissolve well in the monomer, so it can easily separate and create heat transfer problems in a normal heat exchanger. Resultant localized high temperatures and overheating produce a polymer of low molecular weight and relatively little value.

The situation can be improved if the polymerization takes place in an aqueous suspension. The liquid vinyl chloride is kept below its boiling point (-13.4°C at 1 bar, 20°C at 3.5 bar) and vigorously agitated in an aqueous solution containing an initiator. The PVC forms nearly spherical particles varying in size from 100 to 200 mesh up to the size of pearls. The polymerization takes

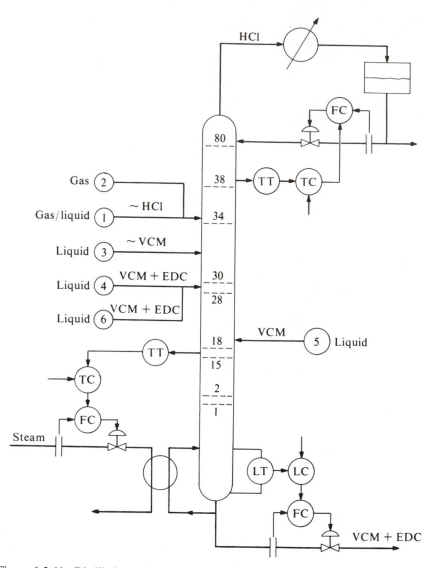

Figure 6.3.11. Distillation column (with controls) for separation of HCl from VCM and EDC.

place in these small particles without their growing together. As a result, one gets a suspension of PVC particles as the product. The heat generated by the reactions is absorbed by the water phase of the suspension. This phase is usually two to four times the volume of vinyl chloride, and because the polymer particles are relatively large, it is quite easy to isolate them by filtration and/or centrifuging.

Emulsion polymerization of vinyl chloride is similar to suspension polymerization except that the process takes place with the monomer dissolved in a special emulsion mixture. Water is also used in the emulsion, thus providing good heat transfer and a medium for temperature control.

A simplifed flowchart for a process for suspension polymerization of vinyl chloride is given in Figure 6.3.12. Three storage tanks for vinyl chloride are shown at the left of the diagram. One of these is for storing vinyl chloride that has been recovered by distillation of unreacted materials from the process. This is a batch process, with the vinyl chloride, water, and other materials for the suspension mixed at the correct ratios in the reactor. The reactor is equipped with a jacket for initial heating of the mixed materials and eventual cooling of the reaction. It also contains piping designed to provide good mixing as well as the possibility for heat exchange inside the reactor. Depending on the process, the optimal reactor size will be in the range of 10 to 20 m^3. The ratio between cooling surface and volume decreases as volume increases, so there is little advantage to be gained from building larger reactors.

Once the reactor is charged, the temperature is increased approximately linearly over a period of about 60 minutes. At the end of this time, the temperature is 50 to 55°C, and it must be held within about 0.5°C of that value for the full reaction period of about 15 hours. The pressure will normally be about 7 to 8 bar.

Following completion of the polymerization reactions, the mass is dumped into an agitated tank where the unreacted vinyl chloride particles are separated

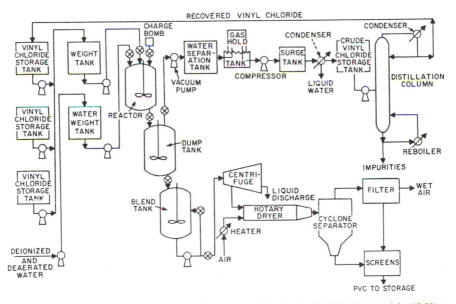

Figure 6.3.12. Suspension polymerization process for vinyl chloride (Sarvetnick, 1969).

and then sent to a special recovery process. This process usually includes a distillation column from which the vinyl chloride is removed as the top product. The PVC particles flow from the dump tank to a mixing tank where particles of different sizes can be blended to give the desired particle size distribution in the final product. This mixture is then centrifuged to separate the particles from the liquid component (see section 3.7). Following the centrifuge, the PVC particles are sent to a rotary dryer (see section 3.11) and then to a cyclone separator for particle size classification.

Figure 6.3.13 shows a somewhat newer process for polymerization of pure vinyl chloride with no water component. This process uses two stages, the first of which converts about 10% of the monomer under intense agitation in a vertical, cylindrical reactor with a heat exchanger jacket. The second stage is a horizontal cylindrical reactor with a piping network over the entire surface area of the reactor. There is less agitation in this reactor, but the mechanical design keeps the heat transfer surfaces clean of gelled high polymer. The total reaction time in this process is 12 to 13 hours.

Temperature control is the most important control problem in the production of PVC. Figure 6.3.14 shows the principles of temperature control used in a batch-type PVC reactor. After the inner vessel is charged with vinyl chloride monomer, it is heated by means of a water-heated jacket. The heating water is brought to temperature in a steam-heated exchanger and is controlled on the basis of temperature measurements following the heat exchanger and on the

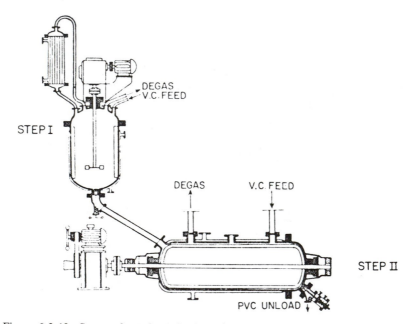

Figure 6.3.13. System for polymerization of pure vinyl chloride (Sarvetnick, 1969).

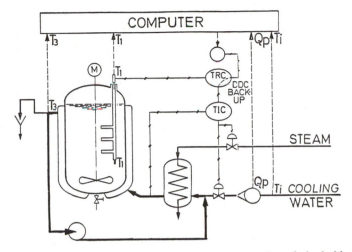

Figure 6.3.14. A typical control system for polymerization of vinyl chloride.

steam supply to the heat exchanger. The set point for this controller is regulated by a second temperature controller in cascade that operates from the temperature of the contents of the reactor (T_1). The polymerization reaction generates a large amount of heat, which must be removed from the reactor by cooling in order to have the temperature in the reactor follow the required, predetermined profile. This means, then, that heating must be replaced by cooling during the progress of the batch reaction. Cooling is provided by passing cold water through the temperature control jacket around the reactor. The two control valves in the steam and water lines are arranged so that one is closed when the other is open, and this is done by a single temperature controller.

Figure 6.3.14 also shows how the simple cascade control system can be replaced with a more sophisticated control system. This system uses a computer where, in addition to the temperature in the reactor (T_1), the temperature (T_i) and flow (Q_p) of the cooling water and the temperature of the water leaving the cooling jacket (T_3) are used. These data are used to compute the heat balance for the system in order to determine the correct set point for the inner temperature control loop. If the computer should fail, the original cascade temperature control system would take over operation of the reactor (this is shown on the figure as "DDC-backup").

6.4. AMMONIA-BASED INDUSTRY

The value of nitrogen as a fertilizer was recognized long ago. The first nitrogenous fertilizer imported into Europe was guano, which came from the islands off Peru. This was followed by sodium nitrate from Chile. As the latter was also in great demand for the manufacture of munitions, it was clear by the end of the nineteenth century that some means of producing fixed nitrogen was necessary in order to avoid a world food shortage. In Norway, Birkeland and Eyde developed such a process in 1905. Their process formed nitric oxide by passing air through an electric arc. The same year, the Frank-Cyro cyanamide process was invented. The former process is now obsolete because of the high cost of electricity, and today the latter provides only a few percent of the world production. The Haber-Bosch process for direct synthesis of ammonia from nitrogen and hydrogen was developed between 1905 and 1913.

The reduction of the Haber-Bosch process to commercial practice at Oppan, Germany in 1913 was one of the major achievements of science and engineering early in this century. This process made it possible to produce ammonia from nitrogen derived from the air and hydrogen taken from coal or hydrocarbons. The synthesis of ammonia provided a cheap source of fixed nitrogen for use in fertilizer and munitions (Vancini, 1971; Stretzoff, 1981).

The original Haber-Bosch process was carried out at 200 bar and 500 to 600°C with an iron catalyst. Under these conditions, the conversion was about 10% per pass, with an overall yield of about 80%. As the technology of reciprocating compressors developed, the process pressure reached a high of 1000 bar (the DuPont process). Recently, however, centrifugal compressors operating at a high of about 150 bar have generally replaced the high pressure processes. This has been the result of process economics: the high pressure processes produce a high yield with no need for recycle, but the lower pressure processes involve much lower capital and maintenance costs, which are balanced against the cost of recycle.

Hydrogen for ammonia production can be obtained from many different sources. The choice is usually based on economics. In some areas of the world, coal is an important source; in others, reformed natural gas, wet gas, propane, and naphtha are used. We will discuss a typical process configuration for production of ammonia from naphtha or propane.

Production of Ammonia from Naphtha or Propane (Torvund, 1971)

Figure 6.4.1 shows a process that produces ammonia from naphtha. The difference between this process and the process for the synthesis of ammonia from propane is primarily the fact that the stages used to remove sulfur in the naphtha process are not needed in the propane process. Desulfurization takes place in two stages. The actual reaction stage following these steps is the same for both naphtha and propane.

The first step in the process is the primary reformer, where the hydrocarbon and recycled synthesis gas are combined with superheated steam. The steam is at about 40 bar and superheated from 360°C to 530°C by heat exchange in the secondary reformer. The steam flow is controlled so that it produces 3.7 (3.0 minimum) kmole H_2O per kmole C in the gas mixture. The steam–hydrocarbon mixture is fed to the primary reformer at 490°C and 35 bar. The reformer has nearly the same function as a cracking unit, which was described in section 6.3. The unit consists of a series of vertical tubes filled with a special catalyst through which the gas is passed. The tubes are heated externally by a fuel consisting partly of either naphtha, propane, or exhaust gas from the synthesis process.

The naphtha–steam mixture is heated to about 830°C in the primary reformer. At a reformer pressure of about 30 bar, the following reactions occur:

$$C_nH_m + nH_2O \rightarrow nCO + (n + m/2)H_2 \tag{6.4.1}$$

$$CO + H_2O \rightleftharpoons CO_2 + H_2 \tag{6.4.2}$$

$$CO + 3H_2 \rightleftharpoons CH_4 + H_2O \tag{6.4.3}$$

Reaction (6.4.1) goes to completion, while reactions (6.4.2) and (6.4.3) are equilibrium reactions where reaction (6.4.2) approaches equilibrium with the outlet temperature of the reformer, and reaction (6.4.3) goes close to equilibrium. The reaction description given by the three reaction equations (6.4.1–6.4.3) is only approximately correct. The reaction in the primary reformer is strongly endothermic and absorbs about half of the heat energy supplied by the fuel. Some of the remaining energy supplied by the fuel is used to increase the temperature of the process gas mixture, and the remainder leaves the reformer with the exhaust gas, which has a temperature of 900 to 950°C. The exhaust gas is cooled to about 180°C by passing it through a series of heat exchangers.

The process gas leaves the primary reformer at a temperature of about 830°C and a pressure of about 29 bar. It is fed to the top of the secondary reformer where it is mixed with 565°C air. The oxygen in the air combines with some of the hydrogen and carbon monoxide in the process gas to form water and carbon dioxide.

$$2H_2 + O_2 \rightarrow 2H_2O \tag{6.4.4}$$

$$2CO + O_2 \rightarrow 2CO_2 \tag{6.4.5}$$

Reactions (6.4.4) and (6.4.5) are exothermic and raise the gas temperature to 1100 to 1200°C. After mixing, the gas goes through two catalytic steps where the following reactions take place:

$$CH_4 + H_2O \rightleftharpoons CO + 3H_2 \tag{6.4.6}$$

$$CO + H_2O \rightleftharpoons CO_2 + H_2 \tag{6.4.7}$$

These reactions are endothermic, and the gas leaves the lower catalyst at a temperature of about 970°C. The air flow added to the secondary reformer is controlled so that the nitrogen flow gives the correct ratio between the nitrogen and hydrogen in the synthesis gas.

After leaving the catalytic section, the gas is cooled by passing through a series of two heat exchangers. The heat removed by the first exchanger pro-

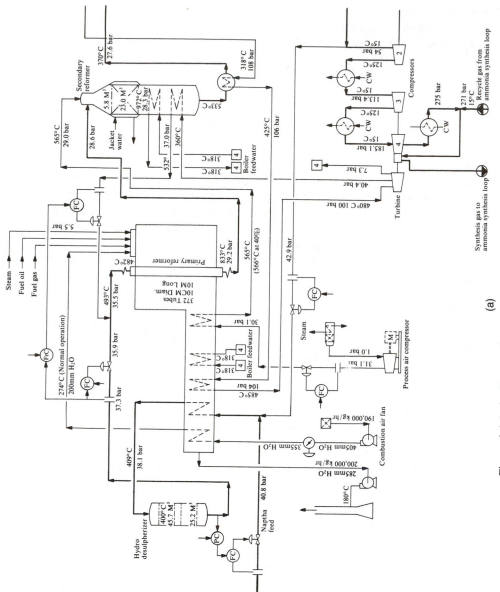

Figure 6.4.1a,b. Process for production of ammonia synthesis gas.

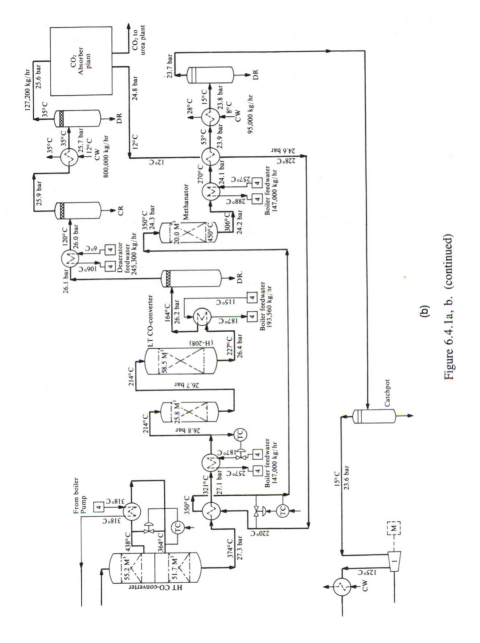

Figure 6.4.1a, b. (continued)

(b)

duces steam, and that from the second is used to superheat another steam flow for use in the primary reformer. When the gas leaves this stage, it has a temperature of about 370°C and a pressure of 28 bar.

The gas, which at this point contains about 14% CO, is next sent to the CO-converter where the CO content is reduced to about 0.33%. The reaction in the converter is:

$$CO + H_2O \rightleftharpoons CO_2 + H_2 \qquad (6.4.8)$$

This is an exothermic equilibrium reaction where the equilibrium is shifted to the right by decreasing the temperature. To get a combination of high reaction rate and efficient conversion, the process is broken into a high temperature (HT) stage and a low temperature (LT) stage. The gas is cooled between and after these stages. The catalyst in the HT converter is iron oxide, and copper is the active component in the LT stage.

The gas temperature must be reduced to 35°C for the CO_2-water scrubbing stage. This is accomplished by sending the gas through three heat exchangers in series. The heat thus recovered is used in various places in the system.

The gas from the CO-converter contains 22 mole% CO_2, which must be removed prior to the ammonia synthesis. Scrubbing, which reduces the content to 0.1 mole%, is accomplished by passing the gas countercurrent to water in an absorption tower. The water temperature in the tower is about 10 to 12°C, and the pressure in the tower is about 25 bar. The absorbed CO_2 is released from the water later by reducing the pressure over two stages. The gas from the first stage is returned to the process gas stream, and that from the second (containing 98% CO_2) is sent to a urea production plant.

The process gas from the scrubber consists mostly of hydrogen and nitrogen at a molar ratio of 3:1. This gas also contains about 0.42% CO and 0.1% CO_2. Because the carbon oxides poison the catalyst in the ammonia synthesis stage, they must be completely removed. This is done by converting them to methane by the following reactions:

$$CO_2 + 4H_2 \rightarrow CH_4 + 2H_2O \qquad (6.4.9)$$

$$CO + 3H_2 \rightarrow CH_4 + H_2O \qquad (6.4.10)$$

Reactions (6.4.9) and (6.4.10) are exothermic and occur in the presence of a nickel catalyst. The carbon oxides in the process gas are reduced by these reactions to a maximum of 5 ppm. Before the process gas enters the methanation reactor, the temperature must be raised from 12°C to 350°C. This is done in two heat exchangers, the first using heat from the output of the methanator and the second using heat from the output flow from the high temperature converter. Following the methanator, the synthesis gas is to be compressed; so the process gas is first cooled from 380°C down to about 15°C by a series of three heat exchanger stages.

Process gas leaves the methanator at a pressure of 24 bar, but it must be at 275 to 290 bar when it enters the ammonia synthesis stage. Compression is obtained by a four-stage centrifugal compressor with gas coolers between the stages. Prior to the first compression stage the gas is passed through a trap to remove all moisture. The first two compressor stages are mounted together and driven by a common electric motor. The other two stages are driven by a steam turbine. The gas pressure is raised to 54 bar and the temperature to 125°C by the first stage; so the temperature must be reduced back to 15°C by a water-cooled heat exchanger. Any moisture still in the gas is easily separated as condensate. In the second compression stage the pressure is raised to 113 bar, and the temperature rises back to 125°C. Once again, the gas is cooled, and any condensate is removed. The third stage raises the pressure to 125 bar, and, as before, the temperature rises to 125°C, again requiring cooling and condensate removal. The last stage of compression produces the final required pressure of 275 bar, the gas is again cooled to 15°C, and then the gas is usually fed to two parallel ammonia synthesis reactors.

The ammonia synthesis takes place in the presence of an iron oxide catalyst according to the reaction:

$$N_2 + 3H_2 \rightleftharpoons 2NH_3 \qquad (6.4.11)$$

This is an exothermic equilibrium reaction. The reaction does not reach completion during a single pass of the gas, but the ammonia produced is condensed out, and the nonreacted gases are recycled to the inlet of the reactor by means of a circulation system. Ultimately, all the gas reacts, and the reaction goes to completion. It should be noted that even though the reaction goes to completion, certain inert substances, including methane, noble gases, and possibly excess nitrogen or hydrogen, still remain. These accumulate in the recirculation system and can be removed only by blowing off a certain amount of gas. When the amount of gas blown off is held at about 5% of the amount of gas fed to the reactor, the inert gases in the recirculation system will be kept down to about 13%.

The synthesis reactor is equipped with heat exchangers for heating and cooling in the catalyst mass (see Figure 5.8.3). The inlet gas is at about 273 bar and 129°C. At the outlet, the gas is at 265 bar and 295°C. In the reaction zone, however, the temperature is in the range of 430 to 500°C.

An ammonia plant has so many instrumentation and control requirements that they cannot all be discussed here. Therefore, we will consider only the most important control loops involved in the system, as shown in Figure 6.4.1.

The most important function of the control systems is to maintain the mass and energy balances. An example of this is the process and surroundings of the primary reformer. The hydrocarbon flow added to the reactor must be controlled precisely, and in order to do this it is essential to hold the pressure upstream of the control valve constant. The precise control needed is obtained with a flow controller in the hydrocarbon (naphtha) feed line, which is updated by a pressure controller in the line after the desulfurization stage. This pressure control loop will present no control problems because the desulfurization stage can be described by a first-order transfer function. In order to maintain the proper stoichiometry, the steam added to the hydrocarbon flow to the primary reformer must be precisely controlled with respect to the hydrocarbon flow. A ratio controller is used to accomplish this. It is also very important that the flow of air added to the top of the secondary reformer be metered precisely, in order to obtain the amount of nitrogen (from the air) required to give the correct ratio relative to the amount of hydrogen obtained from the hydrocarbon.

In the combustion chamber for heating the primary reformer, there are flow controllers on both the fuel supply and the air supply. The air used for combustion is preheated and forced into the chamber by a fan. Neither of these control loops is shown in Figure 6.4.1.

Figure 6.4.1b shows three typical temperature control loops. One loop is responsible for keeping the process gas in the inlet to the lower part of the high temperature CO-converter constant at 364°C. This is done by a control valve located in parallel with the cooler at that point. Similarly, there is a temperature controller on the input to the low temperature CO-converter that varies the flow of cooling water at that point.

The process gas entering the methanator has a temperature of about 350°C. This is obtained by a heat exchanger that gets its heat from the process gas just after the high temperature CO-converter. The temperature in the outlet of the heat exchanger is controlled by means of a parallel valve on the secondary side, as shown in the figure.

This installation is equipped with supervisory control by a computer that processes a large amount of data obtained from many different points in the process. The computer calculates the set points for the most important controllers, especially the flow controllers in the primary reformer system where the requirements of the stoichiometric balance are determined by the particular hydrocarbon used.

Nitric Acid Production

An important intermediate product in the nitrogen-based industries is nitric acid (HNO_3), which can be used for production of a variety of fertilizers. Nitric acid can be made by several different means (Chilton, 1968), but the most important uses nitric oxide (NO). Nitric oxide is converted to nitric acid by the addition of water and oxygen according to the reaction

$$2NO + H_2O + \tfrac{3}{2}O_2 = 2HNO_3$$

Nitric oxide is produced by catalytic combustion of ammonia according to the reaction

$$4NH_3 + 5O_2 = 4NO + 6H_2O$$

This reaction occurs when a mixture of ammonia and air, at exactly the right ratio, flows rapidly through a catalyst of closed-meshed platinum at 600 to 700°C.

There are a number of possible arrangements of this process, two of which are shown in Figure 6.4.2. The essential difference between them is the pressure in the catalytic converter. In Figure 6.4.2a the pressure is nearly atmospheric; in Figure 6.4.2b it is about 7 bar. The temperature in the converter ranges from 750°C to 900°C for the atmospheric and pressurized processes, respectively. There is also a process that uses atmospheric pressure in the converter and high pressure in the absorption stages. Combustion in the converter generates considerable heat in all these processes. This heat is recovered and used in a boiler to produce steam.

The instrumentation and control of a nitric acid plant usually follow conventional principles for the particular unit operations involved. The control systems on the boiler, for example, are based entirely on principles described in section 5.2.

For proper operation of the converter it is especially important to precisely control the ratio of the mass flows of the ammonia and oxygen. This is achieved with a system for ratio control of the ammonia gas and air at the inlet to the catalytic combustion chamber, as shown in Figure 6.4.3. According to equation (4.4.1) the mass flow of the gas is given by:

$$w = K_1 \sqrt{ \Delta p \cdot \frac{p_1}{T_1} }$$

where Δp is the pressure drop across the orifice, p_1 is the upstream pressure, and T_1 is the absolute temperature upstream of the orifice. The instrumentation of Figure 6.4.3 is developed according to this expression in both streams, except that the temperature is measured downstream of the orifice in order to avoid destroying the accuracy of the flow measurement by turbulence due to the measuring unit. The mass flow of ammonia is controlled on the basis of this measurement, and the set point for this controller is set by measurement of the mass flow in the air supply line. All of the calculations in this installation are performed by analog equipment, but obviously they can be done with digital equipment as well. The figure does not indicate how the mass flow of air is controlled. This is done by a flow controller in the air line based upon the calculated mass flow of air. The control valve in the ammonia line has a valve positioner to assure precise valve action and metering by eliminating hysteresis due to friction in the valve mechanism.

Urea Production (Sauchelli, 1964; Powell, 1968)

The synthesis of urea by heating ammonium cyanate, first done in 1828, is considered to have been the beginning of synthetic organic chemistry. In 1920, the

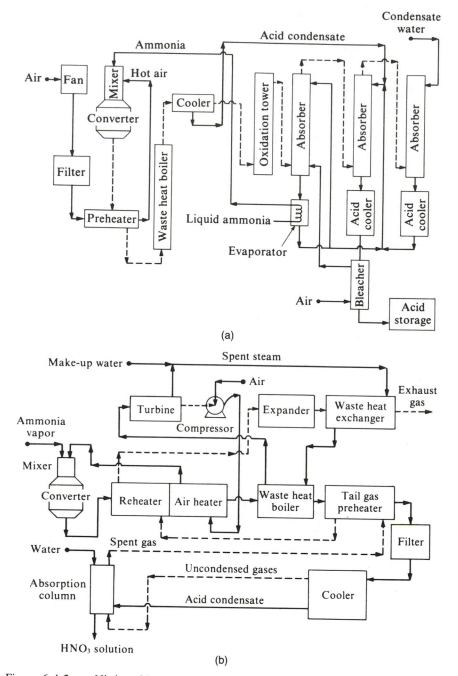

Figure 6.4.2. a. Nitric acid manufacture—atmospheric pressure ammonia oxidation (Sauchelli, 1964). b. Nitric acid manufacture—pressure ammonia oxidation (Sauchelli, 1964).

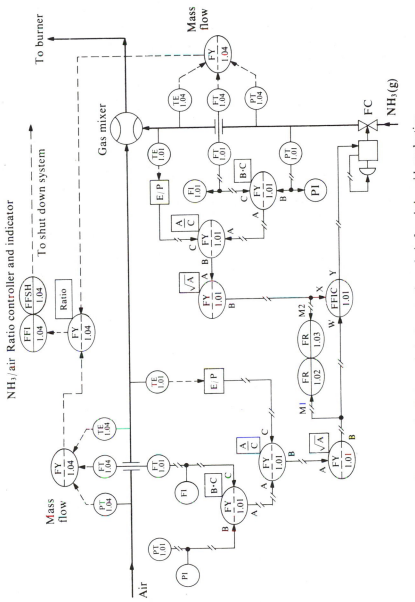

Figure 6.4.3. Ratio control of ammonia and air for nitric acid production.

I. G. Farbenindustrie in Germany first commercially synthesized urea from ammonium carbamate. Important as a nitrogen fertilizer, as a high-protein food supplement for ruminants, and for the production of plastics and glues in combination with formaldehyde and furfural, urea is produced commercially by the combination of ammonia and carbon dioxide to form the intermediate product ammonium carbamate. The carbamate then reacts to form urea and water. Both reactions are reversible:

$$2NH_3 + CO_2 \rightleftharpoons NH_2COONH_4$$

$$NH_2COONH_4 \rightleftharpoons NH_2CONH_2 + H_2O$$

The first reaction is exothermic and easily carried to completion, but the second, which is endothermic, has a conversion rate of only 40 to 70%. The equilibrium of these reactions depends on temperature and pressure, and because urea is formed only in the liquid phase, the second reaction must be maintained with heat under pressure. The combined reactions are highly exothermic, so cooling is usually required (Shreve and Brink, 1977).

Simplified diagrams showing three of many possible process arrangements are given in Figure 6.4.4 (Sauchelli, 1964). A more detailed flowchart for a typical urea plant is shown in Figure 6.4.5. Carbon dioxide from a cooling and scrubbing tower (Figure 6.4.1b) enters a multistage compressor at the left. This compression raises the pressure of the carbon dioxide to about 210 bar. The gas is then fed to two pressure reactors (autoclaves), where it is mixed with ammonia that has been passed through a steam-heated preheating stage by a group of parallel coupled feed pumps from a supply pressure of about 18 bar.

Following the reactor is a series of separation processes designed to separate the urea, carbamate, water, and unreacted gases. The process takes advantage of the fact that urea melts at about 133°C. The carbamate is gradually turned into urea by dehydration in three consecutive stripping and separation stages where the pressure is reduced from about 20 bar to 0.5 bar. In the first stripping stage the urea is superheated to about 145°C.

In each pressure reduction stage a certain amount of residual gas is freed. This gas consists mostly of ammonia which is processed in several cooling and washing stages. The condensed ammonia is returned to the pressure reactors. Any carbamate that has not converted to urea is separated in the scrubber and returned to the autoclaves via high pressure pumps.

The control systems shown for this process are relatively simple, but we see that many of them include multiple process steps. Therefore, they can lead to dynamic difficulties. In particular, the addition of steam to the high pressure splitter (20 bar) is controlled by a temperature controller based on the measurement of temperature in the bottom of the scrubbing tower. The bandwidth for this control system must be very low, and there is a question of whether there should be at least cascade coupling from one or more measurements in the

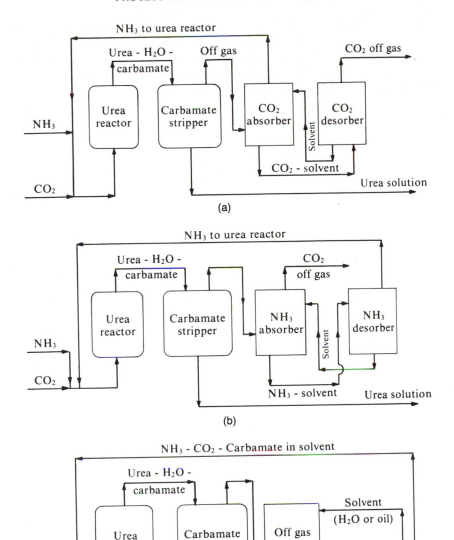

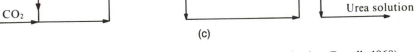

Figure 6.4.4a,b,c. Typical process layouts for urea production (Powell, 1968).

middle of the stage. There are seven other temperature controllers, four pressure controllers, two flow controllers, and five level controllers, all being local loops and free of dynamic problems. Beyond the instrumentation involved with the control systems, there are relatively few measurements of temperature, pressure, or flow supplied to the control room.

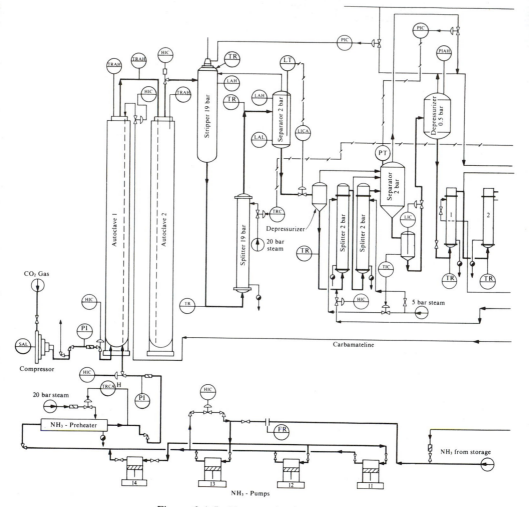

Figure 6.4.5. Urea production process.

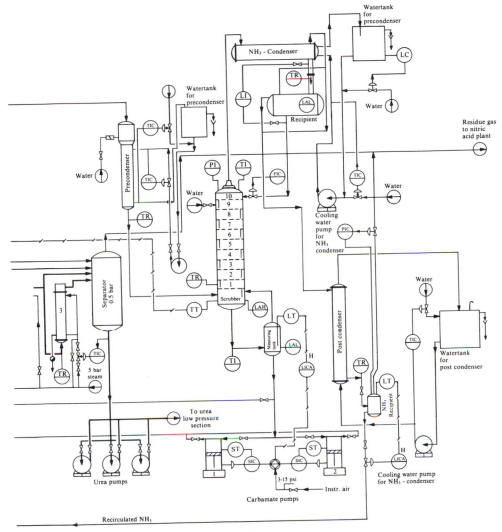

Figure 6.4.5. (continued)

This plant is a typical example of a conventional control structure that works satisfactorily to provide stable operation under normal conditions and small disturbances. The basic principle of the system is simply to maintain mass and energy balances.

Appendix A
ANALYSIS AND DESIGN OF DISCRETE CONTROL SYSTEMS USING THE MODIFIED *w*-TRANSFORMATION

The z-transform is the most commonly used tool in the analysis and synthesis of discrete control systems. This technique dates back to the early control systems literature (for example, James, Nichols, and Phillips, 1947). The technique has many advantages, but also some serious disadvantages. A major advantage is the fact that it bears a strong relationship to the time behavior of the control system. An algorithm to be converted into a computer program for control is easily derived from the z-transform result. One of the major disadvantages is that the relationship between these results and those obtained by using Laplace transform theory and frequency response analysis are not always apparent.

The early literature also introduced the bilinear transformation

$$z = \frac{1 + \tilde{w}}{1 - \tilde{w}} \tag{A.1}$$

where $\tilde{w}$ is a new complex variable. This transformation has advantages, especially with regard to stability analysis. The unit circle in the z-domain, which determines the limit of stability of a closed loop discrete system, is transformed into the imaginary axis in the $\tilde{w}$-domain, so that the standard Routh or Hurwitz stability tests may be used. Another attractive feature of the $\tilde{w}$-transform is that it makes possible the use of convenient Bode plots in the design of discrete control systems.

A slight modification of the $\tilde{w}$-transform was introduced by Balchen (1967), with $\tilde{w}$ replaced by:

$$w = \frac{2}{T}\tilde{w} \tag{A.2}$$

where T is the sampling interval. The advantage of this slight modification is that it makes very apparent the relationships between the behavior of the discrete system and the continuous system resulting from making the sampling time very short. This is seen from the following derivation:

$$w = \frac{2}{T}\frac{z-1}{z+1} = \frac{2}{T}\frac{e^{Ts}-1}{e^{Ts}+1} = \frac{2}{T}\frac{e^{(T/2)s}-e^{-(T/2)s}}{e^{(T/2)s}+e^{-(T/2)s}} = \frac{2}{T}\tanh\left(\frac{T}{2}s\right) \tag{A.3}$$

from which we see that $w \to s$ when $T \to 0$.

The interpretation of equation (A.3) is that the transfer functions [$h(w)$] for discrete systems developed in the w-domain will convert to corresponding transfer functions [$h(s)$] for the continuous system when $T \to 0$. This property is very convenient when comparing the performance of a discrete system with an equivalent continuous system.

Tables of transform pairs, developed by Balchen (1967), give the w-transformed transfer functions $h(w)$ for a number of continuous processes represented by their continuous transfer functions $h(s)$. [*Note:* In the original work (Balchen, 1967) the notation q is used instead of w.]

Table A.1 presents several of the most common w-transformed transfer functions based upon their continuous equivalents. Notice that the w-transformation is applied to a continuous system driven by a sampled signal that has passed through a first-order hold element. This means that the actual signal applied to the continuous system is piecewise constant.

In Table A.1, the factor

$$\gamma_1 = \frac{T}{2T_1} \coth\left(\frac{T}{2T_1}\right) \tag{A.4}$$

is seen to modify the time constants of the w-transfer functions. Figure A.1 is a graphical representation of this factor. From the graph, we can see that when the sampling interval T is small compared with the actual time constant T_1, the factor will have a value of $\gamma_1 \approx 1$.

The w-transformed transfer functions of a discrete system can be seen to be very similar to those of a continuous system. However, one factor is common to all the discrete transfer functions:

$$\left(1 - \frac{T}{2}w\right)$$

which expresses much of the difference between the continuous and discrete systems. It is also seen that a transport delay in the continuous system will appear in the w-transform as the factor

$$\left(\frac{1 - \frac{T}{2}w}{1 + \frac{T}{2}w}\right)^m \tag{A.5}$$

where $mT = \tau =$ transport lag.

It is very convenient to design discrete control system algorithms and filtering algorithms for SISO systems in the w-domain by using transfer functions identical to those used for continuous systems but with s replaced by w. For example, the discrete PI controller will have the transfer function.

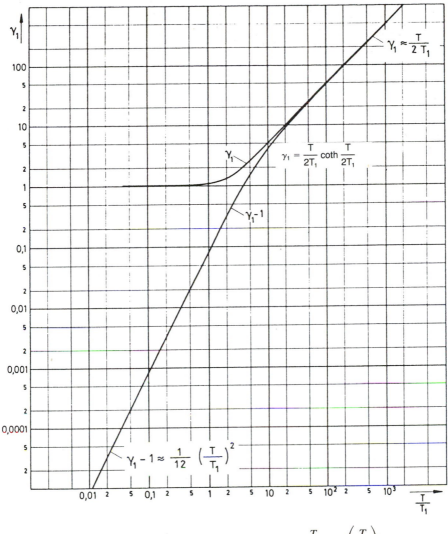

Figure A.1 Graphical representation of $\gamma_1 = \dfrac{T}{2T_1} \coth\left(\dfrac{T}{2T_1}\right)$

$$h_c(w) = K_p \frac{1 + T_i w}{T_i w} \qquad (A.6)$$

where K_p is proportional gain, and T_i is integral time.

The algorithm to be used when writing a computer program for the implementation of the discrete controller is easily derived using the z-transform and the relationships given by equations (A.3) and (A.6):

Table A.1. *w*-transformed transfer functions

	$h(s)$	$h(w) = \mathscr{W}\left(\dfrac{1}{s}(1 - e^{-Ts})h(s)\right)$
	$h_1(s) + h_2(s)$	$h_1(w) + h_2(w)$
	$h_1(s) \cdot h_2(s)$	$h(w) \neq h_1(w) \cdot h_2(w)$
	$h(s)e^{-mTs}$	$h(w)\left(\dfrac{1 - \dfrac{T}{2}w}{1 + \dfrac{T}{2}w}\right)^m$
0.1	1	$\dfrac{1 - \dfrac{T}{2}w}{1 + \dfrac{T}{2}w}$ from 1.2 where $T_1 \to 0$
0.2	$\dfrac{1 + T_2s}{T_2s}$	$\dfrac{\left(1 - \dfrac{T}{2}w\right)\left(1 + \left(T_2 + \dfrac{T}{2}\right)w\right)}{\left(1 + \dfrac{T}{2}w\right)\quad T_2w}$
0.3	$\dfrac{1 + T_2s}{1 + T_1s}$	$\dfrac{\left(1 - \dfrac{T}{2}w\right)\left(1 + \left(\gamma_1 T_2 + \dfrac{T}{2}\left(1 - \dfrac{T_2}{T_1}\right)\right)w\right)}{\left(1 + \dfrac{T}{2}w\right)\quad (1 + \gamma_1 T_1 w)}$
0.4	$\dfrac{T_1s}{1 + T_1s}$	$\dfrac{\left(1 - \dfrac{T}{2}w\right)\left(\gamma_1 T_1 - \dfrac{T}{2}\right)w}{\left(1 + \dfrac{T}{2}w\right)\quad (1 + \gamma_1 T_1 w)}$
1.1	$\dfrac{1}{s}$	$\dfrac{1 - \dfrac{T}{2}w}{w}$
1.2	$\dfrac{1}{1 + T_1s}$	$\dfrac{1 - \dfrac{T}{2}w}{1 + \gamma_1 T_1 w}$
1.3	$\dfrac{1 + T_2s}{s^2}$	$\dfrac{\left(1 - \dfrac{T}{2}w\right)(1 + T_2w)}{w^2}$
1.4	$\dfrac{T_1s}{(1 + T_1s)(1 + T_2s)}$	$\dfrac{\left(1 - \dfrac{T}{2}w\right)\dfrac{\gamma_1 T_1 - \gamma_2 T_2}{T_1 - T_2}T_1w}{(1 + \gamma_1 T_1 w)(1 + \gamma_2 T_2 w)}$
1.5	$\dfrac{1 + T_2s}{s(1 + T_1s)}$	$\dfrac{\left(1 - \dfrac{T}{2}w\right)(1 + (T_2 + (\gamma_1 - 1)T_1)w)}{w(1 + \gamma_1 T_1 w)}$
1.6	$\dfrac{1 + T_3s}{(1 + T_1s)(1 + T_2s)}$	$\dfrac{\left(1 - \dfrac{T}{2}w\right)(1 + Bw)}{(1 + \gamma_1 T_1 w)(1 + \gamma_2 T_2 w)}$ $B = T_3\dfrac{\gamma_1 T_1 - \gamma_2 T_2}{T_1 - T_2} + \dfrac{T_1 T_2}{T_1 - T_2}(\gamma_2 - \gamma_1)$

Table A.1.   *w*-transformed transfer functions (continued)

2.1	$\dfrac{1}{s^2}$	$\dfrac{1 - \dfrac{T}{2}w}{w^2}$
2.2	$\dfrac{1}{s(1 + T_1 s)}$	$\dfrac{\left(1 - \dfrac{T}{2}w\right)(1 + (\gamma_1 - 1)T_1 w)}{w(1 + \gamma_1 T_1 w)}$
2.3	$\dfrac{1}{(1 + T_1 s)(1 + T_2 s)}$	$\dfrac{\left(1 - \dfrac{T}{2}w\right)\left(1 + \dfrac{T_1 T_2}{T_1 - T_2}(\gamma_2 - \gamma_1)w\right)}{(1 + \gamma_1 T_1 w)(1 + \gamma_2 T_2 w)}$
2.4	$\dfrac{1 + T_2 s}{s^2(1 + T_1 s)}$	$\dfrac{\left(1 - \dfrac{T}{2}w\right)(1 + (T_2 + (\gamma_1 - 1)T_1)w - (\gamma_1 - 1)(T_1 - T_2)T_1 w^2)}{w^2(1 + \gamma_1 T_1 w)}$
2.5	$\dfrac{1 + T_3 s}{s(1 + T_1 s)(1 + T_2 s)}$	$\dfrac{\left(1 - \dfrac{T}{2}w\right)(1 + Bw + Cw^2)}{w(1 + \gamma_1 T_1 w)(1 + \gamma_2 T_2 w)}$ $B = T_3 + (\gamma_1 - 1)T_1 + (\gamma_2 - 1)T_2$ $C = \dfrac{T_1 T_2}{T_1 - T_2}[\gamma_1 \gamma_2 (T_1 - T_2) - \gamma_2 (T_1 - T_3) + \gamma_1 (T_2 - T_3)]$
3.1	$\dfrac{1}{s^2(1 + T_1 s)}$	$\dfrac{\left(1 - \dfrac{T}{2}w\right)(1 + (\gamma_1 - 1)T_1 w - (\gamma_1 - 1)T_1^2 w^2)}{w^2(1 + \gamma_1 T_1 w)}$ as 2.4 with $T_2 = 0$
3.2	$\dfrac{1}{s(1 + T_2 s)(1 + T_3 s)}$	as 2.5 with $T_3 = 0$
3.3	$\dfrac{1}{(1 + T_1 s)(1 + T_2 s)(1 + T_3 s)}$	$\dfrac{\left(1 - \dfrac{T}{2}w\right)(1 + Bw + Cw^2)}{(1 + \gamma_1 T_1 w)(1 + \gamma_2 T_2 w)(1 + \gamma_3 T_3 w)}$ $B = a_1(\gamma_2 T_2 + \gamma_3 T_3) + a_2(\gamma_1 T_1 + \gamma_3 T_3) + a_3(\gamma_1 T_1 + \gamma_2 T_2)$ $C = a_1 \gamma_2 T_2 \gamma_3 T_3 + a_2 \gamma_1 T_1 \gamma_3 T_3 + a_3 \gamma_1 T_1 \gamma_2 T_2$ $a_1 = \dfrac{1}{\left(1 - \dfrac{T_2}{T_1}\right)\left(1 - \dfrac{T_3}{T_1}\right)}$ $a_2 = \dfrac{1}{\left(1 - \dfrac{T_1}{T_2}\right)\left(1 - \dfrac{T_3}{T_2}\right)}$ $a_3 = \dfrac{1}{\left(1 - \dfrac{T_1}{T_3}\right)\left(1 - \dfrac{T_2}{T_3}\right)}$

$$z = \frac{1 + \dfrac{T}{2}w}{1 - \dfrac{T}{2}w}, \qquad w = \frac{2}{T}\frac{z - 1}{z + 1}, \qquad \gamma_i = \frac{T}{2T_i}\coth\left(\frac{T}{2T_i}\right)$$

$$h_c(z^{-1}) = K_p \frac{1 + T_i \dfrac{2}{T} \dfrac{1 - z^{-1}}{1 + z^{-1}}}{T_i \dfrac{2}{T} \dfrac{1 - z^{-1}}{1 + z^{-1}}} = \frac{g_0 + g_1 z^{-1}}{1 + f_1 z^{-1}} \tag{A.7}$$

where:

$$g_0 = K_p \left(1 + \frac{T}{2T_i} \right)$$

$$g_1 = -Kp \left(1 - \frac{T}{2T_i} \right)$$

$$f_1 = -1$$

If we let the discrete input to the controller $e(k)$, where k is the sample number, and the associated controller output be $u(k)$, the discrete PI controller determined by equations (A.6) and (A.7) will have the algorithm

$$u(k) = u(k - 1) + g_0 e(k) + g_1 e(k) \tag{A.8}$$

Bode plots, with the concepts of phase and gain margins, or some other algebraic technique may be used to determine the controller parameters of expression (A.7) in the same manner as they would be used for determining the controller parameters for a continuous controller.

The w-transform technique can also be applied to multivariable systems. This makes it possible to design multivariable discrete systems in a manner similar to that described in Chapter 2 and to examine the question of robustness as described in Appendix E. The transfer function matrix $H(w)$ of a discretized continuous system described by $H(s)$ can be found if the continuous system is expressed in the state space form:

$$\dot{\mathbf{x}} = A\mathbf{x} + B\mathbf{u}$$

$$\mathbf{y} = D\mathbf{x} \tag{A.9}$$

It can then be shown (Balchen, 1984a) that:

$$H(w) = \left(1 - \frac{T}{2} w \right) D(wI - \tilde{A})^{-1} \tilde{B} \tag{A.10}$$

where:

$$\tilde{A} = A \left(\frac{AT}{2} \right)^{-1} \tanh \left(\frac{AT}{2} \right) \tag{A.11}$$

$$\tilde{B} = \left(\frac{AT}{2} \right)^{-1} \tanh \left(\frac{AT}{2} \right) B \tag{A.12}$$

Expressions (A.11) and (A.12) are most easily determined from the power series expansions:

$$\tilde{A} = A\left(I - \frac{1}{12}(AT)^2 + \frac{1}{120}(AT)^4 - \frac{1}{1186}(AT)^6 + \frac{1}{11,706}(AT)^8 \ldots\right)$$

(A.13)

$$\tilde{B} = \left(I - \frac{1}{12}(AT)^2 + \frac{1}{120}(AT)^4 \ldots\right)B$$

(A.14)

Appendix B
DERIVATION OF THE EXPRESSION
$(1 + (1 - \hat{M})h_1, h_c)$ IN EQUATION (2.7.11)

Assume that:

$$h_1(s) = \frac{K}{(1 + Ts)} \tag{B.1}$$

and:

$$h_2(s) = e^{-\tau s} \tag{B.2}$$

It is now reasonable to use as feedback in the estimator

$$k(s) = K_1 \frac{(1 + Ts)}{Ts} \tag{B.3}$$

The gain constant $K_0 = K \cdot K_1$ in the estimator loop is chosen such that the loop has a gain margin $\Delta K = 2$ (6 dB). The frequency ω_{180} for this loop is given by $\omega_{180} = \pi/(2\tau)$ and $\omega_c = \pi/(4\tau)$. This gives:

$$K_0 = \frac{\pi T}{4\tau}$$

If we now assume that $\hat{h}_1 = h_1$ and $\hat{h}_2 = h_2$, we can use equation (2.7.10) to write:

$$\hat{M} = M = \frac{\dfrac{\pi}{4} \dfrac{1}{\tau s} e^{-\tau s}}{1 + \dfrac{\pi}{4} \dfrac{1}{\tau s} e^{-\tau s}} \tag{B.4}$$

To get a simpler approximate expression for equation (B.4), we use the approximate transfer function for the transport delay:

$$e^{-\tau s} \approx \frac{1 - \dfrac{\tau}{2} s}{1 + \dfrac{\tau}{2} s} \tag{B.5}$$

When substituted into equation (B.4),

$$M \approx \frac{1 - \dfrac{\tau}{2}s}{1 + \left(\dfrac{4}{\pi} - \dfrac{1}{2}\right)\tau s + \dfrac{2}{\pi}\tau^2 s^2} \tag{B.6}$$

and:

$$N = (1 - M) \approx \frac{\dfrac{4}{\pi}\tau s\left(1 + \dfrac{\tau}{2}s\right)}{1 + 2\zeta\dfrac{s}{\omega_0} + \left(\dfrac{s}{\omega_0}\right)^2} \tag{B.7}$$

where:

$$\omega_0 = \sqrt{\frac{\pi}{2}}\ \frac{1}{\tau} = 1.253\ \frac{1}{\tau}$$

$$\zeta = \sqrt{\frac{\pi}{2}\left(\frac{2}{\pi} - \frac{1}{4}\right)} = 0.485$$

The frequency response of equation (B.7) is shown in Figure B.1 for the case where $|N| < 1$ for $\omega < (\pi/4\tau)$.

If we set this into equation (2.7.4), we find:

$$x = \frac{h_c h_1}{1 + h_c h_1}y_0 + \frac{h_c h_1}{1 + h_c h_1}\left(\frac{1}{h_c} + Nh_1\right)v \tag{B.8}$$

Since the factor

$$\frac{h_c h_1}{1 + h_c h_1} \approx 1$$

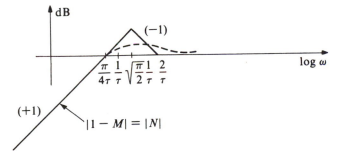

Figure B.1. Frequency response for expression in equation (B.7).

up to fairly high frequencies, and the factor $1/h_c \approx 0$, the response of x resulting from a disturbance v will be dominated by the factor Nh_1. N is introduced to represent the degree of control in the estimator loop. This is the same as one gets in an ordinary control loop for the original process, as shown by the dashed lines in Figure 2.7.1b. Therefore, we see from equation (B.8) the response of x due to the disturbance v will be practically identical to that of the system with ordinary feedback.

By manipulating the block diagram of Figure 2.7.1c, we can develop the diagram of Figure B.2, where we see that the disturbance v now is modified slightly by the transfer function $h_a(s)$, and the controller has a feedback with the transfer function $h_1 s[1 - h_2(s)]$. We find:

$$h_a(s) = \frac{1 + h_c h_1}{1 + h_c h_1 (1 - h_2)} \frac{N}{} \approx 1 \tag{B.9}$$

The so-called Smith predictor has the structure shown in Figure B.2 when $h_a(s) = 1$ and is developed for $h_2(s) = e^{-\tau s}$. The system shown in the figure is, therefore, more generally applicable than the Smith predictor.

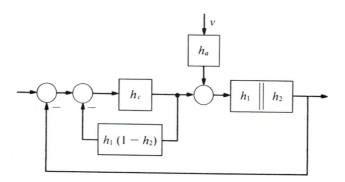

Figure B.2. Modification of block diagram of Figure 2.7.1c.

Appendix C
DERIVATION OF INTERACTION AND PAIRING OF VARIABLES IN MULTIVARIABLE SYSTEMS

Figure C.1 shows the system obtained when the control loop containing u_1 and y_1 is opened, and the rest of the loops are closed. The following notation will be used:

$$H_u = \begin{bmatrix} h_{11} & \vdots & \mathbf{h}^{1T} \\ \cdots & \cdots & \cdots \\ \tilde{\mathbf{h}}^1 & \vdots & H_u^1 \end{bmatrix} \qquad H_c = \begin{bmatrix} h_{c1} & \vdots & 0 \\ \cdots & \cdots & \cdots \\ 0 & \vdots & H_c^1 \end{bmatrix}$$

$$\mathbf{u} = \begin{bmatrix} u_1 \\ \cdots \\ \mathbf{u}^1 \end{bmatrix} \qquad \mathbf{y} = \begin{bmatrix} y_1 \\ \cdots \\ \mathbf{y}^1 \end{bmatrix}$$

The general expressions H_u^{ij} and H_c^{ij} refer to the original matrices H_u and H_c with row i and column j removed. Similarly, we define the reduced vectors such that $\mathbf{u}^j$ refers to the vector $\mathbf{u}$ with element j removed and $\mathbf{y}^i$ refers to the vector $\mathbf{y}$ with element i removed. From Figure C.1 we then find that:

$$y_1 = h_{11}u_1 + \mathbf{h}^{1T}\mathbf{u}^1 \tag{C.1}$$

and:

$$\mathbf{y}^1 = H_u^1\mathbf{u}^1 + \tilde{\mathbf{h}}^1 u_1 = -H_u^1 H_c^1 \mathbf{y}^1 + \tilde{\mathbf{h}}_1 u_1 \tag{C.2}$$

in which we have simplified the notation for the diagonal case as follows:

$$H_u^{ij} \rightarrow H_u^i$$

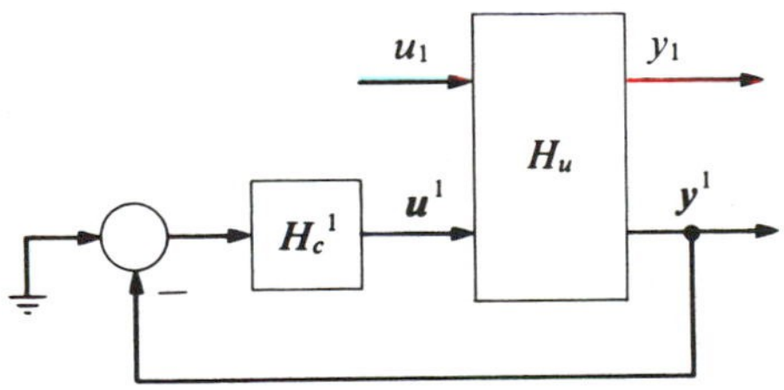

Figure C.1. Multivariable control system with one loop opened.

and:

$$\mathbf{h}^{ij} \rightarrow \mathbf{h}^i$$

The solution of equation (C.2) gives:

$$\mathbf{y}^1 = (I + H_u^1 H_c^1)^{-1} \tilde{\mathbf{h}}^1 u_1 \tag{C.3}$$

Further, we can write:

$$\mathbf{u}^1 = -H_c^1 \mathbf{y}^1 = -H_c^1 (I + H_u^1 H_c^1)^{-1} \tilde{\mathbf{h}}^1 u_1 \tag{C.4}$$

which, when substituted into equation (C.1), gives:

$$y_1 = h_{11} u_1 - \mathbf{h}^1 H_c^1 (I + H_u^1 H_c^1)^{-1} \tilde{\mathbf{h}}^1 u_1 = h_{11}\left(1 - \frac{\mathbf{h}^{1T} H_c^1 (I + H_u^1 H_c^1)^{-1} \tilde{\mathbf{h}}^1}{h_{11}}\right) u_1 \tag{C.5}$$

The last term in the parentheses in equation (C.5) is a scalar quantity. Further manipulation of this term gives:

$$\frac{1}{h_{11}} \mathbf{h}^{1T} H_c^1 (I + H_u^1 H_c^1)^{-1} \tilde{\mathbf{h}}^1 = \frac{1}{h_{11}} \mathbf{h}^{1T} (H_u^1)^{-1} H_u^1 H_c (I + H_u^1 H_c^1)^{-1} \tilde{\mathbf{h}}^1$$

$$= \frac{1}{h_{11}} \mathbf{h}^{1T} (H_u^1)^{-1} M^1 \tilde{\mathbf{h}}^1 \tag{C.6}$$

where we have defined:

$$M^1 = H_u^1 H_c^1 (I + H_u^1 H_c^1)^{-1} = (I + H_u^1 H_c^1)^{-1} H_u^1 H_c^1 \tag{C.7}$$

Equation (C.7) represents the tracking ratio of the remaining part of the multi-variable system. If we can assume that $M^1 \approx I$, which is a good approximation in many cases, then equation (C.6) becomes:

$$\frac{1}{h_{11}} \mathbf{h}^{1T} (H_u^1)^{-1} \tilde{\mathbf{h}}^1 \tag{C.8}$$

Equation (C.5) can be written:

$$y_1 = h_{11}(1 - \Delta_1) u_1 = \overline{h}_{11} u_1 \tag{C.9}$$

where:

$$\Delta_1 = \frac{1}{h_{11}} \mathbf{h}^{1T}(H_u^1)^{-1} M^1 \tilde{\mathbf{h}}^1 \approx \frac{1}{h_{11}} \mathbf{h}_1^T (H_u^1)^{-1} \tilde{\mathbf{h}}^1 \tag{C.10}$$

The factor $(1 - \Delta_1)$ expresses the extent of the influence of the remaining part of the system on the transfer function between u_1 and y_1. In many cases this term is very nearly 1, which means that the design of controller h_{c1} need not take into account the remainder of the system.

If the approximation in equation (C.10) is acceptable, one can use the matrix inversion lemma to get:

$$\overline{h}_{11} = [(H_u)_{11}^{-1}]^{-1} \tag{C.11}$$

where $(H_u)_{11}^{-1}$ is element number $(1, 1)$ in the inverse matrix of H_u. Further manipulation yields:

$$\overline{h}_{11} = \frac{\det(H_u)}{\text{cof}_{11}(H_u)} = \frac{\det(H_u)}{\det_{11}(H_u)} \tag{C.12}$$

where $\text{cof}_{11}(H_u)$ is cofactor number $(1, 1)$ or the subdeterminant number $(1, 1)$, which is found by eliminating row 1 and column 1 in $\det(H_u)$.

The above development can be generalized so that any combination of control loops can be studied. Consideration of output y_i controlled by input u_j leads to the expression

$$y_i = \overline{h}_{ij} u_j = h_{ij}(1 - \Delta_{ij})u_j \tag{C.13}$$

where:

$$\Delta_{ij} = \frac{1}{h_{ij}} \mathbf{h}^{ijT}(H_u^{ij})^{-1} M^{ij}\tilde{\mathbf{h}}^{ij} \approx \frac{1}{h_{ij}} \mathbf{h}^{ijT}(H_u^{ij})^{-1} \tilde{\mathbf{h}}^{ij} \tag{C.14}$$

These general interaction factors can be collected in a matrix:

$$\Delta = \{\Delta_{ij}\} \tag{C.15}$$

which gives an expression for the most desirable pairs of input (control) and output (measurement) variables.

Niederlinski (1971) has shown that when the controllers have integral action (I, PI, or PID controllers), the multiloop system will become unstable for all possible controller parameters if

$$\rho = \frac{\det(H_u(s = 0))}{\prod_{j=1}^{m} h_{jj}(s = 0)} < 0 \tag{C.16}$$

Using equations (C.12) and (C.13), we find:

$$\Delta_{ii}(s) = 1 - \frac{\overline{h}_{ii}(s)}{h_{ii}(s)} = 1 - \frac{1}{h_{ii}(s)} \frac{\det[H_u(s)]}{\det[H_u^{\,ii}(s)]} \tag{C.17}$$

Equation (C.16) must also hold for the system $H_u^{\,ii}(s)$. Thus, a necessary condition for stability is:

$$\rho^i = \frac{\det(H_u^{\,ii}(s = 0))}{\prod\limits_{j=1}^{m-1} h_{jj}(s = 0)} > 0 \tag{C.18}$$

Applying equations (C.16) and (C.18) in (C.17), we find that a necessary (but not sufficient) condition for stability is:

$$\Delta_{ii}(s = 0) = 1 - \frac{\overline{h}_{ii}(0)}{h_{ii}(0)} = 1 - \frac{1}{h_{ii}(0)} \frac{\rho \prod\limits_{j=1}^{m} h_{jj}(0)}{\rho^i \prod\limits_{j=1}^{m-1} h_{jj}(0)} = 1 - \frac{\rho}{\rho^i} < 1$$

$$\tag{C.19}$$

since both ρ and ρ^i must be positive for the system not to be unstable. *Sufficient* conditions for stability are obtained through the application of standard stability criteria.

This development leads to the following rules for the choice of pairs in multiloop control:

1. The direct transfer function $h_{ij}(s)$ defining a pair (assuming no interaction) must be acceptable from a feedback control point of view; that is, the gain–phase–frequency relationships must allow good control.
2. The interaction with other parts of the system should not be excessive as indicated by the following:
   a. $|\Delta_{ij}(s)| \approx 0$: weak interaction, desirable pair.
   b. $0 \gtrless |\Delta_{ij}(s)| \gtrless 0.8$: some interaction, acceptable pair.
   c. $0.8 \gtrless |\Delta_{ij}(s)| < 1$: strong interaction, undesirable pair.
   d. $|\Delta_{ij}(s)| \gtrless 1$: unstable, impossible pair.
3. The final transfer function:

$$\overline{h}_{ij}(s) = h_{ij}(s)[1 - \Delta_{ij}(s)]$$

defining a pair should have acceptable control properties.

When the matrix of equation (C.15) is applied to Figure 5.6.16, the matrix becomes:

$$\Delta(s) \cong \begin{bmatrix} \dfrac{0.8089e^{-0.095}}{(1 + 0.083s)^5} & , & 1.236(1 + 0.083s)^5 e^{0.095} \\[2em] 1.236(1 + 0.083s)^5 e^{0.095}, & & \dfrac{0.8089e^{-0.095}}{(1 + 0.083s)^5} \end{bmatrix} \quad \text{(C.20)}$$

From this we see that the only possible pair is the one chosen. Furthermore, as discussed in section 5.6, the control properties of $\overline{h}_{11}(s)$ and $\overline{h}_{22}(s)$ are acceptable.

The above results were first presented for a 2×2 system in the first edition (1963) of Balchen (1977/84). They are related in the steady state to those obtainable by use of the "relative gain array" introduced by Bristol (1966) and applied by McAvoy (1983) and others. Gagnepain and Seborg (1982) have introduced an alternative measure of interaction to the one discussed above.

Appendix D
ALGORITHMS FOR ADAPTIVE CONTROL OF MONOVARIABLE PROCESSES

Three different structures for adaptive control of a single variable process are shown in Figures 2.14.1.a, b, and c. The solution of Figure 2.14.1c will be discussed here. This adaptive system contains two algorithms:

1. An algorithm for the estimation of the model parameters.
2. An algorithm for adapting the controller parameters on the basis of the estimated model parameters.

Extensive development in each of these areas has led to many solutions. A survey of parameter adaptive control algorithms is given in Isermann (1982).

One of the simplest algorithms for parameter estimation assumes a model of the system using z-transform notation:

$$y(z) = \frac{B(z^{-1})}{A(z^{-1})} z^{-d} u(z) + v(z) \tag{D.1}$$

with the polynomials:

$$A(z^{-1}) = 1 + a_1 z^{-1} + \cdots a_n z^{-n} \tag{D.2}$$

$$B(z^{-1}) = b_1 z^{-1} + \cdots b_n z^{-n} \tag{D.3}$$

where z^{-d} represents a transport delay $\tau = d \cdot T$ with $T = $ the sampling interval.

In the time domain, equation (D.1) is:

$$y(k) = -a_1 y(k - 1) - \cdots a_n y(k - n) + b_1 u(k - d - 1)$$
$$+ \cdots b_n u(k - d - n) + v(k) \tag{D.4}$$

where $v(k)$ is assumed to be a zero mean white noise sequence.

If previous values of the measurements y and the control variable u are stored in a data vector:

$$\boldsymbol{\psi}^T(k) = [-y(k - 1), \ldots, -y(k - n), u(k - d - 1), \ldots, u(k - d - n)] \tag{D.5}$$

and the parameters of equation (D.4) are collected in a vector:

$$\boldsymbol{\theta}^T = [a_1, \ldots, a_n, b_1, \ldots, b_n] \tag{D.6}$$

then equation (D.4) can be written in the compact form

$$y(k) = \boldsymbol{\psi}^T(k)\boldsymbol{\theta} + v(k) \tag{D.7}$$

The system (D.7) can be regarded as a linear system with a state vector $\boldsymbol{\theta}$ observed by the measurements $y(k)$ through a time variable observation matrix $\boldsymbol{\psi}^T(k)$. Therefore, a Kalman filter can be used to estimate $\boldsymbol{\theta}$. If we assume a model of the parameter vector to be

$$\overline{\boldsymbol{\theta}}(k + 1) = \hat{\boldsymbol{\theta}}(k) \tag{D.8}$$

application of the Kalman filter equations given in section 2.13 yields:

$$\hat{\boldsymbol{\theta}}(k) = \overline{\boldsymbol{\theta}}(k) + K(k)(y(k) - \boldsymbol{\psi}^T(k)\overline{\boldsymbol{\theta}}(k)) \tag{D.9}$$

$$K(k) = \overline{\Theta}(k)\boldsymbol{\psi}(k)(\boldsymbol{\psi}^T(k)\overline{\Theta}(k)\boldsymbol{\psi}(k) + v_0)^{-1} \tag{D.10}$$

$$\overline{\Theta}(k + 1) = (I - K(k)\boldsymbol{\psi}^T(k))\overline{\Theta}(k)) \tag{D.11}$$

where:

$$v_0 = \text{cov } v(k) \tag{D.12}$$

$\overline{\Theta}(k)$ is the covariance matrix of the parameter vector $\boldsymbol{\theta}(k)$. The last term in equation (D.10) is a scalar, and the Kalman filter gain matrix $K(k)$ is, in this case, a vector.

Many similar algorithms, with slight modifications, are available for parameter estimation. The most popular control algorithms are the minimum variance controllers and the familiar PID controllers.

The original minimum variance controller proposed by Åstrom (1970) has the transfer function

$$h_c(z^{-1}) = \frac{u(z)}{e(z)} = -\frac{L(z^{-1})}{zB(z^{-1})F(z^{-1})} \tag{D.13}$$

where the polynomials $L(\cdot)$ and $F(\cdot)$ are determined by the identity

$$1 = A(z^{-1})F(z^{-1}) + z^{-(d+1)}L(z^{-1}) \tag{D.14}$$

$B(\cdot)$ and $A(\cdot)$ are the polynomials given in equation (D.1). $F(\cdot)$ and $L(\cdot)$ are given by:

$$F(z^{-1}) = 1 + f_1 z^{-1} + \cdots + f_d z^{-d} \tag{D.15}$$

and:

$$L(z^{-1}) = l_0 + l_1 z^{-1} + \cdots + l_{n-1} z^{-(n-1)} \tag{D.16}$$

Equation (D.14) provides a relationship between the controller parameters

(f_i, l_i) and the parameters of the process model, which are represented by the parameter vector $\boldsymbol{\theta}$.

Other controllers, such as PID controllers, can be adapted in a similar manner as long as a unique relationship exists between the estimated model parameters and the controller parameters.

Appendix E
ROBUSTNESS OF MULTIVARIABLE
CONTROL SYSTEMS

Robustness is a term used to describe the insensitivity of the stability of a control system to changes in the parameters of the system. This consideration is important for systems that are designed and/or tuned on the basis of rather crude model representations of the components. When the system parameters change, or when nonlinearities or multiplicative disturbances are present, the system may become unstable if it is designed without regard for such effects.

The matter of robustness is easily recognized in the classical frequency response design technique for monovariable (SISO) control systems. The concepts of gain margin and phase margin are directly related to robustness (see section 2.2). Of the two systems shown in Figure E.1, system 1 has a higher degree of robustness than system 2 because a small change in the gain of system 2 will lead to a large change in the crossover frequency (ω_c), which is related to a rapidly changing negative phase shift in the loop. Typically, frequency response curves of the type shown for system 2 are found in systems containing elements with nonminimum phase characteristics such as transport delays, transfer functions having zeroes in the right half-plane, and sampling phenomena (discrete time control systems).

It is more difficult to discover the lack of robustness in multivariable systems. In this appendix we will describe a technique for characterizing the robustness of multivariable systems using the concept of singular values (Lehtomaki, Sandell, and Athans, 1981; Doyle and Stein, 1981).

Consider the system shown in Figure E.2. This system consists of a multivariable process with the transfer function matrix $H_u(s)$. The process is controlled by some multivariable control algorithm summarized by the transfer function matrix $H_c(s)$. The product of these two matrices is the loop transfer function matrix $H_0(s)$. Let us now assume that there is a multiplicative perturbation of the elements of the loop transfer function matrix and that these perturbations make up the matrix $P(s) = I + L(s)$. The loop transfer function matrix now becomes:

$$\tilde{H}_0(s) = (I + L(s))H_0(s) \tag{E.1}$$

The perturbations (uncertainties) $L(s)$ are not known precisely. A useful way of expressing a variety of possible perturbations is through the singular value of $L(\cdot)$, defined by:

$$\sigma_i(L(s)) = \sqrt{\lambda_i(L^*(s)L(s))} \tag{E.2}$$

where $\lambda_i(\cdot)$ is the eigenvalue number i of the argument matrix, and $L^*(s)$ is the

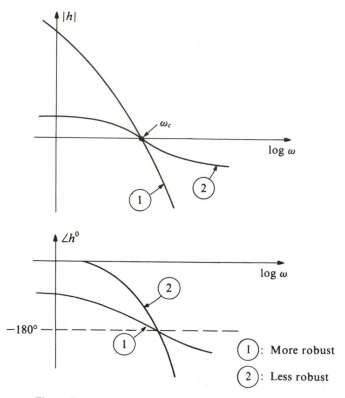

Figure E.1. Comparison of robustness in SISO systems.

complex conjugate transpose matrix of $L(s)$. $\overline{\sigma}(\cdot)$ denotes the maximal singular value and $\underline{\sigma}(\cdot)$ denotes the minimal singular value. Singular values are scalar functions of s and can be compared with transfer functions of monovariable systems. They can also be characterized by their equivalent frequency response functions. Therefore, if the multiplicative perturbations $L(s)$ are known only in terms of a magnitude bound $l_m(\omega)$, this situation may be expressed by $\overline{\sigma}$ with:

$$\overline{\sigma}(L(j\omega)) \leq l_m(\omega) \tag{E.3}$$

In Doyle and Stein (1981) it is shown that a feedback system that is nominally stable will remain stable when subjected to uncertainties of the kind described by equation (E.1) if and only if:

$$\overline{\sigma}(H_0(j\omega)(I + H_0(j\omega))^{-1}) < 1/l_m(\omega) \tag{E.4}$$

for $0 \leq \omega < \infty$.

The expression (E.4) is equivalent to:

$$l_m(\omega) < \underline{\sigma}(I + H_0^{-1}(j\omega)) \tag{E.5}$$

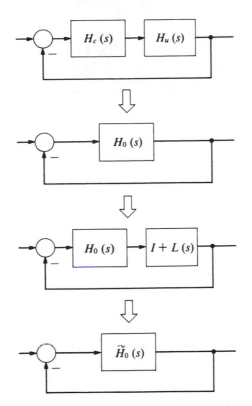

Figure E.2. Multivariable system with controller and perturbations.

Expressions (E.5) and (E.3), taken together, express the result that the perturbed system is stable if the maximum singular value of the perturbations is less than the minimum singular value on the right side of expression (E.5).

This basic technique can be used to shape the controller transfer function matrix, which is part of the total loop transfer function matrix. Because the computation of the singular values requires special computer programs, this is a computer-aided design technique.

For discrete time control systems (sampled data systems) the investigation of robustness can be made using the w-transform technique described in Appendix A and the method outlined above.

Appendix F
COMPRESSION AND DYNAMICS IN GAS SYSTEMS

Consider the system shown in Figure F.1. In this system a fan or compressor with an inlet pressure p_0 works into a volume having a pressure p_1. If we assume adiabatic compression of the gas, we have:

$$pV^k = \text{constant} \tag{F.1}$$

which, after differentiating, gives:

$$V^k \, dp + kV^{k-1}p \, dV = 0 \tag{F.2}$$

so that:

$$dp = -kV^{-1}p \, dV \tag{F.3}$$

Because V represents the volume of a certain amount of gas (mass) being compressed, the volume flow can be expressed as:

$$q = -dV/dt \tag{F.4}$$

which, when inserted into equation (F.3), gives:

$$dp/dt = kV^{-1}pq \tag{F.5}$$

Equation (F.5) also applies to the volume V_0 of the tank in Figure F.1. For the tank, we have:

$$\frac{V_0}{kp_1}\frac{dp_1}{dt} = C\frac{dp_1}{dt} = q \tag{F.6}$$

where $C = V_0/kp_1 =$ gas capacity. For adiabatic compression, the value of k ranges between 1.39 and 1.41 for many gases. For isothermal conditions in

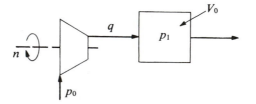

Figure F.1. Compressor working into a fixed volume.

which there is a long holdup time and good heat exchange, $k = 1$. The expression for the gas capacity therefore does not vary more than about 40% between the two extreme cases.

Figure F.2 shows how a volume of gas with capacity C is connected with a compressor having a given characteristic. The compressor is assumed to have an inlet pressure p_0, and the pressure in the tank is p_1. The compressor speed is n, and the gas flow, which can be expressed as $q = f(p_0, p_1, n)$, is determined by the characteristic, as shown in Figure F.3. Within a limited operating range, we can write:

$$\delta q = \frac{\partial q}{\partial p_0} \delta p_0 + \frac{\partial q}{\partial p_1} \delta p_1 + \frac{\partial q}{\partial n} \delta n \tag{F.7}$$

Since the elements of Figure F.2 directly follow equation (F.6), we can take the Laplace transformation and find:

$$\delta p_1(s) = \frac{1}{1 + \tau s} \quad \frac{1}{-\left(\dfrac{\partial q}{\partial p_1}\right)} \left(\frac{\partial q}{\partial p_0} \delta p_0(s) + \frac{\partial q}{\partial n} \delta n(s) \right) \tag{F.8}$$

where:

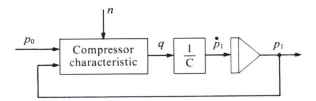

Figure F.2. Elementary block diagram for compressor with gas volume.

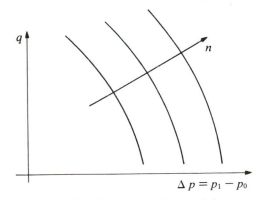

Figure F.3. Compressor characteristics.

$$\tau = C/[-\partial q/\partial p_1] = \text{time constant}$$

and $\partial q/\partial p_1 < 0$.

We see from equations (F.6) and (F.8) that the time constant required for pressure buildup in the system increases with the volume and decreases with absolute pressure.

Appendix G
VELOCITY PROFILE CHARACTERISTICS IN PIPES

Consider the velocity profile shown in Figure G.1. If we let the profile be $v(r)$, a small element with radius r and width dr (area $= 2\pi r\, dr$) will move at a velocity $v(r)$. This element will have a transport time

$$\tau(r) = L/v(r)$$

Therefore, this element can be described by the transfer function $e^{-\tau(r)s}$. If we sum all of these elements over the cross section of the pipe, we will get a total transfer function

$$h(s) = \frac{1}{A} \int_0^R 2\pi r \ e^{-(L/v(r))\cdot s} \cdot dr \tag{G.1}$$

where A is the cross-sectional area of the pipe. If we assume a constant velocity profile, where $v(r) = v_0$, we have:

$$h(s) = e^{-(L/v_0)s} \cdot \frac{1}{A} \int_0^R 2\pi r\, dr = e^{-(L/v_0)s} = e^{-\tau_0 s} \tag{G.2}$$

An integral of the from of equation (G.1) cannot be solved for an arbitrary $v(r)$, but if we know the profile — say, for example, that it is quadratic — then:

$$v(r) = 2v_0[1 - (r/R)^2] \tag{G.3}$$

and it is easy to find the step response of, for example, a concentration change at all points over the cross section at a point measured at a distance L along the pipe. After the step change at point 0, a parabolic front of material moves forward along the pipe. The circular cross section of this parabola, measured at L, can be used as a measure of the step response. If we call the response $c(t)$, we have:

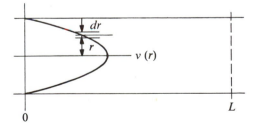

Figure G.1. Velocity profile in a pipe.

$$c(t) = c_0 \quad \pi r^2 \tag{G.4}$$

where r is the radius of the front at point L, as shown in Figure G.2. With a quadratic velocity distribution over the pipe as given in equation (G.3), we can calculate the radius of the front at point L as a function of time from:

$$t = \frac{L}{2v_0\left(1 - \left(\dfrac{r}{R}\right)^2\right)} \tag{G.5}$$

which gives:

$$r^2 = R^2 \quad [1 - L/(2v_0 t)] \tag{G.6}$$

Inserting this into equation (G.4) gives:

$$c(t) = c_0 \pi R^2 \left(1 - \frac{L}{2v_0 t}\right) \cong \tilde{c}_0\left(1 - \frac{L}{2v_0 t}\right) \tag{G.7}$$

The response to a unit step in concentration will be as illustrated in Figure G.3. This response is very much like what would have resulted if the transfer function had been:

$$h(s) = \frac{e^{-\tilde{\tau}_0 s}}{1 + \tilde{\tau}_0 s} \tag{G.8}$$

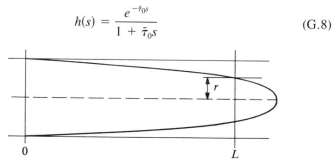

Figure G.2. Derivation of step response in pipe with quadratic velocity profile.

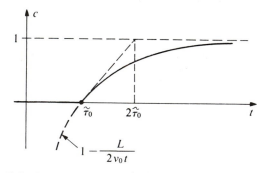

Figure G.3. Step response in pipe with quadratic velocity profile.

where $\bar{\tau}_0 = L/2v_0$, and v_0 is the average flow velocity over the cross section of the pipe.

If this result is compared with that of equation (G.2), we find that the transport delay τ_0 for the case of the quadratic velocity profile relative to that of a constant velocity profile is:

$$\bar{\tau}_0 = (1/2)\tau_0$$

if the average flow velocity in the pipe is the same in both cases.

Appendix H
DEVELOPMENT OF MATHEMATICAL MODELS
FOR HEAT EXCHANGERS

In this appendix we will give an overview of the mathematical model development for these types of heat exchangers:

- Fluid–Fluid heat exchangers
- Condensers
- Reboilers

The major souces used are Thal-Larsen (1960), Hempel (1961), and Buckley (1964).

FLUID–FLUID HEAT EXCHANGERS

In this type of exchanger it is important to differentiate between the cases where one has good (perfect) mixing of the media on the primary and secondary sides and where one has poor or essentially no exchange of heat axially along the contact area of the tubes or surfaces. The situation is simplest when one assumes infinitely good mixing of both media, but it is more realistic to assume a middle ground between perfect and no mixing.

Perfect Mixing of Both Media

Figure H.1 is a heat exchanger with the primary side designated as A and the secondary side as B. The following notation is used:

w_A: flow of primary medium into and out of heat exchanger (kg/s)
w_B: flow of secondary medium into and out of heat exchanger (kg/s)
θ_A: temperature in primary medium (perfect mixing) (°C)
θ_{A1}: inlet temperature of primary medium (°C)

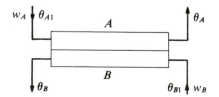

Figure H.1. Simple heat exchanger with perfect agitation.

θ_B: temperature of secondary medium (perfect mixing) (°C)
θ_{B1}: inlet temperature of secondary medium (°C)
A: area of heat transfer surface (m^2)
h_A: heat transfer coefficient of primary medium surface (kJ/s · $m^{2\circ}$C)
h_B: heat transfer coefficient of secondary medium surface (kJ/s · $m^{2\circ}$C)
$h = h_A h_B/(h_A + h_B)$: total heat transfer coefficient between primary and secondary media (kJ/s · $m^{2\circ}$C)
c_A: specific heat of primary medium (kJ/kg°C)
c_B: specific heat of secondary medium (kJ/kg°C)
W_A: mass of primary medium (kg)
W_B: mass of secondary medium (kg)
Δ: symbol for small perturbations around mean value
$^{-}$:   overbar symbol denoting mean values of variables

As perfect mixing is assumed in both the primary and secondary media, it is easy to derive the differential equations and transfer functions on the basis of elementary heat balances. Buckley (1964) does this as follows:

$$\frac{\Delta\theta_B}{\Delta\theta_{B1}}(s) = \frac{c_B\overline{w}_B(c_A\overline{w}_A + hA)\ (1 + T_1 s)}{\alpha(1 + 2\zeta T_0 s + T_0^2 s^2)} \tag{H.1}$$

$$\frac{\Delta\theta_B}{\Delta\theta_{A1}}(s) = \frac{hAc_A\overline{w}_A}{\alpha(1 + 2\zeta T_0 s + T_0^2 s^2)} \tag{H.2}$$

$$\frac{\Delta\theta_B}{\Delta w_B}(s) = \frac{hA(\overline{\theta}_B - \overline{\theta}_A)(c_A\overline{w}_A + hA)(1 + T_1 s)}{\overline{w}_B\alpha(1 + 2\zeta T_0 s + T_0^2 s^2)} \tag{H.3}$$

$$\frac{\Delta\theta_B}{\Delta w_A}(s) = \frac{(hA)^2(\overline{\theta}_A - \overline{\theta}_B)}{\overline{w}_A\alpha(1 + 2\zeta T_0 s + T_0^2 s^2)} \tag{H.4}$$

where:

$$T_1 = \frac{c_A W_A}{c_A\overline{w}_A + hA} \tag{H.5}$$

$$\alpha = c_B\overline{w}_B(c_A\overline{w}_A + hA) + c_A\overline{w}_A hA \tag{H.6}$$

$$T_0 = \sqrt{\frac{c_A W_A c_B W_B}{\alpha}} \tag{H.7}$$

$$\zeta = \frac{c_A W_A c_B\overline{w}_B + c_B W_B c_A\overline{w}_A + hA(c_A W_A + c_B W_B)}{2T_0\alpha} \tag{H.8}$$

The transfer functions in equations (H.1) and (H.3) each have two poles and one zero. When the relative damping factor is larger than 1 ($\zeta > 1$), the quadratic expression in the denominator consists of two first-order terms, each having a time constant. However, we are most interested in the behavior at relatively high frequencies. Equation (H.4) is then:

$$\frac{\Delta\theta_B}{\Delta\theta_{B1}}(s) \approx \frac{c_B\overline{w}_B}{c_B\overline{w}_B + \dfrac{c_A\overline{w}_A hA}{c_A\overline{w}_A + hA}} \quad \frac{1}{1 + T_2 s} \tag{H.9}$$

where:

$$T_2 = \frac{c_B\,W_B}{c_B\overline{w}_B + \dfrac{c_A\overline{w}_A hA}{c_A\overline{w}_A + hA}}$$

At even higher frequencies, the expression given by equation (H.9) will approach:

$$\frac{\Delta\theta_B}{\Delta\theta_{B1}}(s) \approx \frac{1}{\dfrac{W_B}{\overline{w}_B}\,s} \tag{H.10}$$

where $W_B/\overline{w}_B$ is the hydraulic time constant (holdup time) for the secondary side.

Equation (H.3) can be simplified in a similar fashion to yield:

$$\frac{\Delta\theta_B}{\Delta w_B}(s) \approx \frac{hA(\overline{\theta}_B - \overline{\theta}_A)}{\overline{w}_B\left(c_B\overline{w}_B + \dfrac{c_A\overline{w}_A hA}{c_A\overline{w}_A + hA}\right)} \quad \frac{1}{1 + T_2 s} \tag{H.11}$$

where T_2 is defined above.

Equations (H.2) and (H.4) have two real poles characterized by the time constants T_a and T_b, respectively, and these time constants are:

$$T_a, T_b = T_0(\zeta \pm \sqrt{(\zeta^2 - 1)}) \tag{H.12}$$

The ratio between these two time constants is interesting, but it cannot easily be examined without having numerical values.

The following approximations are satisfactory for low frequencies and relatively high frequencies as well:

$$\frac{\Delta\theta_B}{\Delta\theta_{A1}}(s) \approx \frac{hAc_A\overline{w}_A}{\alpha(1 + T_0 s)^2} \tag{H.13}$$

$$\frac{\Delta\theta_B}{\Delta w_A}(s) \approx \frac{(hA)^2(\overline{\theta}_A - \overline{\theta}_B)}{\overline{w}_A\alpha(1 + T_0 s)^2} \tag{H.14}$$

The following comments apply to the above discussion:

- One of the variables θ_{A1}, w_B, or w_A will be the most reasonable control variable if one wishes to control the temperature θ_B. θ_{B1} is usually considered to be a disturbance, although θ_{A1}, w_B, or w_A can also be disturbances. Equations (H.2) or (H.13) and (H.14) show that if we choose either θ_{A1} or w_A as a control variable, the process can be described by a second-order transfer function.
- If w_B is used as a control variable, we see from equation (H.3) or (H.11) that the process can be described by a first-order transfer function.

Heat Exchanger without Axial Mixing

In contrast to the previous case, let us consider a liquid–liquid heat exchanger designed with concentric tubes in which the flows are countercurrent, and there is no heat transfer in either medium in the axial direction. However, we will assume perfect exchange in the radial direction. A heat exchanger of this type is shown in Figure H.2.

In addition to the notation used in the last example, we will add the following:

θ_{A2}: temperature of the primary medium at the outlet of the heat exchanger (°C)

θ_{B2}: temperature of the secondary medium at the outlet of the heat exchanger (°C)

Many investigators have studied this type of heat exchanger and have proposed various simplified models that all seem to be closely related to each other. In this discussion we will refer to the work of Buckley (1964).

Under steady-state conditions the temperature at the outlet of the secondary medium is:

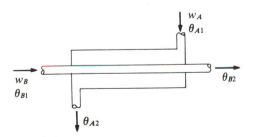

Figure H.2. Countercurrent heat exchanger with no axial mixing.

$$\theta_{B2} = \frac{2c_A\overline{w}_Ac_B\overline{w}_B + hA(c_B\overline{w}_B - c_A\overline{w}_A)}{2c_A\overline{w}_Ac_B\overline{w}_B + hA(c_B\overline{w}_B + c_A\overline{w}_A)} \theta_{B1}$$

$$+ \frac{2hAc_A\overline{w}_A}{2c_A\overline{w}_Ac_B\overline{w}_B + hA(c_B\overline{w}_B + c_A\overline{w}_A)} \theta_{A1} \qquad (H.15)$$

Buckley found that the following transfer functions gave a reasonably good description of the case when the variables were given small perturbations around the steady-state operating point:

$$\frac{\Delta\theta_{B2}}{\Delta\theta_{B1}}(s) \approx K_1 \; e^{-\tau_1 s} \qquad (H.16)$$

$$\frac{\Delta\theta_{B2}}{\Delta\theta_{A1}}(s) \approx K_2 \; \frac{e^{-\tau_2 s}}{(1 + T_1 s)(1 + T_2 s)} \qquad (H.17)$$

$$\frac{\Delta\theta_{B2}}{\Delta w_A}(s) \approx K_3 \; \frac{e^{-\tau_2 s}}{(1 + T_1 s)(1 + T_2 s)} \qquad (H.18)$$

$$\frac{\Delta\theta_{B2}}{\Delta w_B}(s) \approx K_4 \; \frac{e^{-\tau_2 s}}{(1 + T_1 s)(1 + T_2 s)} \qquad (H.19)$$

where:

$$\tau_1 = \frac{W_B}{w_B}, \qquad T_1 = \frac{1}{2}\left(\frac{W_A}{w_A} + \frac{W_B}{w_B}\right), \qquad T_2 = \tau_2 = \frac{1}{8}\left(\frac{W_A}{w_A} + \frac{W_B}{w_B}\right)$$

$$K_1 = \frac{2c_A\overline{w}_Ac_B\overline{w}_B + hA(c_B\overline{w}_B - c_A\overline{w}_A)}{2c_A\overline{w}_Ac_B\overline{w}_B + hA(c_B\overline{w}_B + c_A\overline{w}_A)}$$

$$K_2 = \frac{2hAc_A\overline{w}_A}{2c_A\overline{w}_Ac_B\overline{w}_B + hA(c_B\overline{w}_B + c_A\overline{w}_A)}$$

$$K_3 = \frac{2hAc_Ac_B\overline{w}_B(\overline{\theta}_{A1} - \overline{\theta}_{B1})\left(hA + 2\dfrac{c_A\overline{w}_A{}^2}{h}\dfrac{\partial h}{\partial w_A}\right)}{[2c_A\overline{w}_Ac_B\overline{w}_B + hA(c_B\overline{w}_B + c_A\overline{w}_A)]^2}$$

$$K_4 = \frac{2hAc_Bc_A\overline{w}_A(\overline{\theta}_{A1} - \overline{\theta}_{B1})\left(2\dfrac{\partial h}{\partial w_B}\dfrac{w_B}{h}c_A\overline{w}_A - 2c_A\overline{w}_A - hA\right)}{[2c_A\overline{w}_Ac_B\overline{w}_B + hA(c_B\overline{w}_B + c_A\overline{w}_A)]^2}$$

The transfer functions given by equations (H.16) through (H.19) are only approximations, but they have been shown experimentally to have sufficient precision for control purposes within the frequency range of interest.

We see from the expressions for K_3 and K_4 that the total heat transfer coefficients vary with the liquid flowrates. Therefore, the value of h used in the equations must be considered as the mean value in the region of the operating point.

The following comments apply to these results:

- The response of the temperature at the outlet of the secondary medium to a change of temperature at the inlet of the secondary medium is a pure transport delay with a magnitude equal to the holdup time of the secondary medium in the heat exchanger.
- Except for the gain factors, the transfer functions relating the response of the outlet temperature of the secondary medium to the primary inlet temperature, primary flow changes, and secondary flow changes are all the same. These functions consist of a transport delay and two time constants that have a very simple relationship to the holdup times of the primary and secondary sides.
- The development does not consider the effect of the metal in the tubes separating the primary and secondary media or in the tank surrounding the primary medium. This can be added quite easily by simply increasing the mass of the liquid media by an amount corresponding to the thermal mass of the metal.

CONDENSERS

There are many different condenser configurations in which steam is condensed through cooling by a secondary medium (usually a liquid) flowing in a heat exchanger of some type. To conserve energy, many modern installations use the secondary medium for heating elsewhere in the process. In those cases there can be very strong dynamic interactions that do not exist if the cooling medium is independent of the process, as it would be if it were simply discharged to a sewer.

We will consider dynamic models of two simplified condenser systems based on Figure H.3, which have the following characteristics:

- Perfect mixing in the secondary (cooling) medium due to intense agitation and circulation maintained by pumping.
- No axial mixing of the cooling medium, but perfect radial mixing.

If we assume that the steam entering the primary side contains no inerts, then we can assume that all of the steam will condense. The condensate accumulates outside of the condenser so that the condenser always has the same surface area available for heat transfer. Although the condensate forms a thin film on the inside of the tubes, we assume that the heat transfer coefficient on the steam side is constant and independent of film thickness.

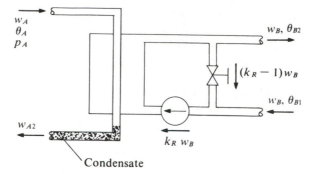

Figure H.3. Condenser configuration.

Buckley (1964) used the results of Hempel (1961) as a basis for the work that we will now review. For convenience, let us introduce the following new notation:

w_{A2}: condensate flow (kg/s)
λ_A: heat of vaporization of the primary medium (kJ/kg)

Under steady-state conditions the flow of condensate produced at a temperature given by the pressure of the saturated steam on the primary side is:

$$w_{A2} = \frac{hAc_Bw_B(\theta_A - \theta_B)}{\lambda_A(hA + c_Bw_B)} \tag{H.20}$$

We can also find the temperature of the cooling medium leaving the condenser (at steady state):

$$\theta_{B2} = \frac{hA\theta_A + c_Bw_B\theta_{B1}}{hA + c_Bw_B} \tag{H.21}$$

When the variables are perturbed around the steady-state operating point, the most important transfer functions are:

$$\frac{\Delta w_{A2}}{\Delta w_B}(s) = K_1 \; \frac{1}{1 + T_1s} \tag{H.22}$$

$$\frac{\Delta w_{A2}}{\Delta \theta_{B1}}(s) = K_2 \; \frac{1}{1 + T_1s} \tag{H.23}$$

$$\frac{\Delta w_{A2}}{\Delta \theta_A}(s) = K_3 \; \frac{1 + T_2s}{1 + T_1s} \tag{H.24}$$

where:

$$T_1 = \frac{c_B W_B}{hA + c_B \overline{w}_B} \tag{H.25}$$

$$T_2 = \frac{W_B}{\overline{w}_B} \tag{H.26}$$

$$K_1 = \frac{c_B(\overline{\theta}_A - \overline{\theta}_{B1})(hA)^2}{\lambda_A(hA + c_B\overline{w}_B)^2} \tag{H.27}$$

$$K_2 = -\frac{c_B\overline{w}_B hA}{\lambda_A(hA + c_B\overline{w}_B)} \tag{H.28}$$

$$K_3 = \frac{c_B\overline{w}_B hA}{\lambda_A(hA + c_B\overline{w}_B)} \tag{H.29}$$

If we are interested in the temperature of the cooling medium (the secondary medium), we need the following transfer functions [which also assume small perturbations around the steady-state values as in equation (H.21)]:

$$\frac{\Delta\theta_{B2}}{\Delta\theta_A}(s) = K_4 \frac{1}{1 + T_1 s} \tag{H.30}$$

$$\frac{\Delta\theta_{B2}}{\Delta w_B}(s) = K_5 \frac{1}{1 + T_1 s} \tag{H.31}$$

$$\frac{\Delta\theta_{B2}}{\Delta\theta_{B1}}(s) = K_6 \frac{1}{1 + T_1 s} \tag{H.32}$$

where:

$$K_4 = \frac{hA}{hA + c_B\overline{w}_B} \tag{H.33}$$

$$K_5 = \frac{c_B(\overline{\theta}_A - \overline{\theta}_{B1})}{(hA + c_B\overline{w}_B)^2} \tag{H.34}$$

$$K_6 = \frac{c_B\overline{w}_B}{hA + c_B\overline{w}_B} \tag{H.35}$$

Except for equation (H.24), all of the transfer functions are first-order with a time constant T_1. This is due to the assumptions that we have perfect mixing of the secondary medium and that we have saturated steam on the primary side. We will deal with a different ideal case in the next section. The transfer functions between changes in steam temperature (or pressure) and the amount of condensate on the primary side shows that we have both a pole and a zero. The ratio between the time constants T_1 and T_2 can be found from equations (H.25) and (H.26):

$$T_2/T_1 = 1 + \frac{hA}{c_B \overline{w}_B} > 1 \qquad (H.36)$$

At high frequencies the amount of condensate follows the temperature of the steam and is inversely proportional to the heat of vaporization. The temperature of the secondary medium changes very little. At low frequencies, the amount of condensate is also proportional to the temperature of the steam, but the amount is lower because the temperature in the secondary medium also changes, thereby giving less heat transfer.

A Condenser with No Axial Mixing in the Secondary Medium

This situation occurs if the cooling (secondary) medium goes through the condenser only once, with no recirculation. In this case we assume there is no axial mixing, but that there is complete mixing in the radial direction. This situation has been studied by many, including Buckley (1964), Thal-Larsen (1960), and Hempel (1961), with somewhat different results. We will discuss Buckley's results, which were based on the work of Thal-Larsen (1960).

Assume that the temperature profile along the secondary side of the heat exchanger increases linearly along the tubes, and that the heat transfer coefficient is based on the mean temperature. Expressions for the steady-state situation can now be found and can be directly compared with those found in equations (H.20) and (H.21):

$$w_{A2} = \frac{hAc_B w_B(\theta_A - \theta_{B1})}{\lambda_A \left(\dfrac{hA}{2} + c_B w_B \right)} \qquad (H.37)$$

$$\theta_{B2} = \frac{hA\theta_A + \left(c_B w_B - \dfrac{hA}{2} \right)\theta_{B1}}{\dfrac{hA}{2} + c_B w_B} \qquad (H.38)$$

The dynamic relationships can be found by applying small perturbations around the steady-state condition. It is worthwhile to first find an expression for the temperature of the secondary medium (Thal-Larsen, 1960):

$$\frac{\Delta \theta_{B2}}{\Delta \theta_A}(s) \approx K_7 \frac{e^{-\tau_3 s}}{(1 + T_3 s)(1 + T_4 s)} \qquad (H.39)$$

$$\frac{\Delta \theta_{B2}}{\Delta w_B}(s) \approx K_8 \frac{e^{-\tau_3 s}}{(1 + T_3 s)(1 + T_4 s)} \qquad (H.40)$$

$$\frac{\Delta\theta_{B2}}{\Delta\theta_{B1}}(s) \approx K_9 \; e^{-\tau_4 s} \tag{H.41}$$

where:

$$\tau_3 = T_3 = \frac{W_B}{4\overline{w}_B} \tag{H.42}$$

$$\tau_4 = T_4 = \frac{W_B}{\overline{w}_B} \tag{H.43}$$

$$K_7 = \frac{2hA}{hA + 2c_B\overline{w}_B} \tag{H.44}$$

$$K_8 = \frac{4c_B A\left(\dfrac{\partial h}{\partial w_B}\overline{w}_B - \overline{h}\right)(\overline{\theta}_A - \overline{\theta}_{B1})}{(\overline{h}A + 2c_B\overline{w}_B)^2} \tag{H.45}$$

$$K_9 = -\frac{hA - 2c_B\overline{w}_B}{hA + 2c_B\overline{w}_B} \tag{H.46}$$

The expressions for the condensate flow can now be written as:

$$\frac{\Delta w_{A2}}{\Delta\theta_A}(s) = K_{10} \; \frac{1 + T_5 s}{(1 + T_3 s)(1 + T_4 s)} \; e^{-\tau_3 s} \tag{H.47}$$

$$\frac{\Delta w_{A2}}{\Delta w_B}(s) = \frac{1}{\lambda_A}\left(c_B\overline{w}_B(1 + T_5 s)\frac{\Delta\theta_{B2}}{\Delta w_B}(s) + (\overline{\theta}_{B2} - \overline{\theta}_{B1})c_B\right) \tag{H.48}$$

$$\frac{\Delta w_{A2}}{\Delta\theta_{B1}}(s) = \frac{c_B\overline{w}_B}{\lambda_A}\left((1 + T_5 s)\frac{\Delta\theta_{B2}}{\Delta\theta_{B1}}(s) - (1 - T_5 s)\right) \tag{H.49}$$

where:

$$T_5 = \frac{W_B}{2\overline{w}_B} \tag{H.50}$$

$$K_{10} = \frac{hAc_B\overline{w}_B}{\lambda_A\left(\dfrac{hA}{2} + c_B\overline{w}_B\right)} \tag{H.51}$$

In equations (H.48) and (H.49) one can use the expressions given by equations (H.40) and (H.41) or the expressions found by Hempel (1961):

$$\frac{\Delta\theta_{B2}}{\Delta\theta_{B1}}(s) = K_{11}e^{-\tau_4 s} \tag{H.52}$$

where:

$$K_{11} = e^{-\tau_4/T} \tag{H.53}$$

$$T = \frac{c_B W_B}{hA} \tag{H.54}$$

and:

$$\frac{\Delta \theta_{B2}}{\Delta w_B}(s) = \frac{K_{12}}{s}(1 - e^{-\tau_4 s}) \tag{H.55}$$

where:

$$K_{12} = \frac{\overline{w_B}}{W_B} K_8 \tag{H.56}$$

K_8 is given by equation (H.45).

REBOILERS

Only the type of reboiler known as a kettle reboiler will be discussed here. Figure H.4 shows the configuration and the variables of the unit. Analysis of this reboiler assumes:

- The process liquid is added to the reboiler at its boiling point, θ_B.
- The primary medium is a saturated vapor (such as steam), and the condensate is collected outside of the reboiler.
- There is no axial temperature gradient on either the liquid or the steam side.
- The heat transfer coefficient between the primary and secondary sides is not significantly affected by perturbations in flowrates or temperatures.

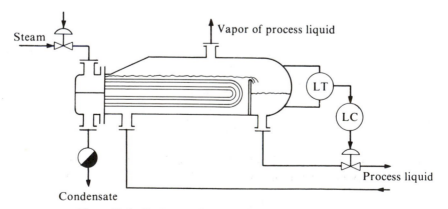

Figure H.4. Typical configuration for kettle type reboiler.

Process Dynamics

From the temperature–pressure diagram for the primary medium used in the reboiler, we find:

$$\Delta\theta_A(s) = (\partial\theta_A/\partial p_A)\Delta p_A(s) \tag{H.57}$$

$$\Delta p_A(s) = \frac{\overline{p_A}}{\rho_A V_A}\frac{1}{s}(\Delta w_{A1}(s) - \Delta w_{A2}(s)) \tag{H.58}$$

where:

$$\Delta w_{A1}(s) = \frac{\partial w}{\partial p_{A0}}\Delta p_{A0}(s) + \frac{\partial w}{\partial p_{A1}}\Delta p_{A1}(s) + \frac{\partial w}{\partial u}\Delta u \tag{H.59}$$

and:

$$\Delta w_{A2}(s) = \Delta Q(s)/\lambda_A \tag{H.60}$$

$$\Delta Q(s) = hA[\Delta\theta_A(s) - \Delta\theta_B(s)] \tag{H.61}$$

where the following new variables have been introduced in addition to those defined earlier:

ρ_A: density of the primary medium (kg/m^3)
p_{A0}: pressure in the primary medium upstream of the control valve
u: stem motion (manipulated variable) in the control valve
Q: heat released by condensation and transfer to the secondary medium (kJ/kg)
λ_A: heat of evaporation of the primary medium (kJ/kg)

Dynamics of the Secondary Side

$$\Delta\theta_B(s) = \frac{1}{\rho_B c_B V_B}\frac{1}{s}(\Delta Q(s) - \lambda_B \Delta w_{B2}(s)) \tag{H.62}$$

$$\Delta p_B(s) = (\partial p_B/\partial\theta_B)\Delta\theta_B(s) \tag{H.63}$$

$$\Delta w_{B2}(s) = (1/Z_B)\Delta p_B(s) \tag{H.64}$$

where the following new variables have been used:

ρ_B: density of the secondary medium (kg/m^3)
c_B: heat content of the secondary medium at temperature θ_B (kJ/kg °C)
V_{B1}: volume of secondary medium in the tank (m^3)
λ_B: heat of evaporation of secondary medium (kJ/kg)

$$Z_B \approx \frac{R_0}{1 + \dfrac{R_0 V_{B2}}{\bar{p}_2} s} : \text{acoustic impedance of steam phase of secondary medium}$$

R_0: equivalent flow resistance in the steam consumer (Ns/m²kg)
V_{B2}: volume of secondary side steam phase (m³)

An elementary block diagram of the system described by equations (H.57) through (H.64) is given in Figure H.5. The figure shows two integrators, which means that the system has two poles. The equation leading to the amount of steam on the secondary side (Δw_{B2}) has a zero. This means that at high frequencies the system will behave approximately the same as a first-order system. Reduction of the block diagram of Figure H.5 leads to:

$$\frac{\Delta w_{B2}}{\Delta u}(s) = K \frac{1 + T_3 s}{(1 + T_1 s)(1 + T_2 s)} \tag{H.65}$$

where:

$$T_1 \approx \frac{\rho_A V_A}{\bar{p}_A} \frac{1}{\dfrac{\partial w_A}{\partial p_{A1}} + \dfrac{\partial \theta_A}{\partial p_A} \dfrac{hA}{\lambda_A}} \tag{H.66}$$

$$T_2 \approx T_3 + \frac{\rho_B c_B V_{B1} R_0}{\dfrac{\partial p_B}{\partial \theta_B} \lambda_B} \tag{H.67}$$

$$T_3 \approx \frac{R_0 V_{B2}}{\bar{p}_B} \tag{H.68}$$

At high frequencies equation (H.65) becomes:

$$\frac{\Delta w_{B2}}{\Delta u}(s) \approx K \frac{1}{1 + T_4 s} \tag{H.69}$$

where:

$$T_4 = T_1 \frac{T_2}{T_3} = T_1 \left(1 + \frac{\rho_B c_B \bar{p}_B}{\dfrac{\partial p_B}{\partial \theta_B} \lambda_B}\right) \tag{H.70}$$

When a reboiler such as that shown in Figure H.4 is used as a stage in a more complex process (perhaps a distillation column), there must be either pressure control (p_A) or flow control (w_{A1}) of the steam added to the primary

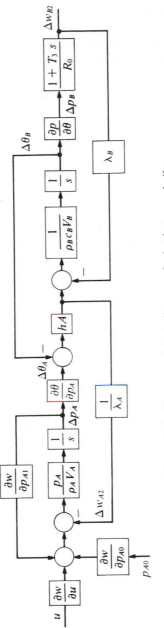

Figure H.5. Elementary block diagram of a kettle type reboiler.

side. Either of these control systems must be fast to keep the time constant T_1 as small as possible.

Regardless of whether the control valve is operated directly by u or by cascade control of the steam pressure or flow, the transfer function to the steam produced from the secondary medium (w_{B2}) will be of the form given in equation (H.65) and therefore presents no significant control problem.

Appendix I
DYNAMIC MODEL OF A BOILER

A great deal of effort has been directed toward development of mathematical models of steam boilers. Much of this work has been useful in the sense that it has provided explanations of many of the inner phenomena occurring in the boiler system. This, in turn, has been useful in the design, construction, and operation of steam boilers. Recently, it has made possible the use of dynamic models as components of control systems in which the models are used for state estimation combined with feedback control.

The following model development is based primarily on the work of Tyssø (1981), although we have changed the notation. The derivation is especially applicable to the case of ships' boilers, which have characteristically small drums in order to provide quick response and ship manueverability. This characteristic absolutely demands precise and rapid control.

The notation is as follows:

V: volume (m^3)
w: mass flow (kg/s)
H: enthalpy (kJ/kg)
ρ: density (kg/m^3)
X: steam quality
Q: heat flow (kJ/s $=$ kW)
λ: heat of vaporization (kJ/kg)

and the subscripts are:

w: water
s: saturated steam
d: drum
r: riser
do: downcomer
o: top of riser
e: economizer
m: mean

The meanings of the various symbols are shown in the schematic diagram of Figure I.1.

A mass balance on the water in the drum leads to the differential equation:

$$\frac{d}{dt}[V_{dw}] = [1/\rho_{dw}][(1 - X_o)w_o + w_w - w_{dow}] \qquad (I.1)$$

503

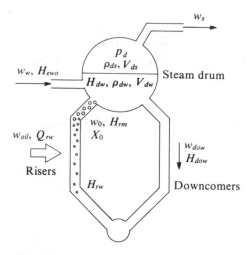

Figure I.1. Simplified diagram of steam–water portion of a boiler.

The mass flow of the steam condensing in the drum can be neglected, and we assume that:

$$\frac{d}{dt}(V_{dw}\,\rho_{dw}) \approx \rho_{dw}\frac{d}{dt}V_{dw}$$

A mass balance on the steam in the drum leads to a dynamic equation for the steam density in that:

$$\frac{d}{dt}(V_{ds}\,\rho_{ds}) = X_0 w_0 - w_s \tag{I.2}$$

However, the volume in the drum is constant, so:

$$\frac{d}{dt}V_{ds} = -\frac{d}{dt}V_{dw}$$

Combining this with equations (I.1) and (I.2) gives:

$$\frac{d}{dt}\rho_{ds} = \frac{1}{V_d - V_{dw}}\left(w_0 + w_w - w_{dow} - w_s + (\rho_{ds} - \rho_{dw})\frac{d}{dt}V_{dw}\right) \tag{I.3}$$

where we have eliminated X_0. After the density of the steam ρ_{ds} has been found, the pressure in the drum (p_d) can be found from the steam tables.

The energy balance on the water in the drum is given by:

$$\frac{d}{dt}(\rho_{dw}\,V_{dw}\,H_{dw}) = w_0(1 - X_0)H_{rw} - w_{dow}H_{dow} + w_w H_{ewo} \tag{I.4}$$

Combining this with equation (I.1), we get:

$$\rho_{dw} V_{dw} \frac{d}{dt} H_{dw} = w_0 (1 - X_0)(H_{rw} - H_{dw}) - w_w(H_{ewo} - H_{dw}) \qquad (I.5)$$

The density of water in the drum (ρ_{dw}) is a function of the drum pressure (p_d).

If we neglect the dynamics in the metal walls of the risers, the heat balance in the steam–water mixture in the risers is:

$$\frac{d}{dt}[\rho_{rm} H_{rm}]V_r = w_{dow} H_{dow} + Q_{rw} + w_0[H_{rw} + X_0 \lambda] \qquad (I.6)$$

Both H_{rw} and λ can be expressed as functions of drum pressure (p_d). The heat flow to the risers is assumed to be proportional to the fuel flow, so:

$$Q_{rw} = k_{rw} w_{oil} \qquad (I.7)$$

The enthalpy of the steam–water mixture can be expressed as a function of the steam quality:

$$H = H_{rw} + (H_{rs} - H_{rw})X \qquad (I.8)$$

H_{rs} is a function of the drum pressure, and if the steam quality is linear with respect to riser length, then:

$$H_{rm} = H_{rw} + (\lambda X_0)/2 \qquad (I.9)$$

We can now rewrite equation (I.6) as:

$$\frac{d}{dt} H_{rm} = \frac{1}{V_r \rho_{rm}}(Q_{rw} - w_{dow}(H_{rm} - H_{dow}) + w_0)\lambda \frac{X_0}{2} \qquad (I.10)$$

The density of the steam–water mixture is:

$$\rho_{rm} = \rho_{rw} - \alpha(\rho_{rw} - \rho_{rs}) \qquad (I.11)$$

where α is approximated by:

$$\alpha = \frac{X \rho_{rw}}{\rho_{rs} + X(\rho_{rw} - \rho_{rs})} \qquad (I.12)$$

The mass flow of water in the downcomers can be calculated from Bernoulli's equation, and a good approximation of this is:

$$w_{dow} = k_c \sqrt{(\rho_{dw} - \rho_{rm})} \qquad (I.13)$$

An approximate expression for the mass flow at the top of the risers is:

$$w_o = w_{dow} - \Delta w_o + k_1 w_{oil} + k_3 w_s \qquad (I.14)$$

where Δw_o is the transient part of w_o and is given by:

$$\frac{d}{dt}\Delta w_o = (1/T)[k_1 w_{\text{oil}} + k_3 w_s - \Delta w_o] \tag{I.15}$$

The quantity k_1 in the above equation is a load-dependent gain factor, T is a load-dependent time constant that is related to the shrink and swell factor of the steam bubbles in the system, and k_3 is a gain factor that is assumed to be constant.

Equations (I.1), (I.3), (I.5), (I.10), and (I.15) provide a set of nonlinear differential equations that can be summarized by the state space form:

$$\dot{\mathbf{x}} = \mathbf{f}(\mathbf{x}, \mathbf{u}, \mathbf{v}, t) \tag{I.16}$$

where the state variables are:

$$x_1 = V_{dw}$$
$$x_2 = H_{dw}$$
$$x_3 = \rho_{ds}$$
$$x_4 = H_{rm}$$
$$x_5 = \Delta w_o$$

The control variables are:

$$u_1 = w_{\text{oil}}$$
$$u_2 = w_w$$

The disturbances are:

$$v_1 = w_s$$
$$v_2 = T_{ewo}:\text{ the temperature of the feedwater from the economizer}$$

Two important process measurements are available, which can be modeled by the form:

$$\mathbf{y} = \mathbf{g}(\mathbf{x}, \mathbf{u}, \mathbf{v}, t) \tag{I.17}$$

where:

$$y_1 = p_d = \text{ the pressure in the drum}$$
$$y_2 = z = \text{ the level of water in the drum}$$

The drum water level is, approximately:

$$z = (V_{dm} - V_{dw})/A_{dm} \tag{I.18}$$

where:

V_{dm} = reference volume (at normal load)
A_{dm} = surface area of water at normal load

Tyssø (1980b) used an augmented (extended) Kalman filter to estimate the five states and four parameters (k_{rw}, V_r, k_c, and k_3) in a discrete version of the model given above. The other parameters in the model were determined from the boiler design data.

Starting with the nonlinear model, it is possible to linearize the model around different steady-state operating points (load demands, in this case). Tyssø (1981) presents the results of this approach. A linearized differential equation model can also be used to find system transfer function matrices similar to those presented by Chien et al. (1958), discussed earlier in section 5.2.

A complete nonlinear model using an augmented Kalman filter is a natural starting point for development of a multivariable, adaptive control system that can work over a wide range of operating and load (demand) conditions.

Appendix J
MATHEMATICAL MODEL OF EVAPORATORS

This development is based essentially on Newell and Fisher (1972) and Seborg and Fisher (1979). This work was developed for a single effect evaporator (Figure J.1) and a double effect evaporator (Figure J.2).

Referring to the diagram and nomenclature of Figure J.1, the mass and energy balances around a single-stage evaporator lead to the following differential equations:

Primary system (steam):

Vapor balance:

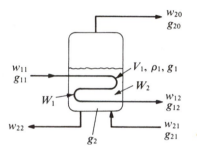

Figure J.1. Schematic diagram of a single effect evaporator.

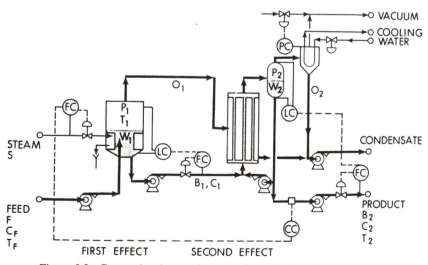

Figure J.2. Conventional control system for a double effect evaporator.

$$V_1(d\rho_1/dt) = w_{11} - w_{12} \tag{J.1}$$

Energy balance:

$$V_1\frac{d(\rho_1 g_1)}{dt} = w_{11}g_{11} - w_{12}g_{12} - Q_1 - Q_{01} \tag{J.2}$$

Heat transfer:

$$Q_1 = h_1 A(\theta_1 - \theta_v) = w_{12}\lambda_1 \tag{J.3}$$

Tube system (metal):

$$W_v c_v(d\theta_v/dt) = h_1 A(\theta_1 - \theta_v) - h_2 A(\theta_v - \theta_2) = Q_1 - Q_2 \tag{J.4}$$

Secondary side (liquid system):

$$dW_2/dt = w_{21} - w_{22} - w_{20} \tag{J.5}$$

$$d(W_2 c_2)/dt = w_{21}c_{21} - w_{22}c_2 \tag{J.6}$$

$$d(W_2 g_2)/dt = w_{21}g_{21} - w_{22}g_2 - w_{20}g_{20} + Q_2 - Q_{02} + L_2 \tag{J.7}$$

$$Q_2 = h_2 A(\theta_v - \theta_2) \tag{J.8}$$

The following notation has been used in the above expressions:

V_1: volume of primary side steam (m^3)
ρ_1: density of primary side steam (kg/m^3)
w_{11}: input flow of primary steam (kg/s)
w_{12}: outflow of primary steam condensate (kg/s)
w_{21}: input of primary thin liquor (kg/s)
w_{22}: outflow of secondary thick liquor (kg/s)
w_{20}: outflow of secondary steam (kg/s)
g_{11}: enthalpy of inlet primary steam (kJ/kg)
g_{12}: enthalpy of primary condensate (kJ/kg)
g_1: enthalpy of steam on primary side (kJ/kg)
g_{21}: enthalpy of incoming thin liquor (secondary side) (kJ/kg)
g_{20}: enthalpy of secondary steam (kJ/kg)
g_2: enthalpy of secondary thick liquor (kJ/kg)
Q_1: heat flow from primary side to metal tubes (kJ/s)
Q_{01}: heat loss from primary side (kJ/s)
Q_2: heat flow from metal to secondary side (kJ/s)
Q_{02}: heat loss from secondary side (kJ/s)
h_1: heat transfer coefficient between primary side and metal (kJ/s°C)
h_2: heat transfer coefficient between metal and secondary side (kJ/s°C)
λ_1: heat of evaporation of primary steam (kJ/kg)
W_v: mass of metal in the tubes (kg)

W_2:   mass of secondary liquid (kg)
c_v: specific heat of metal (kJ/kg°C)
c_{21}:   concentration of nonvolatile components in the secondary feed flow (kg/kg)
c_2:   concentration of nonvolatile components on the secondary side (kg/kg)
A: area of heat transfer surface of the tubes (m^2)
L_2:   heat of solution of the nonvolatile components on the secondary side (kJ/kg · s)

The six differential equations based on the mass and energy balances provide a very reasonable approximate model of the system. When single effect evaporators are connected in cascade, there will be a corresponding set of differential equations for each effect; so, for a double effect system, the model will be twelfth-order. Newell (1971) reduced this somewhat and developed a tenth-order model in which the state variables are as follows:

- First effect: enthalpy of primary steam, metal tube temperature, secondary liquid mass, secondary steam enthalpy, and concentration of the secondary liquid.
- Second effect: mass of the secondary liquid, enthalpy of the secondary steam, concentration of the secondary liquid, temperature of the metal tubes, and temperature in the cooling tubes.

The control and disturbance variables (combined) are: flow, concentration, and enthalpy of the feed to the first effect; flow and enthalpy of the steam to the first effect; liquid flow out of the first effect; liquid flow out of the second effect; and flow and temperature of the cooling water.

Using the tenth-order model as a starting point, fifth-, third-, and second-order models were developed, each having different degrees of approximation. Because it is logical to control the liquid levels in each effect by manipulating the flow of thick liquor out of each effect, the most important control variables and disturbances are: the steam flow, feed rate, feed concentration, and enthalpy of the feed to the first effect. The response quantities having particular interest are as follows: thick liquor concentration out of the second effect, liquid level (mass) in the first effect, liquid level (mass) in the second effect, flow of thick liquor out of the first effect, and flow of thick liquor out of the second effect. Typical response curves for these variables to a 20% step change in steam flow to the first effect are shown in Figure J.3a. The experimental response is compared with those of fifth-order nonlinear and linear models.

We see in Figure J.3a that the response of the concentration of thick liquor out of the first effect is very close to that of a first-order transfer function with a time constant (T_1). The responses of the liquid masses (appearing as levels) both dip down somewhat before they begin to rise toward their final values. This happens because the outlet flows of thick liquor are controlled as described in

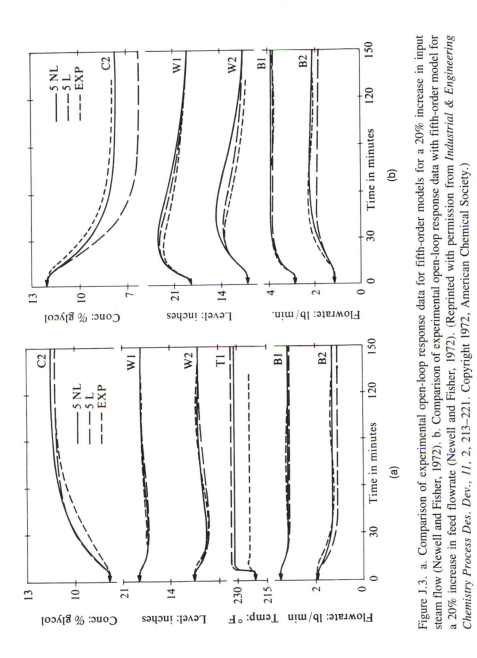

Figure J.3. a. Comparison of experimental open-loop response data for fifth-order models for a 20% increase in input steam flow (Newell and Fisher, 1972). b. Comparison of experimental open-loop response data with fifth-order model for a 20% increase in feed flowrate (Newell and Fisher, 1972). (Reprinted with permission from *Industrial & Engineering Chemistry Process Des. Dev., 11,* 2, 213–221. Copyright 1972, American Chemical Society.)

the discussion given below for the model given by equation (J.9). If these levels had not been adjusted downward as shown, the responses would have been nearly those of pure integrators. The controllers used here were PI types with relatively long integral times (as seen from the tails of the two level response curves, in this case about 100 minutes for each). If these level controllers had been very "tight" (low bandwidth), the system would have shown nearly instantaneous responses of the thick liquor flows (very short time constants).

Figure J.3b gives corresponding response curves to an increase in the feed rate of thin liquor to the first effect. The response of thick liquor concentration follows a trajectory that is nearly first-order, while the two levels respond as integrators with feedback. Both thick liquor flows respond quickly.

Newell and Fisher (1972) developed the following fifth-order model for a laboratory model double effect evaporator:

$$\dot{x} = Ax + Bu + Cv$$

$$y = Dx$$

(J.9)

where:

$$x = \begin{bmatrix} \text{mass of liquid in first effect (kg)} \\ \text{concentration of liquid in first effect} \\ \text{enthalpy of liquid in first effect (kJ/kg)} \\ \text{mass of liquid in second effect (kg)} \\ \text{concentration of liquid in second effect} \end{bmatrix}$$

$$u = \begin{bmatrix} \text{steam flow to first effect (kg/s)} \\ \text{liquid flow out of first effect (kg/s)} \\ \text{liquid flow out of second effect (kg/s)} \end{bmatrix}$$

$$v = \begin{bmatrix} \text{feed rate to first effect (kg/s)} \\ \text{concentration of feed} \\ \text{enthalpy of feed (kJ/kg)} \end{bmatrix}$$

$$A = \begin{bmatrix} 0 & -0.00156 & -0.1711 & 0 & 0 \\ 0 & -0.1419 & 0.1711 & 0 & 0 \\ 0 & -0.00875 & -1.102 & 0 & 0 \\ 0 & -0.00128 & -0.1489 & 0 & 0.00013 \\ 0 & 0.0605 & 0.1489 & 0 & -0.0591 \end{bmatrix}$$

$$B = \begin{bmatrix} 0 & -0.143 & 0 \\ 0 & 0 & 0 \\ 0.392 & 0 & 0 \\ 0 & 0.108 & -0.0592 \\ 0 & -0.0486 & 0 \end{bmatrix}$$

$$C = \begin{bmatrix} 0.2174 & 0 & 0 \\ -0.074 & 0.1434 & 0 \\ -0.036 & 0 & 0.1814 \\ 0 & 0 & 0 \\ 0 & 0 & 0 \end{bmatrix}$$

$$D = \begin{bmatrix} 1 & 0 & 0 & 0 & 0 \\ 0 & 0 & 0 & 1 & 0 \\ 0 & 0 & 0 & 0 & 1 \end{bmatrix}$$

We can use the above matrices to develop the transfer function matrices for the system:

$$\mathbf{y}(s) = H_u(s)\mathbf{u}(s) + H_v(s)\mathbf{v}(s) \tag{J.10}$$

where:

$$H_u(s) = \begin{bmatrix} \dfrac{-1.67 \cdot 10^{-2}(1 + 6.35s)}{s(1 + 0.45s)(1 + 3.49s)} & \dfrac{-3.55 \cdot 10^{-2}(1 + 0.91s)(1 + 7.07s)}{s(1 + 0.45s)(1 + 3.49s)} & 0 \\[3mm] \dfrac{-1.35 \cdot 10^{-2}(1 + 6.97s)}{s(1 + 0.45s)(1 + 3.49s)} & \dfrac{2.7 \cdot 10^{-2}(1 + 6.97s)(1 + 9.09s)}{s(1 + 0.45s)(1 + 3.49s)} & \dfrac{-3.26 \cdot 10^{-4}(1 + 6.97s)(1 + 0.91s)}{s(1 + 16.92s)(1 + 0.45s)(1 + 3.49s)} \\[3mm] \dfrac{0.113(1 + 13.81s)}{(1 + 16.92s)(1 + 0.45s)(1 + 3.49s)} & \dfrac{-0.206(1 + 0.91s)(1 + 6.97s)}{(1 + 16.92s)(1 + 0.45s)(1 + 3.49s)} & 0 \end{bmatrix}$$

Appendix K
MATHEMATICAL MODEL OF A
CRYSTALLIZATION PROCESS

Although there are many models of crystallization processes in the literature, we will base our discussion on the work of Sherwin, Shinnar, and Katz (1967), Han (1969), and Randolph and Larsson (1971).

There are generally five state variables that characterize a crystallization process (see section 3.10):

x_1: degree of supersaturation (kg/m^3)
x_2: number of particles per unit volume in the crystallizer ($1/m^3$)
x_3: average length of particles per unit volume in the crystallizer (m/m^3)
x_4: average surface area of particles per unit volume (m^2/m^3)
x_5: volume of particles per unit volume (m^3/m^3)

The rate of nucleation is a common factor used in the characterization of a crystallization process. It can be expressed as:

$$k_1 \text{ [number of particles/}m^3 \cdot s]$$

Usually, it is assumed to be a function of the degree of supersaturation, $k_1(x_1)$. Another important factor is the linear growth rate of the crystals, which is also a function of the supersaturation, $k_2(x_1)$ (m/s).

The following mathematical model of the crystallization process can be formulated:

$$\dot{x}_1 = \frac{q}{v} \frac{c_1 - x_1}{1 - ax_5} \varepsilon_0 - (\rho - x_1)\left(\frac{3k_2(x_1)ax_4}{1 - ax_5} + k_1(x_1)ar_0^3\right) \quad \text{(K.1)}$$

$$\dot{x}_2 = (1 - ax_5)k_1(x_1) - (q/V)(x_2 - c_2) \quad \text{(K.2)}$$

$$\dot{x}_3 = k_2(x_1)x_2 + (1 - ax_5)k_1(x_1)r_0 - (q/V)(x_3 - c_3) \quad \text{(K.3)}$$

$$\dot{x}_4 = 2k_2(x_1)x_3 + (1 - ax_5)k_1(x_1)r_0^2 - (q/v)(x_4 - c_4) \quad \text{(K.4)}$$

$$\dot{x}_5 = 3k_2(x_1) + (1 - ax_5)k_1(x_1)r_0^3 - (q/V)(x_5 - c_5) \quad \text{(K.5)}$$

where the notation used is:

c_1: concentration of crystallizing material in feed (kg/m^3)
c_2: number of particles in feed, per unit volume ($1/m^3$)
c_3: average length of particles in feed, per unit volume (m/m^3)
c_4: particle surface area in feed, per unit volume (m^2/m^3)
c_5: particle volume in feed, per unit volume (m^3/m^3)

a: formation factor for crystals
ε_0: volumetric filling factor in feed
ρ: density of crystals (kg/m^3)
r_0: size of nucleii (m)
q: volumetric flow of feed (m^3/s)

The growth coefficients k_1 and k_2 can be well described by these approximations:

$$k_1(x_1) = k_{10} \quad \exp[-k_3/(\ln(x_1/c_{1s}))^2] \tag{K.6}$$

$$k_2(x_1) = k_4(x_1 - c_{1s}) \tag{K.7}$$

Figure K.1 illustrates the dependence of the growth functions on the level of supersaturation. Clearly, when the supersaturation is low, the crystal growth rate dominates the nucleation rate. However, at higher supersaturation, the nucleation rate increases very rapidly. When this rate gets too high, there will be little crystallization; so it is important to maintain a proper balance between the two rates. When there are too many nucleii in the crystallizer, the product will consist mostly of very small crystals, with very few large ones. Usually, a desirable product must have a relatively even crystal size distribution, although this depends somewhat on the material being crystallized. In situations where it is difficult to control the rate of nucleation, it may be necessary to add a stage to the process for the purpose of removing the nucleii or to add nucleii when needed. When this is done, it presents a new control variable.

The growth and nucleation rates are both functions of temperature in the sense that the degree of supersaturation is temperature-dependent. The model

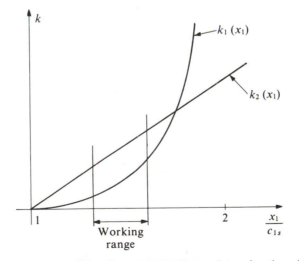

Figure K.1. Nucleation rate (k_1) and crystal growth rate (k_2) as function of supersaturation.

given by equations (K.1) through (K.5) assumes temperature to be constant. Because the temperature is also subject to control (frequently via the pressure in the crystallizer), it is actually a state variable (x_6) that affects the equation for $\dot{x}_1$ and the expressions defining k_1 and k_2. However, temperature is omitted from the model because it is easy to control at a constant value.

Equations (K.3), (K.4), and (K.5) have much in common, and they would have been linearly dependent if the factors in front of the two first terms of each equation were not slightly different. Because of the similarity of the factors, one can say that the degree of controllability for the last three states is relatively low. The degree of controllability is zero when the equations are in fact linearly dependent.

The differential equations given above can be simplified considerably if the feed to the system has no solid material ($\varepsilon_0 = 1$, $c_2 = 0$, $c_3 = 0$, and $c_4 = 0$), and if it can be assumed that the nucleii are so small relative to the average crystal size that they can be neglected (i.e., $r_0 = 0$). Then we have:

$$\dot{x}_1 = \frac{q}{V} \frac{c_1 - x_1}{1 - ax_5} - \frac{3(\rho - x_1)k_2(x_1)ax_4}{1 - ax_5} \tag{K.8}$$

$$\dot{x}_2 = (1 - ax_5)k_1(x_1) - (q/V)x_2 \tag{K.9}$$

$$\dot{x}_3 = k_2(x_1)x_2 - (q/V)x_3 \tag{K.10}$$

$$\dot{x}_4 = 2k_2(x_1)x_3 - (q/V)x_4 \tag{K.11}$$

$$\dot{x}_5 = 3k_2(x_1)x_4 - (q/V)x_5 \tag{K.12}$$

The simplified model is shown in the block diagram of Figure K.2. Since we assumed the temperature to be constant, only the feed rate (q) is available as a control variable. The concentration of the feed stream (c_1) is usually considered to be a disturbance. We see from the diagram that the feed rate (q) affects, in parallel, all five state variables in a multiplicative manner, but because the variations in feed rate take place around a relatively large value, this effect can be approximated as an additive rather than multiplicative influence. The effect of concentration variations (c_1) appears as a disturbance, but only early in the process through x_1.

This process ought to lend itself remarkably well to a control strategy based on state estimation through a model of the type shown in Figure K.2, but apparently this has not been done in practice. However, a scheme has been proposed that uses feedforward of the feed concentration c_1 to the feed flowrate q based on a linearized process model giving the degree of supersaturation x_1 (Han, 1969). These questions were discussed in section 5.4.

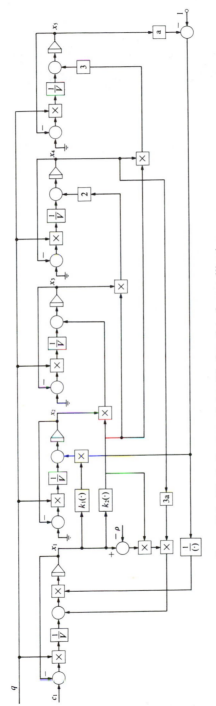

Figure K.2. Block diagram of simplified model of crystallization process.

Appendix L
MATHEMATICAL MODEL OF A
ROTARY DRUM DRYER

Using the background developed in section 3.11 and the basic principles reviewed in Yliniemi, Jutilia, and Uronen (1981), we will now derive a dynamic model for a rotary drum dryer. This model will use the following relatively ideal assumptions:

- Heat and mass transfer coefficients are constant.
- Heat transfer by conduction in the solid and gas phases can be neglected.
- Diffusion of water vapor in the axial direction can be neglected.
- Heat transfer by radiation can be neglected.
- Axial velocity in the gas phase is constant.
- The solid is homogeneous.
- There are no chemical reactions during drying.
- Temperature and moisture of the solid and gas phases are functions only of time and the axial coordinate.

The following notation will be used:

w: moisture in solid phase (kg/kg)
θ_s: temperature in solid phase (°C)
θ_g: temperature in gas phase (°C)
v_s: linear velocity of solid phase (m/s)
v_g: linear velocity of gas phase (m/s)
c_s: specific heat of solid phase (kJ/kg°C)
c_g: specific heat of gas phase (kJ/kg°C)
G_s: specific mass of solid phase (kg/m)
G_g: specific mass of gas phase (kg/m)
V: specific dryer volume (m³/m)
h: specific heat transfer coefficient (kJ/sm³°C)
λ: heat of vaporization of water (kJ/kg)
R: drying rate
z: axial coordinate (m)

We can now formulate three partial differential equations:

$$\frac{\partial w}{\partial t} + v_s \frac{\partial w}{\partial z} = -R \tag{L.1}$$

$$\frac{\partial (c_s \theta_s)}{\partial t} + v_s \frac{\partial (c_s \theta_s)}{\partial z} = \frac{hV}{G_s} (\theta_g - \theta_s) - \lambda R \tag{L.2}$$

$$\frac{\partial(c_g \theta_g)}{\partial t} + v_g \frac{\partial(c_g \theta_g)}{\partial z} = -\frac{hV}{G_g}(\theta_g - \theta_s) - \lambda \frac{G_s}{G_g} R \qquad (L.3)$$

The three state variables are functions of time and the axial coordinate: $w(t, z)$, $\theta_s(t, z)$, and $\theta_g(t, z)$.

Let us divide the dryer into N elements of thickness Δz in the axial direction, and discretize the equations in time by using the interval Δt such that $t = k\Delta t$. If we number the three state variables in each element with an index increasing from 1 to N from the input to the output position of the solid, the original partial differential equations become the following set of discrete equations:

$$w_1(k + 1) = a_1 w_1(k) + b_1 w_0(k) - c_1 \theta_{s1}(k)$$
$$w_2(k + 1) = a_1 w_2(k) + b_1 w_1(k) - c_1 \theta_{s2}(k)$$
$$\vdots \qquad\qquad \vdots \qquad\qquad \vdots$$
$$w_N(k + 1) = a_1 w_N(k) + b_1 w_{N-1}(k) - c_1 \theta_{sN}(k)$$
$$\theta_{s1}(k + 1) = a_2 \theta_{s1}(k) + b_2 \theta_{s0}(k) + c_2 \theta_{g1}(k)$$
$$\theta_{s2}(k + 1) = a_2 \theta_{s2}(k) + b_2 \theta_{s1}(k) + c_2 \theta_{g2}(k)$$
$$\vdots \qquad\qquad \vdots \qquad\qquad \vdots \qquad\qquad (L.4)$$
$$\theta_{sN}(k + 1) = a_2 \theta_{sN}(k) + b_2 \theta_{s,N+1}(k) + c_2 \theta_{gN}(k)$$
$$\theta_{g1}(k + 1) = a_3 \theta_{g1}(k) - b_3 \theta_{g2}(k) + c_3 \theta_{s1}(k)$$
$$\theta_{g2}(k + 1) = a_3 \theta_{g2}(k) - b_3 \theta_{g3}(k) + c_3 \theta_{s2}(k)$$
$$\vdots \qquad\qquad \vdots \qquad\qquad \vdots$$
$$\theta_{gN}(k + 1) = a_3 \theta_{gN}(k) - b_3 \theta_{g,N+1}(k) + c_3 \theta_{sN}(k)$$

where:

$$a_1 = 1 - \frac{v_s \Delta t}{\Delta z}, \qquad b_1 = \frac{v_s \Delta t}{\Delta z}, \qquad c_1 = k_1 \Delta t$$

$$a_2 = 1 - \frac{v_s \Delta t}{\Delta z} - \frac{hV \Delta t}{c_s G_s} - \frac{\lambda k_1 \Delta t}{c_s}, \qquad b_2 = \frac{v_s \Delta t}{\Delta z}, \qquad c_2 = \frac{hV \Delta t}{c_s G_s}$$

$$a_3 = 1 + \frac{v_g \Delta t}{\Delta z} - \frac{hV \Delta t}{c_g G_g}, \qquad b_3 = \frac{v_g \Delta t}{\Delta z}, \qquad c_3 = \left(\frac{hV}{c_g G_g} - \lambda \frac{G_s}{G_g} k_1\right) \Delta t$$

We have assumed that the drying rate in any arbitrary element is given by:

$$R_n = k_1 \quad \theta_{si}(k)$$

where k_1 is constant. The shape of the drying rate curve is given by Figure 3.11.4.

By ordering the elements of w, θ_s, and θ_g as vectors, equation (L.4) can be written as follows:

$$\begin{bmatrix} \mathbf{w} \\ \boldsymbol{\theta}_s \\ \boldsymbol{\theta}_g \end{bmatrix}(k+1) = \Phi \begin{bmatrix} \mathbf{w} \\ \boldsymbol{\theta}_s \\ \boldsymbol{\theta}_g \end{bmatrix}(k) + \Delta \begin{bmatrix} \mathbf{w}_0 \\ \boldsymbol{\theta}_{s0} \\ \boldsymbol{\theta}_{g,N+1} \end{bmatrix}(k) \tag{L.5}$$

For the case where $N = 3$, we have:

$$\Phi = \begin{bmatrix} a_1 & 0 & 0 & -c_1 & 0 & 0 & 0 & 0 & 0 \\ b_1 & a_1 & 0 & 0 & -c_1 & 0 & 0 & 0 & 0 \\ 0 & b_1 & a_1 & 0 & 0 & c_1 & 0 & 0 & 0 \\ 0 & 0 & 0 & a_2 & 0 & 0 & c_2 & 0 & 0 \\ 0 & 0 & 0 & b_2 & a_2 & 0 & 0 & c_2 & 0 \\ 0 & 0 & 0 & 0 & b_2 & a_2 & 0 & 0 & c_2 \\ 0 & 0 & 0 & c_3 & 0 & 0 & a_3 & -b_3 & 0 \\ 0 & 0 & 0 & 0 & c_3 & 0 & 0 & a_3 & -b_3 \\ 0 & 0 & 0 & 0 & 0 & c_3 & 0 & 0 & a_3 \end{bmatrix}$$

$$\Delta = \begin{bmatrix} b_1 & 0 & 0 \\ 0 & 0 & 0 \\ 0 & 0 & 0 \\ 0 & b_2 & 0 \\ 0 & 0 & 0 \\ 0 & 0 & 0 \\ 0 & 0 & 0 \\ 0 & 0 & 0 \\ 0 & 0 & -b_3 \end{bmatrix}$$

We see from equation (L.5) that $w_0(k)$ and $\theta_{s0}(k)$ (the moisture and temperature of the solid at the inlet to the dryer) will normally be system disturbances. $\theta_{g,N+1}(k)$ is the temperature of the gas phase at the inlet, but because the flows of solid and gas are countercurrent, this is at the opposite end of the dryer from the solid phase inlet location. The gas phase inlet temperature is a control variable.

References

Albright, L. F. (1974). *Processes for Major Addition-Type Plastics and Their Monomers*. McGraw-Hill Book Co., New York.

Alevisakis, G., and Seborg, D. E. (1974). Control of multivariable systems containing time delays using a multivariable Smith predictor. *Chem. Eng. Sci., 29*, 373–380.

Al-Shaikh, A. M. (1978). Advances in papermachine control. ATCP-XVIII, Albuquerque, New Mexico.

Al-Shaikh, A. M., and Tu, F. (1978). *Multi-level Control of Continuous Digesters—Applications & Results*. IFAC VII World Congress, Helsinki.

Anderson, B. D. O., and Moore, J. D. (1971). *Linear Optimal Control*. Prentice-Hall, Englewood-Cliffs, N.J.

Arkun, Y., Manousiouthakis, B., and Palazoglu, A. (1984). Robustness analysis of process control systems. A case study of decoupling control in distillation. *Ind. Eng. Chem. Process Des. Dev. 23,* 1, 93–101.

Asbjørnsen, O. A., and Fjeld, M. (1970). Response modes of continuous stirred tank reactors. *Chem. Eng. Sci., 25,* 1627–1636.

Athans, M., and Falb, P. (1966). *Optimal Control*. McGraw-Hill Book Co., New York.

Atherton, D. P. (1982). *Nonlinear Control Engineering*. Van Nostrand Reinhold Co., New York.

Balchen, J. G. (1967). *Reguleringsteknikk*. Bd. II. Tapir Forlag, Trondheim, Norway.

Balchen, J. G. (1979a). Stokastiske systemer. Estimeringsteori. Forelesningsnotater, Institutt for teknisk kybernetikk, Norges Tekniske Høgskole, Trondheim, Norway.

Balchen, J. G. (1979b). The applicability of modern control theory related to dynamic optimization in industry today. In *On-line Optimization Techniques in Industrial Control,* E. J. Kompass and T. J. Williams, eds., Control Engineering.

Balchen, J. G. (1980). Dynamic optimization in the process industry. 5th Int. Congr. in Scandanavia on Chem. Eng. Rapport 80-47-W, Institutt for teknisk kybernetikk, Norges Tekniske Høgskole, Trondheim, Norway.

Balchen, J. G. (1984a). *Reguleringsteknikk,* Bd. I, Tapir Forlag, Trondheim, Norway.

Balchen, J. G. (1984b). A quasi-dynamic optimal control strategy for non-linear multivariable processes based upon non-quadratic objective functionals. *Modeling, Identification, and Control, 5,* 4, 195–209.

Balchen, J. G., and Aune, A. B. (1966). Semidynamic optimal control. In *Theory of Self-Adaptive Control Systems,* P. H. Hammond, ed., 223–233, Plenum Press, New York.

Balchen, J. G., Endresen, T., Fjeld, M., and Olsen, T. O. (1973). Multivariable PID estimation and control with biased disturbances. *Automatica, 9,* 295–307.

Balchen, J. G., Fjeld, M., and Olsen, T. O. (1971). Multivariable control with approximate state estimation of a chemical tubular reactor. IFAC Symposium on Multivariable Systems, Dusseldorf, Oct.

Balchen, J. G., Fjeld, M., and Solheim, O. A. (1970, 1978). *Reguleringsteknikk,* Bd. III, Tapir Forlag, Trondheim, Norway.

Barnes, F. J., and King, C. J. (1974). Synthesis of cascade refrigeration and liquefaction systems. *Ind. Eng. Chem. Process Des. Dev., 13,* 4, 421–433.

Baxley, R. A., Jr. (1969). Local optimizing control for distillation. *Instrumentation Technology* (Oct.), 75–80.

Bell, R. D., Reese, N. W., and Lee, K. B. (1977). Models of a large boiler–turbine plant. IFAC Symposium, Melbourne, Feb. 21–25.

Bierman, G. J. (1977). *Factorization Methods for Discrete Sequential Estimation.* Academic Press, New York.

Bland, W. F., and Davidson, R. L. (1967). *Petroleum Processing Handbook*. McGraw-Hill Book Co., New York.

Borisson, U., and Wittenmark, B. (1974). An industrial application of a self-tuning regulator. 4th IFAC/IFIP Int. Conf. on Dig. Comp. Appl. to Proc. Control, Zurich.

Bristol, E. H. (1966). On a new measure of interaction for multivariable process control. *IEEE Trans. Automatic Control* (Jan.).

Bristol, E. H. (1980). After DDC: Idiomatic control. *Chem. Engr. Prog.* (Nov.), 84–89.

Britt, K. W. (1970). *Handbook of Pulp and Paper Technology*. Van Nostrand Reinhold Co., New York.

Bryson, A. E., and Ho, Y. C. (1969). *Applied Optimal Control*. Blaisdell Pub. Co., Waltham, Mass.

Buchholt, F., and Kummel, M. (1979). Self-tuning control of a pH-neutralization process. *Automatica, 15,* 656–671.

Buckley, P. S. (1964). *Techniques of Process Control*. John Wiley & Sons, New York.

Buckley, P. S. (1982). Control of a heat-integrated distillation column. Chemical Process Control 2, Proc. Eng. Found. Conf., Sea Island, Ga. (347–359).

Buckley, P. S., Luyben, W. L., and Shunta, J. P. (1985). *Design of Distillation Column Control Systems*. Instrument Society of America, Triangle Research Park, N.C.

Bunz, D. and K. Gutschow (1985). CATPAC — An Interactive Software Package for Control System Design. *Automatica, 21:* 2.

Campbell, D. P. (1958). *Process Dynamics*. John Wiley & Sons, New York.

Casey, J. P. (1980). *Pulp and Paper*. Vols. I, II, and III (3rd ed.). John Wiley & Sons, New York.

Cegrell, T., and Hedqvist, T. (1973). Successful adaptive control of papermachines. 3rd IFAC Symp. on Ident. and Sys. Par. Est., The Hague.

Chari, N. C. S. (1973). Integrated control system approach for batch digester control. *Tappi 56,* 7, 65–68.

Chien, K. L., Ergin, E. I., Ling, C., and Lee, A. (1958). The noninteracting controller for a steam-generating system. *Control Engineering, 5* (Oct.), 95–101.

Chilton, Th. H. (1968). *Strong Water. Nitric Acid: Sources, Methods of Manufacture, and Uses.* The MIT Press, Cambridge, Mass.

Clarke, D. W., and Gawthrop, P. J. (1981). Implementation and application of microprocessor based self-tuners. *Automatica, 17,* 233–244.

Cutler, C. R., and Ramaker, B. L. (1980). Dynamic matrix control — A computer control algorithm. Joint Aut. Cont. Conf., 1980, WP5-B.

Dahlqvist, S. A. (1981). Control of a distillation column using self-tuning regulators. *Can. J. Chem. Eng., 59,* 118–127.

Danig, P. (1963). Liquid feed regulation by thermostatic expansion valves. *J. Refrigeration* (May–June), 52–55.

Deshpande, P. B. (1985). *Distillation Dynamics and Control*. Instrument Society of America, Research Triangle Park, N.C.

Deshpande, P. B., and Ashe, R. H. (1981). *Elements of Computer Process Control*. Instrument Society of America, Research Triangle Park, N. C.

Digital Equipment Corp. (1982). Intro. to Local Area Networks. Order no. EB-22714-18, Maynard, Mass.

Douglas, J. M. (1972). *Process Dynamics and Control*. Prentice-Hall, Englewood Cliffs, N.J.

Douglas, J. M., Jafarey, A., and Seeman, R. (1979). Short-cut techniques for distillation column design and control. Part 2: Column operability and control. *Ind. Eng. Chem. Process Des. Dev., 18,* 2, 203–210.

Doyle, J. C., and Stein, G. (1981). Multivariable feedback design: Concepts for a classical/modern synthesis. *IEEE Trans. Automatic Control, 26,* 1.

Driskell, L. (1974). Control valve sizing with ISA formulas. *Instrumentation Technology* (July), 33–48.

Edwards, J. B., and Jassim, H. J. (1977). An analytical study of the dynamics of binary distillation columns. *Trans. Inst. Chem. Engrs.*, *55*, 17–28.

Eklund, K. (1971). Linear drum boiler–turbine models. Report 7117, Inst. för regleringsteknik, Lund Tekniska Högskola.

Eykhoff, P. (1974). *System Parameter and State Estimation*. John Wiley & Sons, London.

Ferson, L. M. (ed.) (1982). *ISA Control Valve Standards*. Instrument Society of America, Research Triangle Park, N.C.

Fisher, D. G., and Seborg, D. E. (1976). *Multivariable Computer Control*. North Holland Pub. Co., New York.

Fjeld, M. (1978). Application of modern control concepts on a kraft papermachine. *Automatica*, *14*, 107–117.

Fjeld, M., and Grimnes, K. (1974). DAPP—a Norwegian co-operation project on multivariable control of a kraft papermachine. 4th IFAC/IFIP Int. Conf. on Dig. Comp. Appln. to Proc. Control, Zurich.

Fjeld, M., and Wilhelm, R. G., Jr. (1981). Self-tuning regulators: The software way. *Control Engineering* (Oct.), 99–102.

Foss, A. S. (1973). Critique of chemical process control theory. *AICHE.*, *19*, 2, 209–214.

Foss, A. S., and Denn, M. (1976). *Chemical Process Control*, AIChE Symposium Series No. 159, Vol. 72. American Institute of Chemical Engineers, New York.

Foss, A. S., Edmunds, J. M., and Kouvaitakis, B. (1980). Multivariable control systems for two-bed reactors by the characteristic locus method. *Ind. Eng. Chem. Fund.*, *19*, 1, 109–117.

Foster, C. C. (1981) *Real Time Programming*. Addison-Wesley, Inc., Reading Mass.

Foust, A. S., Wenzel, L. A., Clump, C. W., Mause, L., and Andersen, L. B. (1980). *Principles of Unit Operations*. John Wiley & Sons, New York.

Franklin, G. F., Powell, J. D., and Emami-Naeini, A. (1986). *Feedback Control of Dynamic Systems*. Addison-Wesley Pub., Reading, Mass.

Franks, R. G. E. (1972). *Modelling and Simulation in Chemical Engineering*. Wiley-Interscience, New York.

Frew, J. A. (1973). Optimal control of batch raw sugar crystallization. *Ind. Eng. Chem. Process Des. Dev.*, *12*, 4, 460–467.

Froment, G. F., Van de Steene, B. O., Van Damme, P. S., Narayanan, S., and Goosens, A. G. (1976). Thermal cracking of ethane and ethane/propane mixtures. *Ind. Eng. Chem. Process Des. Dev.*, *15*, 4, 495–504.

Gagnepain, J. P., and Seborg, D. E. (1982). Analysis of process interactions with application to multiloop control systems design. *Ind. Eng. Chem. Process Des. Dev.*, *21*, 1, 5–11.

Gault, G. A., Howarth, W. J., Lynch, A. J., and Whiten, W. J. (1979). Automatic control of flotation circuits: Theory and operation. *World Mining* (Dec.), 54–57.

Geankopolis, C. J. (1983). *Transport Processes and Unit Operations, 2nd ed.* Allyn and Bacon, Newton, Mass.

Gibson, J. E. (1963). *Non-Linear Automatic Control*. McGraw-Hill, New York.

Gilles, E. D., and Retzbach, B. (1980). Reduced models and control of distillation columns with sharp temperature profiles. 19th CDC–IEEE Conf., Albuquerque, N. Mex., 865–870.

Goodwin, G. C., and Payne, R. L. (1977). *Dynamic System Indentification: Experimental Design and Data Analysis*. Academic Press, New York.

Gould, L. A. (1969). *Chemical Process Control. Theory and Applications*. Addison-Wesley Pub. Co., London.

Gould, L. A., Evans, L. B., and Kurihara, H. (1970). Optimal control of fluid catalytic cracking processes. *Automatica*, *6*, 695–703.

Govind, R., and Powers, G. J. (1978). Synthesis of process control systems. *IEEE Trans. Systems, Man, and Cybernetics*, SMC-8, 11, 792–795.

Gupta, G., and Timm, D. C. (1971). Predictive/corrective control for continuous crystallization. *Chem. Eng. Prog. Symp. Series No. 110*, *67*, 121–128.

Gustavsson, I. (1972). Comparison of different methods for identification of industrial processes. *Automatica, 8,* 127–142.

Gustavsson, I., Ljung, L., and Søderstrom, T. (1976). Identification of processes in closed loop. 4th IFAC Symp. on Ident., Tbilisi, USSR.

Haglund, S., Wallin, G., and Karlsson, T. (1980). A new digester control method for the manufacture of sulphite viscose pulps. TAPPI Conf. on Dissolving Pulps, Vienna, Oct. 1, 1980 (31–36).

Han, C. D. (1969). A control study of isothermal mixed crystallizers. *Ind. Eng. Chem. Process Des. Dev., 8,* 150–158.

Hempel, A. (1961). On the dynamics of steam–liquid heat exchangers. *Trans. ASME, J. Basic Eng.* (June), 244–252.

Henley, E. J., and Seader, J. D. (1981). *Equilibrium-Stage Separation Operations in Chemical Engineering.* John Wiley & Sons, New York.

Henriksen, R., Kaggerud, I., and Olsen, T. O. (1982). Application of online estimation to a rougher flotation process. 6th IFAC Symp. on Ident. and Syst. Parameter Estimation, Washington, D.C.

Himmelblau, D. M. (1970). *Process Analysis by Statistical Methods.* John Wiley & Sons, New York.

Himmelblau, D. M., and Bischoff, K. B. (1968). *Process Analysis and Simulation. Deterministic Systems.* John Wiley & Sons, New York.

Irving, M. R., Boland, F. M., and Nicholson, H. (1979). Optimal control of argon–oxygen decarburising steelmaking process. *Proc. IEE, 127,* 2, 198–202.

Isermann, R. (1982). Parameter adaptive control algorithms — a tutorial. *Automatica, 18,* 5, 513–528.

Jafarey, A., Douglas, J. M., and McAvoy, T. J. (1979). Short-cut techniques for distillation column design and control. Part 1. *Ind. Eng. Chem. Process Des. Dev., 18,* 2, 197–202.

James, H. M., Nichols, N. B., and Phillips, R. S. (1947). *Theory of Servomechanisms.* McGraw-Hill Book Co., New York.

Kaggerud, I. (1983). Experience with optimal control in a chalcopyrite flotation circuit. Preprints, IFAC Symp. on Automation in Mining, Mineral, and Metal Processing, Helsinki.

Kalman, R. E., and Bucy, R. S. (1961). New results in linear filtering and prediction theory. *Trans. ASME J. Basic Engineering, 82,* 95.

Keey, R. B. (1972). *Drying. Principles and Practices.* Pergamon Press, Oxford.

Kelsall, Reid, K. J., and Restarick, C. J. (1967-70). "Continuous Grinding in a Small Wet Ball Mill: Part I — A Study of the Influence of Ball Diameter", *Powder Technology 1,* pg. 291. "Part II — A Study of Holdup Weight" *Powder Technology 2,* pg. 162. "Part III — A Study of Distribution of Residence Time", *Powder Technology 3,* pg. 170.

Kline, J. P. (1982). *Paper and Paperboard.* Miller Freeman Pub., San Francisco.

Koivo, H. N., and Cojocariu, R. (1977). An optimal control for a flotation circuit. *Automatica, 13,* 37–45.

Latour, P. R. (1979). Use of steady-state optimization for computer control in the process industries. In *On-line Optimization Techniques in Industrial Control,* E. J. Kompass and T. J. Williams (eds.), Control Engineering.

Laub, A. J. (1979). A Schur method for solving Riccati equations. *IEEE Trans. Automatic Control, 24,* 6, 913–921.

Lavigne, J. R. (1977). *An Introduction to Paper Industry Instrumentation.* Miller Freeman Pub., San Francisco.

Lee, W., and Kugelman, A. M. (1973). Number of steady-state operating points and local stability of open loop fluid catalytic cracker. *Ind. Eng. Chem. Process Des. Dev., 12,* 2, 197–204.

Lee, W., and Weekman, W. (1976). Advanced control practice in the chemical process industry: A view from industry. *AIChE J., 22,* 1, 27–38.

LeGuea, M. J. A. F. (1975). The control of a flotation process. *CIM Bulletin* (Apr. 1975), 113–128.

Lehtomaki, N. A., Sandell, N. R., and Athans, M. (1981). Robustness results in linear-quadratic gaussian based multivariable control designs. *IEEE Trans. Automatic Control, 26,* 1, 75–93.

Liptak, B. G. (1969/70). *Instrument Engineers Handbook.* Vol. I: *Process Measurements.* Vol. II: *Process Control.* Chilton Book Co., Philadelphia.

Liss, B., and Shinnar, R. (1976). The dynamic behavior of continuous crystallizers in which nucleation and growth depend on properties of the crystal magna. *AIChE Symp. Series, 72,* 153, 28–40.

Ljung, L. (1978). Convergence analysis of parametric identification methods. *IEEE Trans. Automatic Control, 23,* 770–783.

Luyben, W. L. (1969). Distillation feedforward control with intermediate feedback control trays. *Chem. Eng. Science, 24,* 997–1001.

Luyben, W. L. (1970). Distillation decoupling. *AIChE J., 16,* 2, 198–203.

Luyben, W. L. (1973). *Process Modeling, Simulation, and Control for Chemical Engineers.* McGraw-Hill Book Co., New York.

MacFarlane, A. G. J. (1979). *Frequency Response Methods in Control Systems.* IEEE Press, John Wiley & Sons, New York.

MacFarlane, A. G. J. (1980). *Complex Variable Methods for Linear Multivariable Feedback Systems.* Taylor & Francis Ltd., London.

MacFarlane, A. G. J., and Postlethwaite, I. (1977). A generalized Nyquist stability criterion and multivariable root loci. *Int. J. Control, 25,* 81–127.

MacPherson, J. (1985). Optimal process control—it pays. *Instrumentation Technology* (Apr.), 65–67.

Magliozzi, T. L. (1967). Control system prevents surging in centrifugal compressors. *Chem. Eng.,* 74(May), 139–142.

Marchetti, J. L., Mellichamp, D. A., and Seborg, D. E. (1983). Predictive control based on discrete convolution models. *Ind. Eng. Chem. Process Des. Dev., 22,* 488–495.

Marshall, S. A., and James, R. W. (1975). Dynamic analysis of industrial refrigeration system to investigate capacity control. *Proc. Inst. Mech. Engrs., 189,* 44/75, 437–444.

Maybeck, P. S. (1979–1982). *Stochastic Models, Estimation, and Control,* Vols. 1, 2, and 3. Academic Press, New York.

Mayne, D. Q. (1973). The design of multivariable linear control systems. *Automatica, 9,* 201–207.

McAvoy, T. J. (1983). *Interaction Analysis.* Instrument Society of America, Research Triangle Park, N.C.

McCabe, W. L., and Smith, J. C. (1956). *Unit Operations of Chemical Engineering.* McGraw-Hill Book Co., New York.

McCabe, W. L., Smith, J. C., and Harriott, P. (1985). *Unit Operations of Chemical Engineering, 4th Ed.* McGraw-Hill Book Co., New York.

McMillan, G. K. (1984). pH control: A magical mystery tour. *Instrumentation Technology* (Sept.), 69–76.

Michelsen, M. L., Vakil, H. B., and Foss, A. S. (1973). State space formulation of fixed bed reactor dynamics. *Ind. Eng. Chem. Fund., 12,* 3, 323–328.

Miller, S. A. (1969). *Ethylene and Its Industrial Derivatives.* Ernest Benn Ltd., London.

Mitchell and Gauthier, Associates (1981). ACSL: Advanced Continuous Simulation Language—Users' Guide/Reference Manual. Concord, Mass.

Moore, C. F., Smith, C. L., and Murrill, P. W. (1970). Improved algorithms for direct digital control. *Instruments and Control Systems,* 70–74.

Morari, M., Arkun, Y., and Stephanopoulous, G. (1980). Studies in the synthesis of control structures for chemical processes. Part I, II, III. *AIChE J., 26,* 2, 220–260.

Mummé, K. I. (1983). *Papermaking.* Monograph, University of Maine Press, Orono.

Myklestad, O. (1963a). Heat and mass transfer in rotary dryers. *Chem. Eng. Prog. Symp. Series, 59,* 41, 129–137.

Myklestad, O. (1963b). Moisture control in rotary dryers. *Chem. Eng. Prog. Symp. Series, 59,* 41, 138–144.

Nass, L. I. (1976). *Encyclopedia of PVC.* Marcel Dekker, Inc., New York.

Newell, R. B. (1971). Multivariable computer control of an evaporator. PhD thesis, Univ. of Alberta, Canada.

Newell, R. B., and Fisher, D. G. (1972). Model development, reduction, and experimental evaluation for an evaporator. *Ind. Eng. Chem. Process Des. Dev., 11,* 2, 213–221.

Niederlinski, A. (1971). A heuristic approach to the design of linear multivariable interacting control systems. *Automatica, 7,* 691–701.

Niemi, A. J. (1966). A study of dynamic and control properties of industrial flotation processes. *Acta Polytechnica Scandanavica,* no. 8.

Niemi, A. J. (1983). Dynamic models of ore dressing processes, their steady state counterparts, and their use for control. IFAC Symposium on Automation in Mining, Mineral and Metal Processing. Helsinki.

Nilsen, R. N. (1984). The CSSL-IV Simulation Language, Reference Manual. Simulation Services, Chatsworth, California.

Nisenfeld, A. E., and Seeman, R. C. (1981). *Distillation Columns.* Instrument Society of America, Research Triangle Park, N.C.

Olsen, T. O. (1972). Modelling and control of ball mill grinding. Rapport 72-86-W, Institutt for teknisk kybernetikk, Norges Tekniske Høgskole (Dr. ing.-avhandling). Part III: A study of distribution residence time. *Powder Technology, 3,* 170.

Olsen, T. O. (1975). Matematiske modeller for flotasjons-prosesser. Rapport 75-42-U, Institutt for teknisk kybernetikk, Norges Tekniske Høgskole, Trondheim, Norway.

Olsen, T. O., and Henriksen, R. (1976). An innovation approach to parameter estimation in flotation processes. 4th IFAC Symp. on Ident. and Syst. Parameter Estimation. Tbilisi, USSR.

Palmenberg, R. E., and Ward, T. J. (1976). Derivative decoupling control. *Ind. Eng. Chem. Process Des. Dev., 15,* 1, 41–47.

Perlmutter, D. D. (1972). *Stability of Chemical Reactors.* Prentice-Hall, Englewood Cliffs, N.J.

Perry, R. H., and Chilton, C. H. (1973). *Chemical Engineers Handbook.* McGraw-Hill Book Co., New York.

Pike, D. H., and Thomas, M. E. (1974). Optimal control of a continuous distillation column. *Ind. Eng. Chem. Process Des. Dev., 134,* 2, 97–102.

Piovoso, M. J., and Williams, J. M. (1985). Self-tuning pH control. A difficult problem, an effective solution. *Instrumentation Technology* (May), 45–49.

Pontryagin, L. S., Boltyanskii, W. G., Gamkredlidze, R. W., and Mischenko, E. F. (1962). *The Mathematical Theory of Optimal Processes.* John Wiley & Sons, New York.

Powell, R. (1968). *Urea Process Technology.* Noyes Development Corp., Park Ridge, N.J.

Proudfoot, C. G., Gawthrop, P. J., and Jacobs, O. L. K. (1983). Self-tuning PI-control of a pH neutralization process. *Proc. IEE, 130,* 5, 167–272.

Rademaker, O., and Rijnsdorp, J. E. (1959). Dynamics and control of continuous distillation columns. Proc. 5th World Petr. Cong., New York.

Rademaker, O., Rijnsdorp, J. E., and Maarleveld, A. (1975). *Dynamics and Control of Continuous Distillation Units.* Elsevier Scientific Pub. Co., Amsterdam.

Randolf, A. D., and Larsson, M. A. (1971). *Theory of Particulate Processes.* Academic Press, New York.

Ray, W. H. (1981). *Advanced Process Control.* McGraw-Hill Book Co., New York.

Ray, W. H. (1983). Multivariable process control—a survey. *Computers and Chemical Engineering (UK), 7,* 367–394.

Richalet, J., Rault, A., Testud, J. L., and Papon, J. (1978). Model predictive heuristic control: Application to industrial processes. *Automatica, 14,* 413–428.

Rijnsdorp, J. E., and Seborg, D. E. (1976). A survey of experimental applications of multivariable control to process control problems. *AIChE Symp. Series, No. 159, 72,* 112–123.

Rosenbrock, H. H. (1969). Design of multivariable control systems using the inverse Nyquist array. *Proc. IEE, 116,* 1929–1936.

Rouhani, R., and Mehra, R. K. (1982). Model algorithmic control (MAC); basic theoretical properties. *Automatica, 18,* 4, 401–414.

Safonov, M. G., and Athans, M. (1977). Gain and phase margin for multiloop LQG regulators. *IEEE Trans. Automatic Control, 22,* 173–179.

Sage, A. P., and Melsa, J. L. (1971). *System Identification.* Academic Press, New York.

Saltman, D. (1979). *Paper Basics.* Van Nostrand Reinhold Co., New York.

Sarvetnick, H. A. (1969). *Polyvinylchloride.* Reinhold Plastics Appl. Ser., Van Nostrand Reinhold Co., New York.

Sastry, W. A., Seborg, D. E., and Wood, R. K. (1977). Self-tuning regulator applied to a binary distillation column. *Automatica, 13,* 417–420.

Sauchelli, V. (1964). *Fertilizer Nitrogen. Its Chemistry and Technology.* Reinhold Pub. Co., New York.

Schuldt, S. B., and Smith, F. B., Jr. (1971). An application of quadratic performance synthesis techniques to a fluid cat cracker. Joint Auto. Contr. Conf. (Washington University), paper no. 3-E4 (270–276).

Schulz, R. (1973). The drum water level in the multivariable control system of a steam generator. *IEEE Trans. Indus. Electronics and Control Instrumentation, IECI-20,* No. 3, 164–169.

Seborg, D. E., and Fisher, D. G. (1979). Experience with experimental applications of multivariable computer control. *Trans. ASME J. Dynamic Systems, Measurement, and Control, 101,* 108–116.

Seborg, D. E., Edgar, T. F., and Shah, S. L. (1986). Adaptive Control Strategies for Process Control: A Survey. *AIChE J. 32,* 6, 881.

Shale, C. C. and Faschig, G. E. (1969) Operating characteristics of a high temperature electrostatic precipitator. U.S. Bureau of Mines report 7276.

Sherwin, M. B., Shinnar, R., and Katz, S. (1967). Dynamic behavior of a well mixed isothermal crystallizer. *Chem. Eng. Prog. Symp. Series, No. 95, 65,* 59–74.

Shinnar, R. (1981). Chemical reactor modelling for purposes of controller design. *Chem. Eng. Commun., 9,* 73–99.

Shinskey, F. G. (1973). *pH and pION Control in Process and Waste Streams.* John Wiley & Sons, New York.

Shinskey, F. G. (1978). *Energy Conservation through Control.* Academic Press, New York.

Shinskey, F. G. (1979). *Process Control Systems.* McGraw-Hill Book Co., New York.

Shinskey, F. G. (1977/84). *Distillation Control.* McGraw-Hill Book Co., New York.

Shreve, R. N., and Brink, J. A., Jr. (1977). *Chemical Process Industries* (4th ed.). McGraw-Hill Book Co., New York.

Silva, J. M., Wallman, P. H., and Foss, A. S. (1979). Multi-bed catalytic reactor control system: Configuration development and experimental testing. *Ind. Eng. Chem. Fund., 18,* 4, 383–391.

Simon, J. D. and Mitter, S. K. (1968). A Theory of Modal Control. *Information and Control 13,* 316.

Sittig, M. (1976). *Polyolefin Production Processes. Latest Developments.* Noyes Data Corp., Park Ridge, N.J.

Smith, O. J. M. (1957). Closer control of loops with dead time. *Chem. Eng. Prog., 53,* 5, 217–219.

Solheim, O. A. (1972). Design of Optimal Control System with Prescribed Eigenvalues. *Int'l J. Cont. 15,* 143–160.

Solheim, O. A. (1976). *Optimalregulering.* Tapir Forlag, Trondheim, Norway.

Speight , J. G. (1980). *The Chemistry and Technology of Petroleum.* Marcel Dekker, New York.

Stephanopoulous, G. (1982). Synthesis of control systems for chemical plants—a challenge for creativity. Int. Symp. on Process Systems Engineering, Kyoto, Japan.

Stephanopoulos, G. (1984). *Chemical Process Control: An Introduction to Theory and Practice.* Prentice Hall, Englewood Cliffs, N.J.

Stretzoff, S. (1981). *Technology and Manufacture of Ammonia.* John Wiley & Sons, New York.

Sælid, S. (1976). Modeling, estimation, and control of a rotary cement kiln. Rapport 76-157-W, Institutt for teknisk kybernetikk, Norges Tekniske Høgskole (Dr. ing.-avhandling), Trondheim, Norway.

Sælid, S. (1982). Modellering av dynamiske systemer. Forelesningsnotater, Institutt for teknisk kybernetikk, Norges Tekniske Høgskole, Trondheim, Norway.

Sælid, S., Jenssen, N. A., and Balchen, J. G. (1983). Design and analysis of a dynamic positioning system based on Kalman filtering and optimal control. *IEEE Trans. Automatic Control, 28* (special issue on Kalman filtering).

Taiwo, O. (1980). Application of the method of inequalities to the multivariable control of binary distillation columns. *Chem. Eng. Sci., 35,* 4, 847–858.

Teager, H. M. (1955). An analysis of transients in mass transfer operations. D. Sc. thesis, MIT, Cambridge, Mass.

Thal-Larsen, H. (1960). Dynamics of heat exchangers and their models. *Trans. ASME J. Basic Eng.* (June), 489–504.

Toijala, K., and Fagervik, K. (1972). A digital simulation study of two point feedback control of distillation columns. *Kemian Teollisuus,* (Finland) No. 1, 5–16.

Torvund, L. (1971). Prosessbeskrivelse for Ammoniakkfabrik NII. ved Eidanger Saltpeterfabriker, Norsk Hydro.

Tyssø, A. (1980a). CYPROS — cybernetic program packages. *Modeling, Identification and Control, 1,* 4, 215–229.

Tyssø, A. (1980b). Modeling, parameter estimation, and control of a ship boiler. Rapport 80-13-W, Institutt for teknisk kybernetikk, Norges Tekniske Høgskole, Trondheim, Norway.

Tyssø, A. (1981). Modeling and parameter estimation of a ship boiler. *Automatica, 17,* 1, 157–166.

Tyssø, A., and Brembo, J. C. (1978). Installation and operation of a multivariable ship boiler control system. *Automatica, 14,* 213–221.

Tyssø, A., Brembo, J. C., and K. Lind (1976). The design of a multivariable control system for a ship boiler. *Automatica, 12,* 3, 211–224.

Uhran, J. J. and W. I. Davisson (1984). The Structure of NDTRAN — A Systems Simulation Language. *IEEE Trans. Sys, Man, Cyb.* SMC-14:6.

Vakil, H. B., Michelsen, M. L., and Foss, A. S. (1973). Fixed-bed reactor control with state estimation. *Ind. Eng. Chem. Fund., 12,* 3, 328–335.

Vancini, C. A. (1971). *Synthesis of Ammonia,* MacMillan Press, Ltd., London.

Viswanadham, N., Patnaik, L. M., and Sarma, I. G. (1979). Robust multivariable controllers for a tubular ammonia reactor. *Trans. ASME J. Dynamic Systems, Measurements, and Control, 110,* 290–298.

Vroom, K. E. (1957). The "H" factor: A means of expressing cooking times and temperatures as a single variable. *Pulp and Paper Magazine of Canada (PPMC),* No. 3, 228.

Wahl, E. F., and Harriott, P. (1970). Understanding and prediction of the dynamic behavior of distillation columns. *Ind. Eng. Chem. Process Des. Dev., 9,* 3, 396–407.

Waller, K. W., and Gustafsson, T. K. (1983). Fundamental properties of continuous pH control. *ISA Trans., 22,* 1, 25–34.

Wallman, P. H., and Foss, A. S. (1981). Experiences with dynamic estimators for fixed-bed reactors. *Ind. Eng. Chem. Fund., 20,* 3, 234–239.

Wallman, P. H., Silva, J. M., and Foss, A. S. (1979). Multivariable integral controls for fixed-bed reactors. *Ind. Eng. Chem. Fund., 18,* 4, 392–399.

Wells, C. H., Johns, E. C., and Chapman, F. L. (1975). Computer control of batch digesters using a Kappa number model. *Tappi, 58,* 8, 177–181.

Whalley, R. (1978). The mathematical model for a naval boiler. *Inst. MC Trans., 11,* No. 9.

White, M. H. (1972). Surge control for centrifugal compressor. *Chem. Eng., 79* (Dec).

Wilhelm, R. B., and Fjeld, M. (1983). Control algorithms for cross-directional control: The state of the art. IFAC/IMACS PRP Conf., 1983.

Williams, T. J., and Harnett, R. T. (1957). Automatic control in continuous distillation. *Chem. Eng. Prog., 53,* 5, 220–225.

Wilson, J. T., and Belanger, P. R. (1971). Application of dynamic optimization techniques to reactor control in the manufacture of ethylene. Joint Auto. Control Conf. (IEEE), paper no. 3-E5.

Yliniemi, L., Jutila, E. A. A., and Uronen, P. (1981). Modeling and control of a pilot plant rotary drier used for drying of industrial ore concentrates. Proc. IFAC — 8th Triennial World Congress, Kyoto, Japan (2837–2843).

Yliniemi, L. and Uronen, P. (1983). A comparison of different control strategies of a rotary kiln. IFAC 4th Symposium on Automation in Mining, Mineral and Metal Processing, Helsinki.

Åstrøm, K. J. (1964). Control problems in papermaking. IBM Scientific Computing Symposium: Control Theory and Applications, New York.

Åstrøm, K. J. (1970). *Introduction to Stochastic Control Theory*. Academic Press, New York.

Åstrøm, K. J., and Bohlin, T. (1966). Numerical identification of linear dynamic systems from normal operating records. *Proc. 2nd IFAC Symp. on Theory of Self-tuning Regulators*, P. H. Hammond (ed.). Plenum Press, New York.

Åstrøm, K. J., and Eykhoff, P. (1971). System identification — a survey. *Automatica, 7,* 123–162.

Åstrøm, K. J., and Haggman, B. (1974). Papermachine dynamics. 4th IFAC/IFIP Int. Conf. on Dig. Comp. Appln. to Proc. Control, Zurich (463).

Åstrøm, K. J., and Wittenmark, B. (1973). On self-tuning regulators. *Automatica, 9,* 185–199.

Åstrøm, K. J., Borisson, U., Ljung L., and Wittenmark, B. (1977). Theory and applications of self-tuning regulators. *Automatica, 13,* 457–476.

INDEX